Lecture Notes in Electrical Engineering

Volume 504

**** Indexing: The books of this series are submitted to ISI Proceedings, EI-Compendex, SCOPUS, MetaPress, Springerlink ****

Lecture Notes in Electrical Engineering (LNEE) is a book series which reports the latest research and developments in Electrical Engineering, namely:

- Communication, Networks, and Information Theory
- Computer Engineering
- Signal, Image, Speech and Information Processing
- Circuits and Systems
- Bioengineering
- Engineering

The audience for the books in LNEE consists of advanced level students, researchers, and industry professionals working at the forefront of their fields. Much like Springer's other Lecture Notes series, LNEE will be distributed through Springer's print and electronic publishing channels.

For general information about this series, comments or suggestions, please use the contact address under "service for this series".

To submit a proposal or request further information, please contact the appropriate Springer Publishing Editors:

Asia:

China, *Jessie Guo, Assistant Editor* (jessie.guo@springer.com) (Engineering)

India, *Swati Meherishi, Senior Editor* (swati.meherishi@springer.com) (Engineering)

Japan, *Takeyuki Yonezawa, Editorial Director* (takeyuki.yonezawa@springer.com) (Physical Sciences & Engineering)

South Korea, *Smith (Ahram) Chae, Associate Editor* (smith.chae@springer.com) (Physical Sciences & Engineering)

Southeast Asia, *Ramesh Premnath, Editor* (ramesh.premnath@springer.com) (Electrical Engineering)

South Asia, *Aninda Bose, Editor* (aninda.bose@springer.com) (Electrical Engineering)

Europe:

Leontina Di Cecco, Editor (Leontina.dicecco@springer.com)
(Applied Sciences and Engineering; Bio-Inspired Robotics, Medical Robotics, Bioengineering; Computational Methods & Models in Science, Medicine and Technology; Soft Computing; Philosophy of Modern Science and Technologies; Mechanical Engineering; Ocean and Naval Engineering; Water Management & Technology)

Christoph Baumann (christoph.baumann@springer.com)
(Heat and Mass Transfer, Signal Processing and Telecommunications, and Solid and Fluid Mechanics, and Engineering Materials)

North America:

Michael Luby, Editor (michael.luby@springer.com) (Mechanics; Materials)

More information about this series at http://www.springer.com/series/7818

Ali Boyaci · Ali Riza Ekti
Muhammed Ali Aydin · Serhan Yarkan
Editors

International Telecommunications Conference

Proceedings of the ITelCon 2017, Istanbul

İSTANBUL TİCARET ÜNİVERSİTESİ

CARIT

Springer

Editors
Ali Boyaci
Department of Electrical-Electronics Engineering
Istanbul Commerce University
Istanbul
Turkey

Ali Riza Ekti
Department of Information Technology
Balıkesir University
Balıkesir
Turkey

Muhammed Ali Aydin
Computer Engineering Department
Istanbul University
Istanbul
Turkey

Serhan Yarkan
Istanbul Commerce University
Istanbul
Turkey

ISSN 1876-1100 ISSN 1876-1119 (electronic)
Lecture Notes in Electrical Engineering
ISBN 978-981-13-0407-1 ISBN 978-981-13-0408-8 (eBook)
https://doi.org/10.1007/978-981-13-0408-8

Library of Congress Control Number: 2018940400

Cover design: INF Technology, Istanbul, Türkiye (http://www.inf.com.tr)
LaTeX editor: Serhan Yarkan

Printed on acid-free paper

This Springer imprint is published by the registered company Springer Nature Singapore Pte Ltd.
The registered company address is: 152 Beach Road, #21-01/04 Gateway East, Singapore 189721, Singapore

Preface

ITelCon is an organization which is supported by both pioneering academicians in their fields along with experts and professionals from industry. Under the patronage of ITelCon organization, ITelCon 2017 is the First International Telecommunications Conference, which was held in Teknopark Kurtköy, İstanbul, Türkiye, on December 28–29, 2017. The conference was sponsored by several institutions and companies in different categories. Istanbul Commerce University, the foundation sponsor, hosted the conference in a great cooperation with Istanbul Teknopark. Besides, Spark Measurement Technologies and Turkish Aerospace Industries (TAI), Inc. supported ITelCon 2017 as platinum and gold sponsor, respectively.

ITelCon 2017 came true with great efforts and collaboration of several committees. Honorary Chair, Nazım Ekren, Rector, along with İbrahim Baz, Vice Rector, Istanbul Commerce University, supported the organizing committee with immense enthusiasm. Executive Committee, Gülsüm Zeynep Gürkaş Aydın and Serap Tepe, handled and run the conference seamlessly covering publicity, logistics and announcements. Technical Program Chair, Özgür Can Turna, administered the entire manuscript submission, delegation, and review processes in collaboration with technical program committee consisting of more than 50 members representing 15 countries across the globe.

The conference program started with seven outstanding keynote speeches delivered by prominent academicians and experts in the field. Opening speech was given by Temel Kotil from TAI, which focused on concurrent and near-future "research and development" trends in the industry. Second talk, which was given by Tolga M. Duman from Bilkent University, was on the general research perspectives and opportunities for next-generation wireless communications. BİLGEM Vice President, Ali Görçin, delivered a talk on behalf of BİLGEM Informatics and Information Security Research Center, President, Hacı Ali Mantar, outlining the comprehensive near-future research and development options and cutting-edge research facilities and infrastructures at TÜBİTAK. Next, Erchin Serpedin from Texas A&M University at College Station emphasized the importance of green communications and networking in the upcoming era. In what follows, Hüseyin

Arslan from University of South Florida underlined the physical layer security concerns, possible solutions, and emerging security and privacy requirements for 5G and beyond. After that Tülay Yıldırım delivered an enlightening speech regarding artificial intelligence, machine learning, and their use in the field of telecommunications. Khalid A. Qaraqe from Texas A&M University at Qatar put an emphasis on telehealth applications, how they would foster the daily life, and very sensitive nature of the data to be carried over wireless networks in conjunction with the security, privacy, and authenticity perspectives. Finally, Spark Measurement Technologies, platinum sponsor of ITelCon 2017, introduced the state-of-the-art measurement and instrumentation devices and software for 5G and beyond.

On behalf of the ITelCon organizing committee, we would like to thank all the people, companies, and organizations who contributed to making ITelCon 2017 successful. First and foremost, we acknowledge contributions of the authors who considered submitting their valuable research outcomes to ITelCon 2017. Second, ITelCon organizing committee appreciate the program committee members and the reviewers across the globe whose sincere efforts in reviewing the manuscripts improved both individual and overall scientific quality of the works and the conference, respectively. Special thanks are extended to the invited speakers, Temel Kotil, Tolga M. Duman, Hacı Ali Mantar, Erchin Serpedin, Hüseyin Arslan, Tülay Yıldırım, Khalid A. Qaraqe, and Çağan Irmak, for their sincere enthusiasm to take an active part in ITelCon 2017. Under the patronage of Istanbul Chamber of Commerce, we would like to thank Istanbul Commerce University, who made ITelCon 2017 possible by providing all sorts of permissions, access, and man power whenever needed along with Bilal Macit and his team from Istanbul Teknopark who let organizing committee use their facilities throughout the conference. Last but not least, ITelCon 2017 would like to extend special thanks select students of the EE Engineering Department at Istanbul Commerce University, for their great assistance in running the workshop.

Istanbul, Turkey — Ali Boyaci
Balıkesir, Turkey — Ali Riza Ekti
Istanbul, Turkey — Muhammed Ali Aydin
Istanbul, Turkey — Serhan Yarkan
ITelCon 2017—Organizing Committee

Organization

Organizing Committee

General Chairs

Ali Boyaci, Istanbul Commerce University, Türkiye
Ali Riza Ekti, Balikesir University, Türkiye
Muhammed Ali Aydin, Istanbul University, Türkiye
Serhan Yarkan, Istanbul Commerce University, Türkiye

Executive Chairs

Gülsüm Zeynep Gürkaş Aydın, Istanbul University, Türkiye
Serap Tepe, Uskudar University, Türkiye

Program Committee

Program Chair

Özgür Can Turna, Istanbul University, Türkiye

Steering Committee

Erchin Serpedin, Texas A&M University at College Station, USA
Hüseyin Arslan, University of South Florida, USA
Khalid A. Qaraqe, Texas A&M University at Qatar Station, Qatar
Tülay Yıldırım, Yıldız Technical University, Türkiye

Local Committee

Gamze Kirman, Istanbul Commerce University, Türkiye
Sezer Can Tokgöz, Istanbul Commerce University, Türkiye
Özgür Alaca, Istanbul Commerce University, Türkiye

Sponsoring Institution

Istanbul Commerce University

Powered by

INF Technology, İstanbul, Türkiye

Contents

Part V Security

Part VI Signal Processing

Part I
xG Networks

Performance Analysis of Relaying FSO System over $\mathcal{M}$-Distributed Turbulent Channel with Variable Gain AF Protocol

V. K. Jagadeesh, Vineeth Palliyembil, Imran Shafique Ansari, P. Muthuchidambaranathan and Khalid A. Qaraqe

Abstract In this work, the performance of an all optical relaying free space optical (FSO) system over Málaga ($\mathcal{M}$)-distributed turbulent channel is analyzed with variable gain amplify and forward (AF) protocol. For the relay system, the approximate cumulative distribution function (CDF) expression of the output signal to noise ratio (SNR) is derived and corresponding outage probability, channel capacity and asymptotic bit error rate (BER) are analyzed for different modulation schemes in the presence of pointing errors.

Keywords Free space optics · Amplify and forward · Outage probability · Bit error rate · Channel capacity · Pointing errors

1 Introduction

Free space optics (FSO) is an emerging technology that uses light emitting diodes (LED) or laser diodes to transmit the data through wireless medium. The main advantages of FSO communication systems include large bandwidth with high data rate, license free spectrum, easy installation, low power consumption, and greater security [1, 2]. Due to these features, FSO communication systems have found some

V. K. Jagadeesh (✉)
Department of Electronics and Communication Engineering,
NMAMIT, Mangalore, India
e-mail: vkjagadeesh@gmail.com

V. Palliyembil · P. Muthuchidambaranathan
Department of Electronics and Communication Engineering,
National Institute of Technology, Tiruchirappalli, India

I. S. Ansari · K. A. Qaraqe
Department of Electrical and Computer Engineering,
Texas A & M University at Qatar (TAMUQ), Doha, Qatar

I. S. Ansari · K. A. Qaraqe
Education City, Ar-Rayyan, Qatar

A. Boyaci et al. (eds.), *International Telecommunications Conference*,
Lecture Notes in Electrical Engineering 504,
https://doi.org/10.1007/978-981-13-0408-8_1

applications in the area of indoor communications, underwater communications, chip to chip communications, inter-satellite communications etc.

Even though, there are some challenges limiting the full potential of FSO communication systems. The atmospheric turbulence induced fading and pointing errors due to misalignment are the main impairments affecting an optical signal when propagating through the turbulence channel and can degrade the performance of the system, if the link range is more than 1 km [3–5]. Spatial diversity [6] and relay techniques [7] are promising approaches to overcome these problems in FSO systems. The relay techniques create a virtual multiple aperture system and realize the advantages of multiple-input multiple-output (MIMO) techniques.

Relay based FSO systems is a well known technique that is utilized to improve the performance of FSO systems by mitigating the limiting effects of scintillation and misalignment fading [7]. It provides higher capacity with wider and more energy efficient coverage. The performance analysis of relay based FSO systems has been been analyzed with amplify-and-forward (AF) and decode-and-forward (DF) relays over different channel models [8, 9]. The AF relays have attracted a lot of research interest due to less complexity. The variable gain AF relays use the channel state information (CSI) of the preceding hop to control the relay gain; whereas, a fixed-gain AF relay does not need the CSI. In [10, 11], the capacity performance of AF based FSO systems have been discussed. The outage probability and the average bit error rate (BER) performance analysis of DF based FSO system over Gamma-Gamma channel analyzed in [12]. The outage and error performances of bi-directional decode-XOR-forward (DXF) relay FSO systems were analyzed over Málaga ($\mathcal{M}$)-turbulence channel in [13].

In this work, the performance of an all optical relaying FSO system over $\mathcal{M}$-distributed turbulent channel is analyzed with variable gain AF protocol. For the relay system, the approximate cumulative distribution function (CDF) expression of the output signal-to-noise ratio (SNR) is derived and the corresponding outage probability, channel capacity, and asymptotic BER are analyzed for different modulation schemes.

2 System and Channel Model

A multi hop FSO system with variable AF gain relay under $\mathcal{M}$-distribution [14] is considered. The system model consists of a source (S), a destination (D), and a single relay (R) node as shown in Fig. 1.

With the help of the relay, signals are transmitted from the transmitter to the target user through the turbulence induced fading channels and are corrupted at the receiver by the additive white Gaussian noise (AWGN) noise of zero mean and variance N_0. The received signal at the relay R can be expressed as

$$y_{s,r} = I_{s,r}x + n_{s,r}, \tag{1}$$

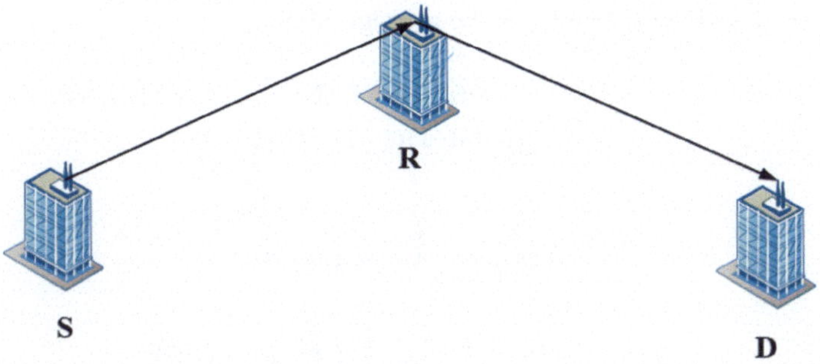

Fig. 1 Relay assisted dual-hop transmission system

where x is the transmitted information by source, $y_{s,r}$ is the received signal at the relay, $I_{s,r}$ represents the irradiance from source to the relay and $n_{s,r}$, is the AWGN at relay. The received signal at the destination is expressed as

$$y_{r,d} = I_{r,d}G[1 + \eta_{r,d}(I_{s,r}x + n_{s,r})] + n_{r,d}, \tag{2}$$

where $y_{r,d}$ is the received signal at the destination, $I_{r,d}$ represents the irradiance from relay to destination, G is the normalized amplifying gain, $\eta_{r,d}$ is the electrical-to-optical conversion coefficient, and $n_{r,d}$ is the AWGN at the destination. Most of the previous relaying FSO networks, G is assumed as a constant [15]. Here the value of G is assumed as

$$G = \sqrt{\frac{P_r}{I_{s,r}^2 P_{s,r}^2 \eta_{r,d}^2 + N_0}}, \tag{3}$$

where P_r is the power received at the destination and $P_{s,r}$ is the transmitted power at the source. After removing the direct current (DC) component, the expression for the received signal can be written as

$$y_{r,d} = I_{r,d}G\eta_{r,d}(I_{s,r}x + n_{s,r}) + n_{r,d}. \tag{4}$$

The unconditional PDF for the $\mathcal{M}$-distributed turbulence channel can be expressed in terms of the SNR [16] as

$$f_\gamma(\gamma) = \frac{g^2 A}{4\gamma} \sum_{m=1}^{\beta} b_m G_{1,3}^{3,0}\left(B\sqrt{\frac{\gamma}{\mu}}\,\middle|\, \begin{matrix} g^2+1 \\ g^2, \alpha, m \end{matrix}\right), \tag{5}$$

where

$$\mu = \frac{g^2(g^2+1)^{-2}(g^2+2)(\xi_g + \Omega')}{\alpha^{-1}(\alpha+1)[2\xi_g(\xi_g + 2\Omega') + \Omega'^2(1+1/\beta)]}\bar{\gamma},$$

is the electrical SNR of (5) and $\bar{\gamma}$ is the average SNR,

$$B = \frac{g^2\alpha\beta(\xi_g + \Omega^{'})}{[(g^2+1)(\xi_g\beta + \Omega^{'})]},$$

$$A = \frac{2\alpha^{\frac{\alpha}{2}}}{\xi_g^{1+\frac{\alpha}{2}}\Gamma(\alpha)}\left(\frac{\xi_g\beta}{\xi_g\beta + \Omega^{'}}\right)^{\beta+\frac{\alpha}{2}},$$

$$a_k = \binom{\beta-1}{k-1}\frac{(\xi_g\beta + \Omega^{'})^{1-\frac{k}{2}}}{(k-1)!}\left(\frac{\Omega^{'}}{\xi_g}\right)^{k-1}\left(\frac{\alpha}{\beta}\right)^{\frac{k}{2}},$$

$$b_m = a_k\left(\frac{\alpha\beta}{\xi_g\beta + \Omega^{'}}\right)^{-\frac{\alpha+k}{2}}$$

g is the pointing error parameter. α is a positive parameter related to the effective number of large-scale cells of the scattering process, and β represents the amount of fading parameter. ξ_g is denoted as $\xi_g = 2b_0(1-\rho)$. b_0 is the average power of the total scatter components and ρ is the amount of scattering power coupled [17]. $\Omega^{'}$ represents the average power from the coherent contributions [14].

When $g^2(\xi_g + \Omega^{'})/(g^2+1) = 1$, this PDF given in (5) comes in agreement with [18, Eq. (19)].

Utilizing (3) and (4), the end-to-end SNR can be obtained and is tightly upper bounded at high SNR values as

$$\gamma_{eq} = \frac{\gamma_{s,r}\gamma_{r,d}}{\gamma_{s,r} + \gamma_{r,d} + 1} \approx \frac{\gamma_{s,r}\gamma_{r,d}}{\gamma_{s,r} + \gamma_{r,d}}, \tag{6}$$

where $\gamma_{s,r} = \frac{P_{s,r}^2 I_{s,r}^2}{N_0}$ and $\gamma_{r,d} = \frac{\eta_{r,d}^2 I_{r,d}^2}{N_0}$.

3 Cumulative Distribution Function

When using the upper bound of γ_{eq}, it is well known that the CDF of γ_{eq} is given as

$$\begin{aligned} F_{\gamma_{eq}}(\gamma) &= 1 - (1 - F_{\gamma_{s,r}}(\gamma))(1 - F_{\gamma_{r,d}}(\gamma)) \\ &= F_{\gamma_{s,r}}(\gamma) + F_{\gamma_{r,d}}(\gamma) - F_{\gamma_{s,r}}(\gamma)F_{\gamma_{r,d}}(\gamma), \end{aligned} \tag{7}$$

where $F_{\gamma_{s,r}}(\gamma)$ and $F_{\gamma_{r,d}}(\gamma)$ are the CDF of $\gamma_{s,r}$ and $\gamma_{r,d}$, respectively. From Eq. (5), the CDF of $\gamma_{s,r}$ is obtained as

$$F_{\gamma_{s,r}}(\gamma) = \frac{g_{s,r}^2 A_{s,r}}{8\pi} \sum_{m_{s,r}=1}^{\beta_{s,r}} b_{m_{s,r}} 2^{\alpha_{s,r}+m_{s,r}-1}$$
$$\times G_{3,7}^{6,1}\left(\frac{B_{s,r}^2 \gamma}{16\mu_{s,r}} \middle| \begin{matrix} 1, \frac{g_{s,r}^2+1}{2}, \frac{g_{s,r}^2+2}{2} \\ k_1 \end{matrix} \right), \tag{8}$$

where $k_1 = \frac{g_{s,r}^2}{2}, \frac{g_{s,r}^2+1}{2}, \frac{\alpha_{s,r}}{2}, \frac{\alpha_{s,r}+1}{2}, \frac{m_{s,r}}{2}, \frac{m_{s,r}+1}{2}, 0$. A similar expression is obtained for $F_{\gamma_{r,d}}(\gamma)$ as

$$F_{\gamma_{r,d}}(\gamma) = \frac{g_{r,d}^2 A_{r,d}}{8\pi} \sum_{m_{r,d}=1}^{\beta_{r,d}} b_{m_{r,d}} 2^{\alpha_{r,d}+m_2-1}$$
$$\times G_{3,7}^{6,1}\left(\frac{B_{r,d}^2 \gamma}{16\mu_{r,d}} \middle| \begin{matrix} 1, \frac{g_{r,d}^2+1}{2}, \frac{g_{r,d}^2+2}{2} \\ k_2 \end{matrix} \right), \tag{9}$$

where the parameters $A_{r,d}$, $\alpha_{r,d}$, $m_{r,d}$ and $\beta_{r,d}$ have the same definitions as previously mentioned, but they are related to the link condition between the relay and the destination.

4 System Performance Analysis

In this section, the performance of the relay based FSO system over $\mathcal{M}$-distributed turbulence conditions is analyzed. More specifically, the expressions for the asymptotic outage probability and BER are derived.

4.1 Outage Probability

Outage probability is an important performance metric in communication system and is defined as the probability that the instantaneous SNR γ_{eq} falls below a given threshold value γ_{th}. From this definition, the outage probability is given as

$$P_{out} = F_{eq}(\gamma_{th}). \tag{10}$$

For high SNR regime, the last term in Eq. (7) is neglected and the accurate expression for the outage probability is given as

$$P_{out} \approx F_{s,r}(\gamma_{th}) + F_{r,d}(\gamma_{th}). \tag{11}$$

Table 1 Parameters (p and q) for various binary modulation techniques

Type of modulation	p	q
Coherent binary frequency shift keying (CBFSK)	0.5	0.5
Non-coherent binary frequency shift keying (NBFSK)	1	0.5
Coherent binary phase shift keying (CBPSK)	0.5	1
Differential binary phase shift keying (DBPSK)	1	1

4.2 BER Analysis

The average BER of the proposed system can be expressed in terms of the CDF as [19]

$$P_e = \frac{q^p}{2\Gamma(p)} \int_0^{\infty} \exp(-q\gamma)\gamma^{p-1} F_{\gamma}(\gamma) \mathrm{d}\gamma. \tag{12}$$

In the above equation p and q are the BER parameters corresponding to the four modulation schemes as given in Table 1 [15].

On substituting Eq. (7) into (12) and utilizing [20, Eq. (7.813.1)], the average BER performance of a relaying FSO network for different modulation schemes, at high SNR regime, can be derived as

$$P_e = P_{s,r} + P_{r,d}, \tag{13}$$

where

$$P_{s,r} = \frac{g_{s,r}^2 A_{s,r}}{16\pi\Gamma_p} \sum_{m_{s,r}=1}^{\beta_{s,r}} b_{m_{s,r}} 2^{\alpha_{s,r}+k_1-1} \times G_{4,7}^{6,2}\left(\frac{B_{s,r}^2}{16q\mu_{s,r}} \middle| \begin{matrix} k_3 \\ k_4 \end{matrix} \right), \tag{14}$$

where $k_3 = 1-p, \frac{g_{s,r}^2+1}{2}, \frac{g_{s,r}^2+2}{2}$ and $k_4 = \frac{g_{s,r}^2}{2}, \frac{g_{s,r}^2+1}{2}, \frac{\alpha_{s,r}}{2}, \frac{\alpha_{s,r}+1}{2}, \frac{m_{s,r}}{2}, \frac{m_{s,r}+1}{2}, 0$. $P_{r,d}$ has a similar expression as $P_{s,r}$.

4.3 Channel Capacity

The ergodic channel capacity can be defined as $\overline{C} = \mathbb{E}[\log_2(1 + c\gamma)]$, where $c =$ 1for heterodyne detection and $c = e/2\pi$ for IM/DD [21, Eq. (7.43)]. This can be represented as

$$\overline{C} = \int_{0}^{\infty} \log_2(1 + e/2\pi\gamma) f_\gamma(\gamma) d\gamma. \tag{15}$$

Now substituting Eq. (5) into (15) and utilizing [22], the ergodic channel capacity of a relaying FSO network, at high SNR regime, is obtained as

$$\overline{C} = C_{s,r} + C_{r,d}, \tag{16}$$

where

$$C_{s,r} = \frac{g_{s,r}^2 A_{s,r}}{16\pi ln2} \sum_{m_{s,r}=1}^{\beta_{s,r}} b_{m_{s,r}} 2^{\alpha_{s,r}+k_1} \times G_{3,7}^{7,1}\left(\frac{B_{s,r}^2 \pi}{8e\mu_{s,r}} \middle| \begin{matrix} 0, 1, \frac{g_{s,r}^2+2}{2} \\ k_5 \end{matrix} \right), \tag{17}$$

where $k_5 = \frac{g_{s,r}^2}{2}, \frac{\alpha_{s,r}}{2}, \frac{\alpha_{s,r}+1}{2}, \frac{m_{s,r}}{2}, \frac{m_{s,r}+1}{2}, 0, 0$. $C_{r,d}$ has a similar expression as $C_{s,r}$.

5 Results and Discussion

In this section, the analysis is conducted with some numerical results for the channel model under consideration with variable gain AF relay. In all cases, the following parameters are considered in the numerical results: the transmitted power is normalized as $\Omega + 2b_0 = 1$, the channel parameters $\alpha_1 = 10$, $\alpha_2 = 4$, $\beta_1 = 5$, $\beta_1 = 2$, and $\gamma_{th} = 6$dB. The average and electrical SNRs of both the links are assumed to be equal. The performance of the system is analyzed with outage probability, the channel capacity, and the asymptotic BER of coherent binary frequency shift keying (CBFSK), non-coherent binary frequency shift keying (NBFSK), coherent binary phase shift keying (CBPSK), differential binary phase shift keying (DBPSK) modulation schemes.

In Fig. 2, the outage performance is analyzed for different values of ρ and g. From the figure, it is observed that increasing ρ can improve the outage performance of the system. It is also observed that the outage performance is also improved with increase in g parameter because the definition of g is the ration between equivalent beam radius at the receiver and the pointing error displacement standard deviation at the receiver.

In Figs. 3 and 4, the asymptotic BER of relay based FSO system is analyzed under the effect of pointing errors with turbulence for different modulation schemes. Figure 3 shows the asymptotic BER performance of the system with $\rho = 0$ and Fig. 4 shows the same analysis for $\rho = 1$. From the figures it is observed that the CBFSK modulation outperforms NBFSK because it is the most power-efficient modulation schemes. Coherent modulation schemes perform better than their non coherent

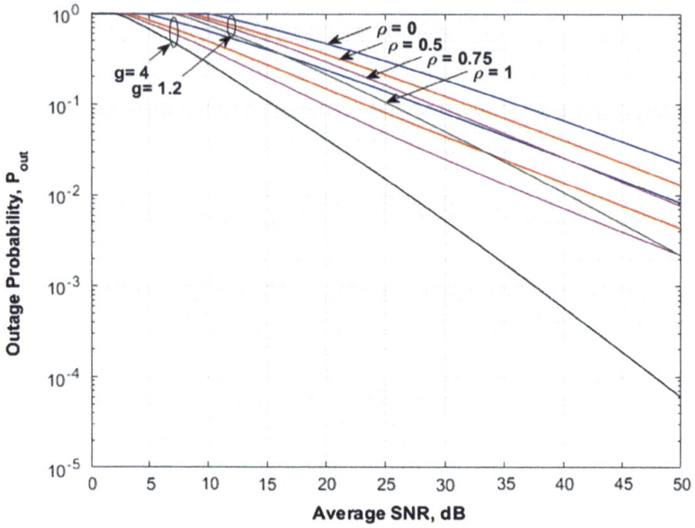

Fig. 2 Outage performance of a relaying FSO network with pointing errors

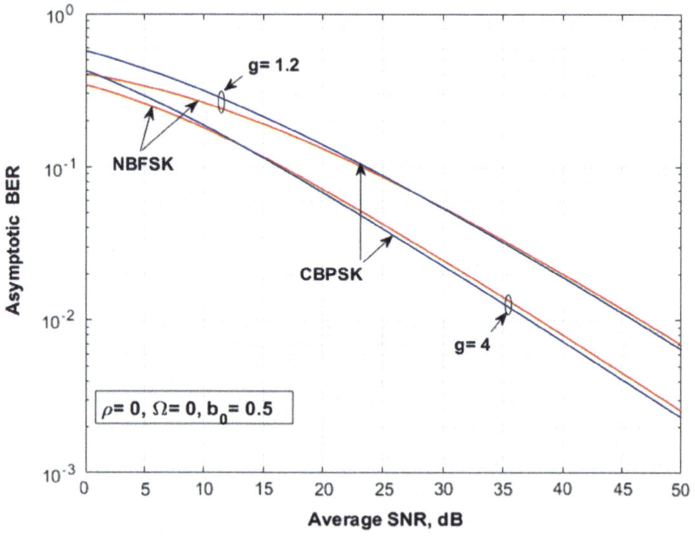

Fig. 3 BER performance of a relaying FSO network for $\rho = 0$ with pointing errors

scheme. It is also observed that the performance for $\rho = 1$ is much better. And it is also observed that as the effect of pointing errors (as the value of g increases, the effect of pointing error decreases) increases, the BER deteriorates and vice versa.

In Fig. 5, the ergodic channel capacity of the system is analyzed under the combined effect of atmospheric turbulence and pointing errors for different values of

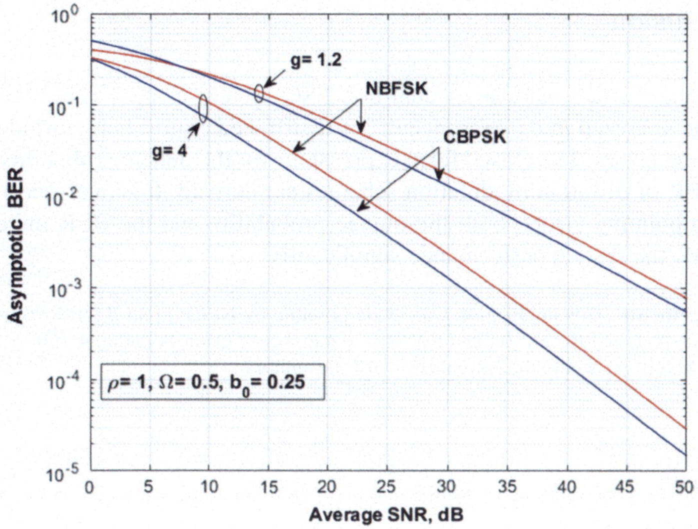

Fig. 4 BER performance of a relaying FSO network for $\rho = 1$ with pointing errors

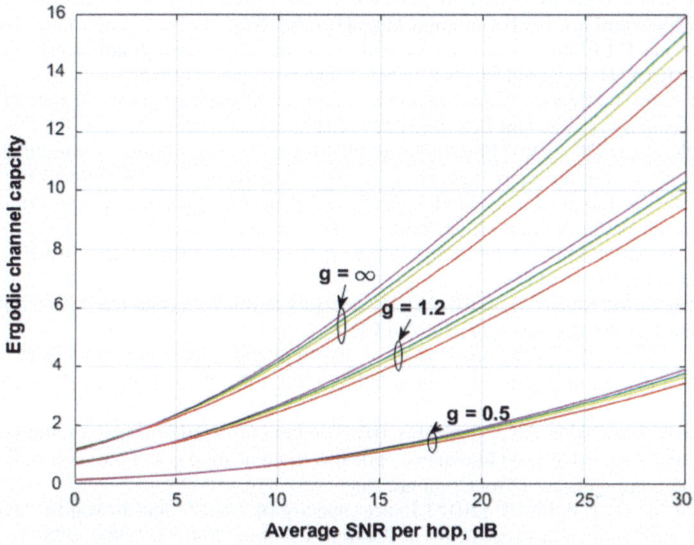

Fig. 5 Channel capacity of a relaying FSO network with pointing errors

$\rho = 0, 0.5, 0.75, 1$ and $g = 0.5, 1.2$. The cases of $\rho = 0$ and $\rho = 1$ correspond to the special cases of the K and the Gamma-Gamma distributions, respectively. From the figure, it is observed that the channel capacity improves with g parameter. The better channel capacity is obtained for $g = \infty$ (zero pointing error). Also, it is observed that the better channel capacity is obtained for $\rho = 1$ because the turbulence intensity, σ_I^2 of the FSO link, decreases with increase in ρ.

6 Conclusion

The outage and BER analysis of dual-hop FSO system using a variable gain AF protocol are analyzed in the presence of $\mathcal{M}$-distributed atmospheric turbulence and misalignment errors. The closed form expressions for the outage probability and the average BER of different modulation schemes are derived. It is analyzed that with increase in pointing errors, both the outage probability and the BER increase and thus degrade the system performance significantly.

Acknowledgements This publication was made possible by NPRP grant # [NPRP8-648-2-273] from the Qatar National Research Fund (QNRF) (a member of Qatar Foundation (QF)). The statements made herein are solely the responsibility of the authors

References

1. Majumdar AK, Ricklin JC (2010) Free-space laser communications: principles and advances. Springer Science & Business Media, Berlin
2. Kedar D, Arnon S (2004) Urban optical wireless communication networks: the main challenges and possible solutions. IEEE Commun Mag 42(5):S2–S7
3. Zhu X, Kahn JM (2002) Free-space optical communication through atmospheric turbulence channels. IEEE Trans Commun 50(8):1293–1300
4. Arnon S (2003) Effects of atmospheric turbulence and building sway on optical wireless-communication systems. Opt Lett 28(2):129–131
5. Borah DK, Voelz DG (2009) Pointing error effects on free-space optical communication links in the presence of atmospheric turbulence. J Lightwave Technol 27(18):3965–3973
6. Navidpour SM, Uysal M, Kavehrad M (2007) BER performance of free-space optical transmission with spatial diversity. IEEE Trans Wirel Commun 6(8)
7. Safari M, Uysal M (2008) Relay-assisted free-space optical communication. IEEE Trans Wirel Commun 7(12):5441–5449
8. Karimi M, Nasiri-Kenari M (2010) Outage analysis of relay-assisted free-space optical communications. IET Commun 4(12):1423–1432
9. Bhatnagar MR (2012) Average BER analysis of differential modulation in DF cooperative communication system over gamma-gamma fading FSO links. IEEE Commun Lett 16(8):1228–1231
10. Peppas KP, Stassinakis AN, Nistazakis HE, Tombras GS (2013) Capacity analysis of dual amplify-and-forward relayed free-space optical communication systems over turbulence channels with pointing errors. J Opt Commun Netw 5(9):1032–1042
11. Aggarwal M, Garg P, Puri P (2015) Exact capacity of amplify-and-forward relayed optical wireless communication systems. IEEE Photonics Technol Lett 27(8):903–906
12. Bhatnagar MR (2012) Performance analysis of decode-and-forward relaying in gamma-gamma fading channels. IEEE Photonics Technol Lett 24(7):545–547
13. Sharma PK, Garg P (2014) Bi-directional decode-XOR-forward relaying over$\mathcal{M}$-distributed free space optical links. IEEE Photonics Technol Lett 26(19):1916–1919
14. Jurado-Navas A, Garrido-Balsells JM, Paris JF, Puerta-Notario A (2011) A unifying statistical model for atmospheric optical scintillation. arXiv:1102.1915
15. Ansari IS, Yilmaz F, Alouini MS (2013) Impact of pointing errors on the performance of mixed RF/FSO dual-hop transmission systems. IEEE Wirel Commun Lett 2(3):351–354
16. Ansari IS, Yilmaz F, Alouini MS (2016) Performance analysis of free-space optical links over Málaga ($\mathcal{M}$) turbulence channels with pointing errors. IEEE Trans Wirel Commun 15(1):91–102

17. Jurado-Navas A, Balsells JMG, Paris JF, Castillo-Vzquez M, Puerta-Notario A (2011) General analytical expressions for the bit error rate of atmospheric optical communication systems. Opt Lett 36(20):4095–4097
18. Samimi H, Uysal M (2013) End-to-end performance of mixed RF/FSO transmission systems. J Opt Commun Netw 5(11):1139–1144
19. Ansari IS, Al-Ahmadi S, Yilmaz F, Alouini MS, Yanikomeroglu H (2011) A new formula for the BER of binary modulations with dual-branch selection over generalized-K composite fading channels. IEEE Trans Commun 59(10):2654–2658
20. Gradshteyn IS, Ryzhik IM (2000) Table of integrals, series, and products. Academic Press, San Diego
21. Arnon S, Barry J, Karagiannidis G (eds) (2012) Advanced optical wireless communication systems. Cambridge University Press, Cambridge
22. Adamchik VS, Marichev OI (1990) The algorithm for calculating integrals of hypergeometric type functions and its realization in reduce system. In: Proceedings of the international symposium on Symbolic and algebraic computation, ACM, pp 212–224

Performance Analysis of Relay Assisted Mixed Dual-Hop RF-FSO Systems with Pointing Errors

V. K. Jagadeesh, Vineeth Palliyembil, Imran Shafique Ansari, P. Muthuchidambaranathan and Khalid A. Qaraqe

Abstract In this work, a fixed-gain amplify-and-forward (AF)- relay assisted dual-hop RF-FSO communication system with pointing errors is analyzed by considering Rayleigh/Málaga ($\mathcal{M}$)- distributed turbulence channel. For this mixed RF-FSO cooperative system, a novel closed form analytical expressions are obtained for moment generating function (MGF), outage probability and average bit error rate (BER) for various binary and M-ary modulation techniques. Additionally, by using the asymptotic expansion of the Meijer's G function at the high SNR regime, we derive new asymptotic expressions for the CDF, outage probability, and the average BER in terms of simple elementary functions.

Keywords RF-FSO · Rayleigh fading · Málaga distribution · Amplify-and-forward relay · Bit error rate · Outage probability

1 Introduction

Wireless communication is so pervasive that it has even changed the way we live and radio frequency (RF) spectrum scarcity has become one of the major concerns in the area of wireless communications. Optical wireless communication (OWC) has been emerging as a promising alternative due to its advantages including higher bandwidth

V. K. Jagadeesh (✉)
Department of Electronics and Communication Engineering, NMAMIT, Mangalore, India
e-mail: vkjagadeesh@gmail.com

V. Palliyembil · P. Muthuchidambaranathan
Department of Electronics and Communication Engineering,
National Institute of Technology, Tiruchirappalli, India

I. S. Ansari · K. A. Qaraqe
Department of Electrical and Computer Engineering, Texas A & M University at Qatar (TAMUQ), Doha, Qatar

I. S. Ansari · K. A. Qaraqe
Education City, Ar-Rayyan, Qatar

A. Boyaci et al. (eds.), *International Telecommunications Conference*,
Lecture Notes in Electrical Engineering 504,
https://doi.org/10.1007/978-981-13-0408-8_2

and higher capacity compared to the traditional RF communication systems. The main advantages of free-space optical (FSO) communication systems include larger bandwidth, low power consumption, cost effectiveness, greater security, and license free spectrum [1]. In addition to all this, FSO links are highly secure and higher amount of data can be transferred over short distances with high speed through the use of lasers [1, 2]. The atmospheric turbulence and pointing errors degrade the performance of the FSO system, if the link range extends more than 1 km [3]. To minimize the effect of turbulence and pointing errors various methods are studied. Diversity [4–6] and relay techniques [7] are the main techniques to improve the performance of the system. Relay techniques provides wider and energy efficient coverage. The performance of relay systems were analyzed under different turbulence conditions with different modulation schemes [8–11]. Recently some research are concentrating on asymmetric relay system based on both RF and FSO technology [12–18] to solve the problems arising due to RF scarce spectrum.

In [13], the exact closed form for the outage probability of the RF-FSO relay link was analyzed using amplify-and-forward (AF) relay by taking RF as Rayleigh channel model and FSO as Gamma-Gamma distributed. The effect of pointing errors on the performance of RF-FSO system was analyzed in [14]. The authors derived exact closed form equations for the bit error rate (BER) and channel capacity in terms of Meijer's G function. In [15], the performance of a dual-hop RF-FSO link was analyzed using both fixed and variable AF relay in the presence of pointing errors. Here the RF link assumed as Nakagami- m and FSO link assumed as Gamma-Gamma distributed. The outage probability was analyzed in [16] using AF relay by assuming RF link as Rayleigh and FSO link as newly proposed Málaga ($\mathcal{M}$) distributed turbulence channel [19–21]. The multi-user AF relay based RF-FSO system was studied and analyzed under the effect of pointing errors in [17]. In [18, 22], the performance of a dual hop RF-FSO link was analyzed with fixed gain AF and DF relay using power series method. The RF link was taken as Nakagami-m and FSO as Gamma-Gamma distributed. In [23], exact analytical expressions of channel capacity are derived for fixed AF and DF relay mixed RF-FSO system.

2 Paper Contributions and Organization

In this paper, we analyze the end-to-end BER and outage performance of an asymmetric dual-hop RF-FSO system with fixed gain AF relay. The relay utilizes subcarrier intensity modulation (SIM) technique to get optical signals for retransmission from input RF signals. In the proposed system, RF link follows Rayleigh distribution and FSO link follows $\mathcal{M}$-distribution. The $\mathcal{M}$-distribution is a generalized channel as K-distribution, Gamma-Gamma and the negative exponential distribution can be derived from this. We derive the cumulative distribution function (CDF), probability density function (PDF), and moment generating function (MGF) of the asymmetric RF-FSO system, under the effect of pointing errors. Using these parameters, the analytical expressions are obtained for the outage probability, the average error rate for various binary and M-ary modulation techniques.

Remaining paper is organized as follows. The channel and system models are presented in Sect. 3. In Sect. 4, the channel characteristics of the system are presented. Section 5 explains the performance of RF-FSO system. The numerical results are discussed in Sect. 6. The paper is finally concluded in Sect. 7.

3 Channel and System Model

As the model is shown in Fig. 1, there is no direct link between the source and destination. The source (S) communicates with the destination (D) through the fixed gain AF relay (R). Source to relay (S-R) is based on RF technology and assumed as Rayleigh faded and relay to destination (R-D) is based on FSO technology and is modeled as $\mathcal{M}$ distributed. The relay converts the RF signal to optical using SIM method [24]. The final signal to noise ratio (SNR) can be expressed as $\gamma = \frac{\gamma_1 \gamma_2}{\gamma_1 + C}$, where γ_1 is the SNR corresponding to S-R hop and γ_2 is the corresponding SNR of R-D hop ($\mathcal{M}$-distributed). C is the fixed relay gain defined in [13]. The Rayleigh fading SNR follows the exponential distribution and its PDF is given as [25]

$$f_{\gamma_1}(\gamma_1) = \frac{1}{\bar{\gamma_1}} \exp\left(-\frac{\gamma_1}{\bar{\gamma_1}}\right), \tag{1}$$

where $\bar{\gamma_1}$ corresponds to the average SNR of Rayleigh fading channel. In $\mathcal{M}$-distributed fading, while modeling small scale fluctuations new scattering component U_S^C is coupled to already existing U_L (line of sight) component and U_S^G represents classic scatter field component independent of LOS contribution. Considering Ω represents average power of LOS component and represented as $\Omega = E[|U_L|^2]$. The average power of total scatter components is denoted by $2b_0 = E[|U_S^C|^2 + |U_S^G|^2]$. Every single scatter component has an average power and is given as $E[|U_S^C|^2] = \rho 2b_0$ and $E[|U_S^G|^2] = (1-\rho)2b_0$, where ρ is the amount of scattering power coupled to the LOS component and ranges from 0 to 1.

The $\mathcal{M}$-distributed fading SNR PDF with pointing errors is given by [20]

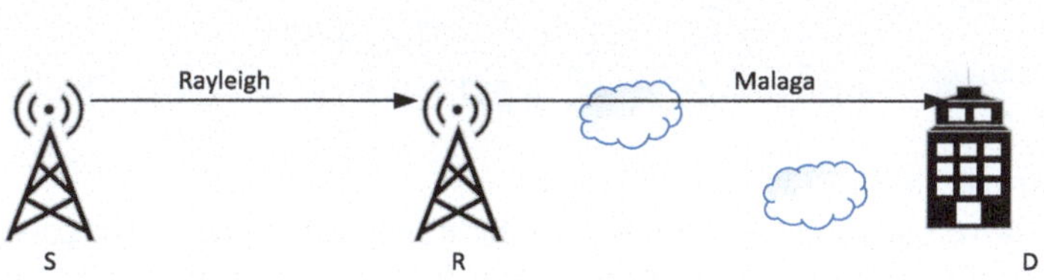

Fig. 1 Dual hop RF-FSO system with a source (S), relay (R), and a destination (D)

$$f_{\gamma_2}(\gamma_2) = \frac{g^2 A}{2^r \gamma_2} \sum_{m=1}^{\beta} b_m G_{1,3}^{3,0}\left(B\left(\frac{\gamma_2}{\mu_r}\right)^{\frac{1}{r}} \,\middle|\, \begin{matrix} g^2+1 \\ g^2, \alpha, m \end{matrix}\right), \tag{2}$$

where

$$B = \frac{g^2 \alpha \beta (\xi_g + \Omega^{'})}{[(g^2+1)(\xi_g \beta + \Omega^{'})]},$$

$$A = \frac{2\alpha^{\frac{\alpha}{2}}}{\xi_g^{1+\frac{\alpha}{2}} \Gamma(\alpha)} \left(\frac{\xi_g \beta}{\xi_g \beta + \Omega^{'}}\right)^{\beta + \frac{\alpha}{2}},$$

$$b_m = a_m \left(\frac{\alpha \beta}{\xi_g \beta + \Omega^{'}}\right)^{-\frac{\alpha+m}{2}},$$

$$a_m = \binom{\beta - 1}{m - 1} \frac{(\xi_g \beta + \Omega^{'})^{1-\frac{m}{2}}}{(m-1)!} \left(\frac{\Omega^{'}}{\xi_g}\right)^{k-1} \left(\frac{\alpha}{\beta}\right)^{\frac{m}{2}}.$$

When $g^2(\xi_g + \Omega^{'})/(g^2+1) = 1$, this PDF given in (2) comes in agreement with [16, Eq. (19)]

In Eq. (2), r is the parameter defining the type of detection technique (i.e. r = 1 represents heterodyne detection and r = 2 represents intensity modulation/direct detection (IM/DD)).

$$f_{\gamma_{heterodyne}}(\gamma) = \frac{g^2 A}{2\gamma_2} \sum_{m=1}^{\beta} b_m G_{1,3}^{3,0}\left(B \frac{\gamma}{\mu_{heterodyne}} \,\middle|\, \begin{matrix} g^2+1 \\ g^2, \alpha, m \end{matrix}\right), \tag{3}$$

where $\mu_{heterodyne}$ is the average SNR of (3), and

$$f_{\gamma_{IM/DD}}(\gamma) = \frac{g^2 A}{4\gamma_2} \sum_{m=1}^{\beta} b_m G_{1,3}^{3,0}\left(B\left(\frac{\gamma}{\mu_{IM/DD}}\right)^{\frac{1}{2}} \,\middle|\, \begin{matrix} g^2+1 \\ g^2, \alpha, m \end{matrix}\right), \tag{4}$$

where

$$\mu_{IM/DD} = \frac{g^2(g^2+1)^{-2}(g^2+2)(\xi_g + \Omega^{'})}{\alpha^{-1}(\alpha+1)[2\xi_g(\xi_g + 2\Omega^{'}) + \Omega^{'2}(1+1/\beta)]} \bar{\gamma}_{IM/DD},$$

is the electrical SNR of (4),

In the above equations α is related to the effective number of large-scale cells of the scattering process and the parameter for amount of fading is represented by β. In short, $E[|U_S^G|^2]$ is denoted as $\xi_g = (1-\rho)2b_0$. The parameter $\Omega^{'}$ is the average power from the coherent contributions and is given as $\Omega^{'} = \Omega + \rho 2b_0 + 2\sqrt{\rho 2b_0 \Omega} cos(\phi_A - \phi_B)$, ϕ_A and ϕ_B are the deterministic phases of the U_L and U_S^C, respectively, and g is the pointing error parameter [26].

4 Statistical Characteristics

4.1 Cumulative Distribution Function

Exact Cumulative Distribution Function The expression of the CDF for the AF based RF-FSO system can be given as

$$F_{\gamma}(\gamma) = P_r\left[\frac{\gamma_1\gamma_2}{\gamma_2 + C} < \gamma\right], \tag{5}$$

which can be written as

$$F_{\gamma}(\gamma) = \int_0^{\infty} P_r\left[\frac{\gamma_1\gamma_2}{\gamma_2 + C} < \gamma|\gamma_2\right] f_{\gamma_2}(\gamma_2)\mathrm{d}\gamma_2. \tag{6}$$

$$F_{\gamma}(\gamma) = \int_0^{\infty}\left(\int_0^{\gamma\left(1+\frac{C}{\gamma_2}\right)} f_{\gamma_1}(\gamma_1)\mathrm{d}\gamma_1\right) f_{\gamma_2}(\gamma_2)\mathrm{d}\gamma_2. \tag{7}$$

By substituting (1) and (2) in (7), leads to

$$\begin{aligned} F_{\gamma}(\gamma) = \quad & 1 - \frac{g^2 A}{2^r}\sum_{m=1}^{\beta} b_m \int_0^{\infty}\frac{1}{\gamma_2}\exp\left(-\frac{\gamma}{\bar{\gamma}_1}\right)\exp\left(\frac{-C\gamma}{\bar{\gamma}_1\gamma_2}\right) \\ & \times G_{1,3}^{3,0}\left(B\left(\frac{\gamma_2}{\mu_r}\right)^{\frac{1}{r}} \middle| \begin{matrix} g^2+1 \\ g^2, \alpha, m \end{matrix}\right)\mathrm{d}\gamma_2. \end{aligned} \tag{8}$$

The exponential term $\exp(\frac{-C\gamma}{\bar{\gamma}_1\gamma_2})$ can be written as $G_{0,1}^{1,0}\left(\frac{C\gamma}{\bar{\gamma}_1\gamma_2}\middle| \begin{matrix} - \\ 0 \end{matrix}\right)$ [27, Eq. (11)]. In addition to this, using [27, Eq. (21)], the CDF is obtained as

$$\begin{aligned} F_{\gamma}(\gamma) = 1 - \frac{g^2 A}{2^r(2\pi)^{r-1}}\exp\left(-\frac{\gamma}{\bar{\gamma}_1}\right)\sum_{m=1}^{\beta} b_m r^{\alpha+m-1} \\ G_{r,3r+1}^{3r+1,0}\left(\left(\frac{B}{r^2}\right)^r \frac{C\gamma}{\bar{\gamma}_1\mu_r}\middle| \begin{matrix} K_1 \\ K_2 \end{matrix}\right), \end{aligned} \tag{9}$$

where $K_1 = \frac{g^2+1}{r}, \ldots, \frac{g^2+r}{r}$ comprises of r terms and
$K_2 = \frac{g^2}{r}, \ldots, \frac{g^2+r-1}{r}, \frac{\alpha}{r}, \ldots, \frac{\alpha+r-1}{r}, \frac{m}{r}, \ldots, \frac{m+r-1}{r}, 0$ comprises of 3r terms.

Asymptotic Cumulative Distribution Function The arguments of the Meijer's G function in (9) can be inverted using [28, Eq. (6.2.2)] and applying [20, Eq. (41)], the asymptotic expression of the CDF at high SNR is derived in terms of basic elementary functions as

$$F_\gamma(\gamma) = 1 - \frac{g^2 A}{2^r (2\pi)^{r-1}} \exp\left(-\frac{\gamma}{\bar{\gamma}_1}\right) \sum_{m=1}^{\beta} b_m r^{\alpha+m-1}$$
$$\times \left\{ \sum_{k=1}^{3r+1} \left(\left(\frac{r^2}{B}\right)^r \frac{\bar{\gamma}_1 \mu_r}{C\gamma} \right)^{-K_{2,k}} \frac{\prod_{l=1;l\neq k}^{3r+1} \Gamma(K_{2,l} - K_{2,k})}{\prod_{l=1}^{r} \Gamma(K_{1,l} - K_{2,k})} \right\}. \tag{10}$$

4.2 Probability Density Function

By differentiating (9) with γ, utilizing product rule, and using [29, Eq. 07.34.20.0001.01], the closed form expression for the PDF is obtained as

$$f_\gamma(\gamma) = \frac{g^2 A}{2^r (2\pi)^{r-1}} \exp\left(\frac{-\gamma}{\bar{\gamma}_1}\right) \sum_{m=1}^{\beta} b_m r^{\alpha+m-1}$$
$$\left\{ \frac{1}{\bar{\gamma}_1} G_{r,3r+1}^{3r+1,0}\left(\left(\frac{B}{r^2}\right)^r \frac{C\gamma}{\bar{\gamma}_1 \mu_r} \,\middle|\, \begin{matrix} K_1 \\ K_2 \end{matrix} \right) - \left(\frac{1}{\gamma}\right)^{-1} G_{r+1,3r+2}^{3r+1,1}\left(\left(\frac{B}{r^2}\right)^r \frac{C\gamma}{\bar{\gamma}_1 \mu_r} \,\middle|\, \begin{matrix} 0, K_1 \\ K_2, 1 \end{matrix} \right) \right\}. \tag{11}$$

4.3 Moment Generating Function

Moment generating function is defined as [14]

$$M_\gamma(s) \cong E(e^{-\gamma s}), \tag{12}$$

which can be expressed in terms of the CDF as

$$M_\gamma(s) \cong s \int_0^\infty e^{-\gamma s} F_\gamma(\gamma) \mathrm{d}\gamma. \tag{13}$$

The MGF with pointing errors is derived by substituting (9) in (13) and the closed form expression is obtained as

$$M_{\gamma}(s) = 1 - s\left(s + \frac{1}{\bar{\gamma}_1}\right)^{-1} \frac{g^2 A}{2^r (2\pi)^{r-1}} \sum_{m=1}^{\beta} b_m r^{\alpha+m-1}$$
$$G_{r+1,3r+1}^{3r+1,1}\left(\left(\frac{B}{r^2}\right)^r \frac{C}{\bar{\mu}_r} \frac{1}{s\bar{\gamma}_1 + 1} \,\middle|\, \begin{matrix} 0, K_1 \\ K_2 \end{matrix}\right). \tag{14}$$

5 Performance of RF-FSO Cooperative System

In this section, the expressions are derived for the outage probability and average BER of the dual-hop asymmetric fixed gain AF relay based RF-FSO systems.

5.1 *Outage Probability*

Exact Outage Probability The outage probability (P_{out}) is defined as the probability that the instantaneous SNR, γ, falls below a specified threshold γ_{th}. The end-to-end exact outage probability for the system can be derived by using (9) as

$$P_{out} = 1 - \frac{g^2 A}{2^r (2\pi)^{r-1}} \exp\left(-\frac{\gamma_{th}}{\bar{\gamma}_1}\right) \sum_{m=1}^{\beta} b_m r^{\alpha+m-1}$$
$$G_{r,3r+1}^{3r+1,0}\left(\left(\frac{B}{r^2}\right)^r \frac{C\gamma_{th}}{\bar{\gamma}_1 \mu_r} \,\middle|\, \begin{matrix} K_1 \\ K_2 \end{matrix}\right). \tag{15}$$

Asymptotic Outage Probability By using (10), the asymptotic outage probability is expressed as

$$P_{out} = 1 - \frac{g^2 A}{2^r (2\pi)^{r-1}} \exp\left(-\frac{\gamma_{th}}{\bar{\gamma}_1}\right) \sum_{m=1}^{\beta} b_m r^{\alpha+m-1}$$
$$\times\left\{\sum_{k=1}^{3r+1}\left(\left(\frac{r^2}{B}\right)^r \frac{\bar{\gamma}_1 \mu_r}{C\gamma_{th}}\right)^{-K_{2,k}} \frac{\prod_{l=1;l\neq k}^{3r+1} \Gamma(K_{2,l} - K_{2,k})}{\prod_{l=1}^{r} \Gamma(K_{1,l} - K_{2,k})}\right\}. \tag{16}$$

Equation (16) is asymptotic in high SNR regime with respect to the FSO link i.e. μ_r.

5.2 Average Bit Error Rate

Exact Average Bit Error Rate The average BER of the proposed system can be expressed as [30]

$$P_e = \frac{q^p}{2\Gamma p} \int_0^{\infty} \exp(-q\gamma)\gamma^{p-1} F_\gamma(\gamma) \mathrm{d}\gamma. \tag{17}$$

On substituting (9) in (17) and utilizing [31, Eq. (7.813.1)], the average BER is derived as

$$P_e = \frac{1}{2} - \frac{q^p A g^2}{2^{2r}(\pi)^{r-1}\Gamma(p)} \sum_{m=1}^{\beta} b_m r^{\alpha+m-1} \left(q + \frac{1}{\bar{\gamma}_1}\right)^{-p}$$
$$G_{r+1,3r+1}^{3r+1,1}\left(\left(\frac{B}{r^2}\right)^r \frac{C}{\mu_r} \frac{1}{q\bar{\gamma}_1 + 1} \,\middle|\, \begin{matrix} 1-p, K_1 \\ K_2 \end{matrix}\right). \tag{18}$$

Asymptotic Average Bit Error Rate The arguments of the Meijer's G function in (18) can be inverted using [28, Eq. (6.2.2)] and applying [20, Eq. (41)], the asymptotic expression of the BER at high SNR is derived in terms of basic elementary functions as

$$P_e = \frac{1}{2} - \frac{q^p A g^2}{2^{2r}(\pi)^{r-1}\Gamma(p)} \sum_{m=1}^{\beta} b_m r^{\alpha+m-1} \left(q + \frac{1}{\bar{\gamma}_1}\right)^{-p}$$
$$\times \left\{ \sum_{k=1}^{3r+1} \left(\left(\frac{r^2}{B}\right)^r \frac{\mu_r(q\bar{\gamma}_1 + 1)}{C}\right)^{-K_{2,k}} \right.$$
$$\left. \frac{\left(\prod_{l=1;l\neq k}^{3r+1} \Gamma(K_{2,l} - K_{2,k})\right)\Gamma(p + K_{2,k})}{\prod_{l=2}^{r+1} \Gamma(K_{1,l} - K_{2,k})} \right\}. \tag{19}$$

Equation (19) is asymptotic in high SNR regime with respect to the FSO link i.e. μ_r.

In the above equation p and q are the parameters corresponding to the four modulation schemes as given in Table 1 [14].

Table 1 Parameters (p and q) for various binary modulation techniques

Type of modulation	p	q
Coherent binary frequency shift keying (CBFSK)	0.5	0.5
Non-coherent binary frequency shift keying (NBFSK)	1	0.5
Coherent binary phase shift keying (CBPSK)	0.5	1
Differential binary phase shift keying (DBPSK)	1	1

6 Numerical Results and Discussion

In this section, the numerical results for the performance of the average BER and outage probability of the proposed RF-FSO system under different turbulence conditions is presented. Here we considered fixed gain AF relay based dual-hop asymmetric RF-FSO communication system. The RF link experiences Rayleigh fading, and the FSO link experiences $\mathcal{M}$-distributed turbulence channel. In all cases, the transmitted power is normalized as $\Omega + 2b_0 = 1$. The link distance is taken as 1 km, wavelength, $\lambda = 785nm$, and optical wave number $k = 2\pi/\lambda$. The refraction structure parameter C_n^2 is considered as $C_n^2 = 2.8 \times 10^{-14} m^{-2/3}$. Using these parameters the Rytov variance can be calculated as $\sigma_l^2 = 1.23 C_n^2 k^{7/6} L^{11/6}$. The channel parameters α and β are obtained as 8 and 4 [32]. The relay gain factor $C = 1$. The average and electrical SNR of both the links are assumed to be equal.

In Figs. 2 and 3, the outage probability performance of mixed RF-FSO transmission system is investigated under the combined effects of the turbulence and pointing errors. Figure 2 shows the outage performance of the system for both IM/DD and heterodyne detection schemes with different value of $\rho = 0, 0.5, 1$ with fixed $g = 1$. The corresponding scintillation index is calculated using the equation in [33] and are 1.2, 0.98, 0.32. It is seen that the scintillation index parameter decreases with increasing ρ value. So, from the figure it is observed that the outage performance is

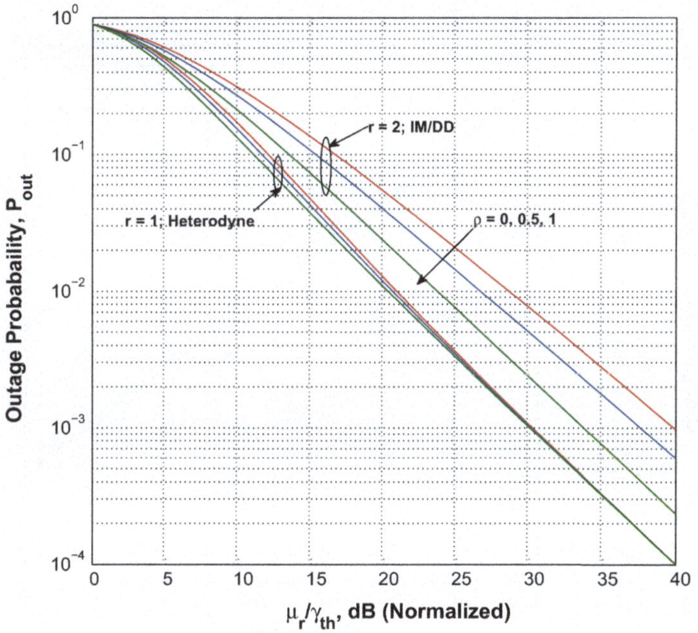

Fig. 2 Outage probability versus normalized SNR per hop for different values of ρ with $g = 1.2$

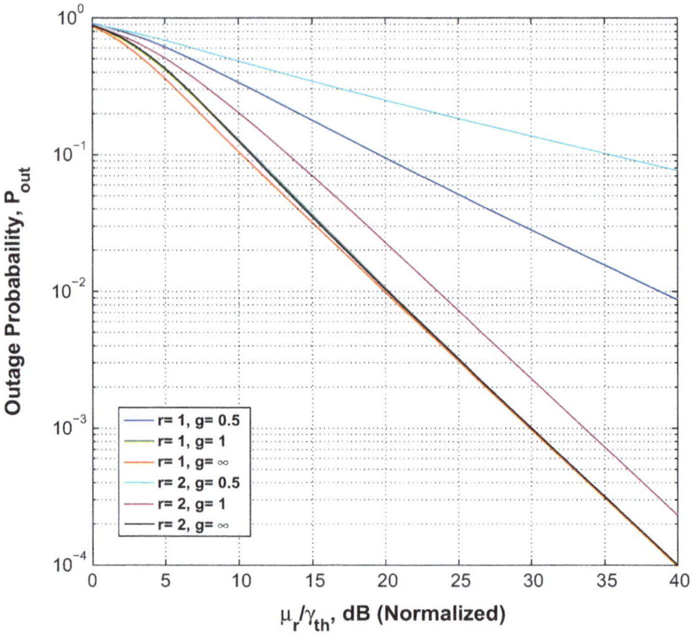

Fig. 3 Outage probability versus normalized SNR per hop for different values of g with $\rho = 1$

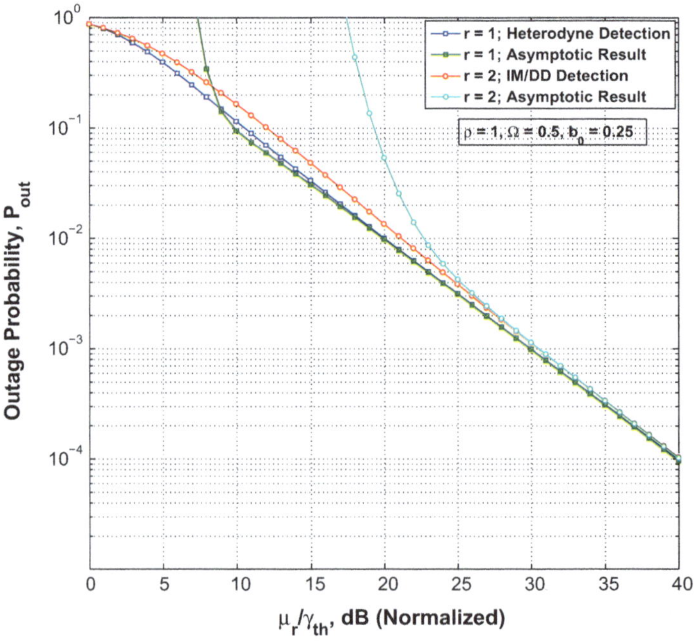

Fig. 4 Exact and asymptotic outage performance with $\rho = 1$ and $g = 1.2$

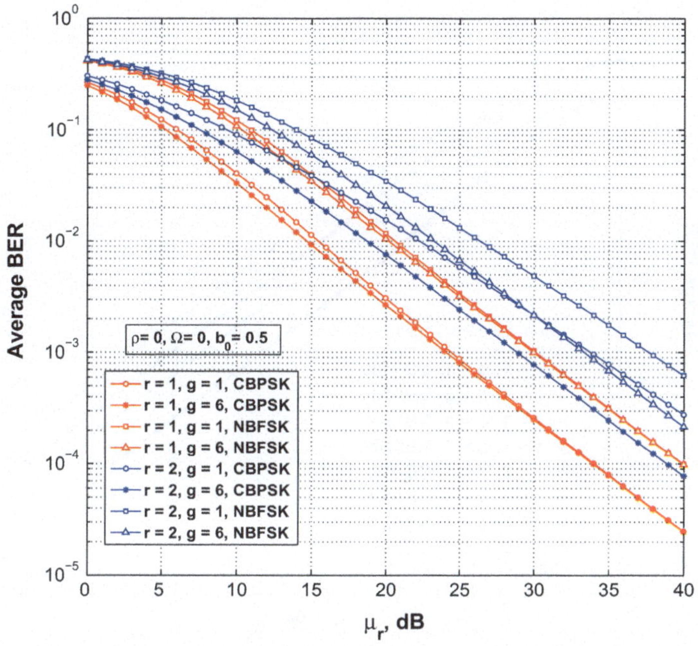

Fig. 5 Average BER versus average SNR per hop of different modulations for $\alpha = 8, \beta = 4, \rho = 0,$ $\Omega = 0, b_0 = 0.5$ with different pointing error parameters; $g = 1$ and 6

getting improved with ρ value. Figure 3 shows the outage performance of the system for varying pointing error parameter, $g = 0.5, 1, \infty$ with constant ρ parameter. With the increase in g, it is noted that there is an improvement in the outage performance of the system. As g parameter increases, the impact of g on pointing error decreases because the definition of g is the ration between equivalent beam radius at the receiver and the pointing error displacement standard deviation at the receiver. Figure 4 shows the comparison between the exact outage probability and the asymptotic probability for AF relay based RF-FSO system. From the figure it is observed that at high SNR, the asymptotic expression converges to the exact outage probability.

In Figs. 5 and 6, the BER performance of the RF-FSO system for different modulation schemes are analyzed for $\rho = 0$ and $\rho = 1$ with different values of $g = 1$ and 6. Here we consider both types of detection schemes for the analysis. For convenience, only the CBPSK and NBFSK are demonstrated in plot. From the figures, it is observed that the CBPSK demonstrates better performance than NBFSK because it is the power efficient modulation technique. Comparing with Figs. 5 and 6 shows better BER performance. This is because when $\rho = 1$ the overall scattering power is entirely coupled to LOS component and reduces its adverse effects. Figure 7 shows the comparison between the exact BER performance and the asymptotic BER performance of the system. In this case, we consider only CBPSK modulation scheme.

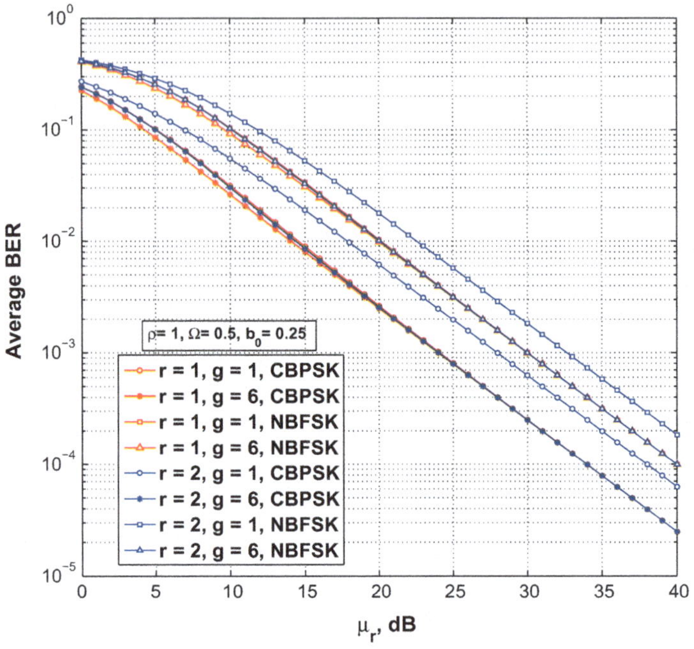

Fig. 6 Average BER versus average SNR per hop of different modulations for $\alpha = 8, \beta = 4, \rho = 1$, $\Omega = 0.5$, $b_0 = 0.25$ with different pointing error parameters; $g = 1$ and 6

From the figure it is observed that at high SNR, the asymptotic expression converges to the exact BER performance.

7 Conclusion

In this paper, the mathematical expressions have been derived for the CDF, the PDF, and the MGF of the equivalent SNR of a fixed gain AF-based mixed dual hop RF-FSO system. Using these parameters, exact analytical expressions are derived for the outage probability and the BER for the proposed system and the performance is analyzed. Here the RF link is considered as Rayleigh distributed and FSO is considered as $\mathcal{M}$ distributed. Furthermore, the effect of fading, turbulence, jitter, and pointing error has been investigated on the performance of the proposed RF-FSO link. In addition, the asymptotic expressions for the CDF, outage probability, and the average BER are derived. Furthermore, the asymptotic outage and BER performance for RF-FSO system is analyzed.

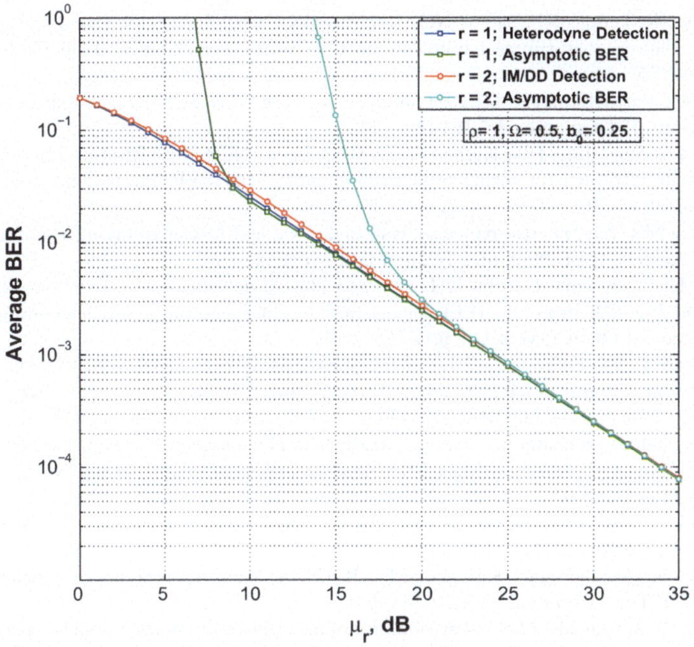

Fig. 7 Exact and asymptotic BER performance of CBPSK modulation for $g = 1.2$

Acknowledgements This publication was made possible by NPRP grant # [NPRP8-648-2-273] from the Qatar National Research Fund (QNRF) (a member of Qatar Foundation (QF)). The statements made herein are solely the responsibility of the authors

References

1. Majumdar AK, Ricklin JC (2010) Free-space laser communications: principles and advances. Springer Science & Business Media, Berlin, 5 May 2010
2. Kedar D, Arnon S (2004) Urban optical wireless communication networks: the main challenges and possible solutions. IEEE Commun Mag 42(5):S2–S7
3. Arnon S (2003) Effects of atmospheric turbulence and building sway on optical wireless-communication systems. Opt Lett 28(2):129–131
4. Tsiftsis TA, Sandalidis HG, Karagiannidis GK, Uysal M (2009) Optical wireless links with spatial diversity over strong atmospheric turbulence channels. IEEE Trans Wirel Commun 8(2):951–957
5. Navidpour SM, Uysal M, Kavehrad M (2007) BER performance of free-space optical transmission with spatial diversity. IEEE Trans Wirel Commun 6(8)
6. Wang Z, Zhong WD, Fu S, Lin C (2009) Performance comparison of different modulation formats over free-space optical (FSO) turbulence links with space diversity reception technique. IEEE Photonics J 1(6):277–285
7. Safari M, Uysal M (2008) Relay-assisted free-space optical communication. IEEE Trans Wirel Commun 7(12):5441–5449

8. Datsikas CK, Peppas KP, Sagias NC, Tombras GS (2010) Serial free-space optical relaying communications over gamma-gamma atmospheric turbulence channels. J Opt Commun Netw 2(8):576–586
9. Karimi M, Nasiri-Kenari M (2010) Outage analysis of relay-assisted free-space optical communications. IET Commun 4(12):1423–1432
10. Tsiftsis TA, Sandalidis HG, Karagiannidis GK, Sagias NC (2006) Multihop free-space optical communications over strong turbulence channels. In: IEEE international conference on communications, ICC 2006, vol. 6, pp 2755–2759
11. Bhatnagar MR (2012) Performance analysis of decode-and-forward relaying in gamma-gamma fading channels. IEEE Photonics Technol Lett 24(7):545–547
12. Moradi H, Falahpour M, Refai HH, LoPresti PG, Atiquzzaman M (2010) On the capacity of hybrid FSO/RF links. In: IEEE international conference on global telecommunications conference (GLOBECOM 2010), pp 2755–2759
13. Lee E, Park J, Han D, Yoon G (2011) Performance analysis of the asymmetric dual-hop relay transmission with mixed RF/FSO links. IEEE Photonics Technol Lett 23(21):1642–1644
14. Ansari IS, Yilmaz F, Alouini MS (2013) Impact of pointing errors on the performance of mixed RF/FSO dual-hop transmission systems. IEEE Wirel Commun Lett 2(3):351–354
15. Zedini E, Ansari IS, Alouini MS (2015) Performance analysis of mixed Nakagami-m and gammagamma dual-hop FSO transmission systems. IEEE Photonics J 7(1):1–20
16. Samimi H, Uysal M (2013) End-to-end performance of mixed RF/FSO transmission systems. J Opt Commun Netw 5(11):1139–1144
17. Miridakis NI, Matthaiou M, Karagiannidis GK (2014) Multiuser relaying over mixed RF/FSO links. IEEE Trans Commun 62(5):1634–1645
18. Anees S, Bhatnagar MR (2015) Performance of an amplify-and-forward dual-hop asymmetric RFFSO communication system. J Opt Commun Netw 7(2):124–135
19. Jurado-Navas A, Garrido-Balsells JM, Paris JF, Puerta-Notario A (2011) A unifying statistical model for atmospheric optical scintillation. arXiv:1102.1915
20. Ansari IS, Yilmaz F, Alouini MS (2016) Performance analysis of free-space optical links over Málaga ($\mathcal{M}$) turbulence channels with pointing errors. IEEE Trans Wirel Commun 15(1):91–102
21. Jurado-Navas A, Garrido-Balsells JM, Paris JF, Castillo-Vzquez M, Puerta-Notario A (2012) Impact of pointing errors on the performance of generalized atmospheric optical channels. Opt Express 20(11):12550–12562
22. Anees S, Bhatnagar MR (2015) Performance evaluation of decode-and-forward dual-hop asymmetric radio frequency-free space optical communication system. IET Optoelectron 9(5):232–240
23. Vellakudiyan J, Ansari IS, Palliyembil V, Muthuchidambaranathan P, Qaraqe KA (2016) Channel capacity analysis of a mixed dual-hop radio-frequencyfree space optical transmission system with Málaga distribution. IET Commun 10(16):2119–2124
24. Popoola WO, Ghassemlooy Z (2009) BPSK subcarrier intensity modulated free-space optical communications in atmospheric turbulence. J Lightwave Technol 27(8):967–973
25. Simon MK, Alouini MS (2005) Digital communication over fading channels, vol 95. Wiley, New York
26. Borah DK, Voelz DG (2009) Pointing error effects on free-space optical communication links in the presence of atmospheric turbulence. J Lightwave Technol 27(18):3965–3973
27. Adamchik VS, Marichev OI (1990) The algorithm for calculating integrals of hypergeometric type functions and its realization in reduce system. In: Proceedings of the international symposium on symbolic and algebraic computation, ACM, pp 212–224
28. Springer MD (1979) The algebra of random variables. Wiley, New York (No. 519.2 S67)
29. Wolfram I (2010) Mathematica edition: version 8.0. Wolfram Research Inc., Illinois, Champaign
30. Ansari IS, Al-Ahmadi S, Yilmaz F, Alouini MS, Yanikomeroglu H (2011) A new formula for the BER of binary modulations with dual-branch selection over generalized-K composite fading channels. IEEE Trans Commun 59(10):2654–2658

31. Gradshteyn IS, Ryzhik IM (2000) Table of integrals, series, and products. Academic press Inc., San Diego
32. Jurado-Navas A, Balsells JMG, Paris JF, Castillo-Vzquez M, Puerta-Notario A (2011) General analytical expressions for the bit error rate of atmospheric optical communication systems. Opt Lett 36(20):4095–4097
33. Al-Habash MA, Andrews LC, Phillips RL (2001) Mathematical model for the irradiance probability density function of a laser beam propagating through turbulent media. Opt Eng 40(8):1554–1562

Coverage Probability Analysis by Fractional Frequency Reuse Scheme

Joydev Ghosh, Dushanta Nalin K. Jayakody, Marwa Qaraqe and Theodoros A. Tsiftsis

Abstract In this work, we present the cell edge coverage probability (CECP) performance of cellular networks under the composite multi-path fading environment, where Rayleigh fading is superimposed on log-normal shadowing, by the fractional frequency reuse (FFR) scheme. We demonstrate that our analytical results and the simulation results are in line with the analysis presented in Novlan et al. (IEEE Trans Wirel Commun 10(12):4294–4305, 2011, [12]). In addition, we highlight that the superiority of the proposed analytical results over Novlan et al. (IEEE Trans Wirel Commun 10(12):4294–4305, 2011, [12]) in relation to the computational complexity.

Keywords Cell Edge Coverage Probability (CECP) · Fractional Frequency Reuse (FFR) Scheme · Composite Multi-path Fading

1 Introduction

Drastically growing energy costs and rising rigid scenario standards have led to an increased demand of addressing the energy efficiency (EE) aspect of next generation networks [1]. In early wireless networks, the radio access portion consumes more

J. Ghosh · D. N. K. Jayakody
School of Computer Science and Robotics, National Research Tomsk Polytechnic University, Tomsk, Russia
e-mail: joydev.ghosh.ece@gmail.com

D. N. K. Jayakody · M. Qaraqe (✉)
College of Science and Engineering Hamad Bin Khalifa University, Qatar Foundation, Education City, Doha, Qatar
e-mail: mqaraqe@hbku.edu.qa

D. N. K. Jayakody
e-mail: nalin.jayakody@ieee.org

T. A. Tsiftsis
School of Engineering, Nazarbayev University, Astana, Kazakhstan
e-mail: theodoros.tsiftsis@nu.edu.kz

A. Boyaci et al. (eds.), *International Telecommunications Conference*, Lecture Notes in Electrical Engineering 504,
https://doi.org/10.1007/978-981-13-0408-8_3

than 2/3 of the total energy consumption [2]. Hence, improving the energy efficiency (EE) of wireless networks is very important in order to fulfill the constraint issues that arise by the large requirement of traffic and energy consumption. The femtocell is an emerging and non-discriminatory issue in current and next generation radio access networks [3]. Owing to its small transmission path between transmitter and receiver and low transmit power feature, the femtocell can reduce the consumption of energy, increase the battery life of low-power smart devices, and increase the network coverage in terms of the quality of experience (QoE) [4]. As of today, a great deal of work has been dedicated to resource distribution and interference reduction techniques in future generation networks with small cell. The majority of the work articulate on the orthogonal frequency division multiple access (OFDMA) resource allocations [5] and sub-channel allocations [6] to reduce the cross tier interference. The fractional frequency reuse (FFR) scheme is considered as one of the efficient interference mitigation techniques in cellular networks. This can reduces cross-tier interference and optimize spatial reuse. In the the conventional approach, where the $FFR = 1$, the entire frequencies to each macrocell provides large spatial reuse, but less coverage. In the this approach, where the $FRF = 3$, (1/3) of whole spectrum resources in individual macrocell can provide a less spatial reuse, but large coverage. We treat coverage probability (CP) as our performance metric. This can be defined as the probability that the achieved signal to interference plus noise ratio (SINR) higher than the required threshold for better quality of experience (QoE) [7]. Coverage probability is an important parameter to compute the communication reliability. Below, we outline the contribution given in this work.

1. We analytically derive the expressions for the CECP metric of a FFR scheme by explicitly considering the independent composite multi-path fading power for inner cell users (ICUs) and outer cell users (OCUs). We extend the analysis to obtain an expression for the CP metric of this network.
2. We show that the analytical results for the CECP converge to the corresponding simulation for all values of path loss exponents, α (Fig. 1).

2 System Model

We consider OFDMA based multi-cellular DL wireless networks where the macro base stations (MBSs) are distributed as a spatial Poisson point process (PPP) and users are distributed uniformly [8]. For the sake of simplicity of analysis, we assume k number of interfering MBSs with density ρ. The SINR of a user associated to the macrocell centre can be expressed as [9]:

$$\gamma_{cc} = \frac{P_0 h_0 ||d_0||^{-\alpha}}{\sum_{j \in k} P_j \tau_j h_j ||D_j||^{-\alpha} + N_0}, \tag{1}$$

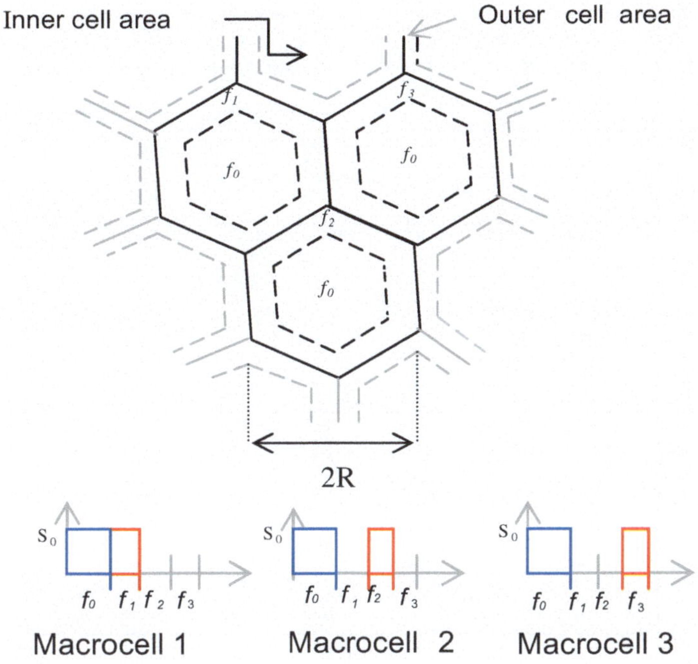

Fig. 1 FFR with three macrocell clusters in size

where P_0 indicates transmit power by the reference user located at random position (x_0, y_0), h_0 and α indicates the composite multi-path fading and path loss exponent, respectively. The probability density function (PDF) of the SINR can be expressed as in [10]

$$p_{\gamma}\gamma = \int_0^{\infty} \frac{1}{\bar{\gamma}\Gamma(1)} \exp\left(-\frac{\gamma}{\bar{\gamma}}\right) \frac{\xi}{\sqrt{2\pi}\sigma\gamma} \exp\left[-\frac{(10\log_{10}\gamma - \mu)^2}{2\sigma^2}\right] d\gamma, \qquad (2)$$

where $\bar{\gamma}$ denotes the average SINR per symbol, γ is distributed according to the gamma distribution $\Gamma(1) = \int_0^{\infty} e^{-t} dt$, $\xi = 10/ln10 = 4.343$, μ(dB), and σ(dB) are the mean and standard deviation of $10\log_{10}\gamma$ respectively. $P_j\tau_j h_j ||D_j||^{-\alpha}$, $\forall j = 0, 1, 2, ..., J$ stands for the received power from a base station (BS) positioned at (x_j, y_j). $||D_j|| = R\sqrt{3}\sqrt{(x_j - \phi x_0)^2 + (y_j - \phi y_0)^2 + (x_j - \phi x_0)(y_j - \phi y_0)}$ indicates jth BS and the user positioned at (x_0, y_0) where $\phi = \frac{R_t}{R} \in [0, 1]$ indicates dividing condition between middle and edge of the macrocell. $\tau_j = 0, 1$ represents an indicator parameter to find the status, i.e, inactive or busy, of each channel and N_0 is the additive white Gaussian noise (AWGN) power [11].

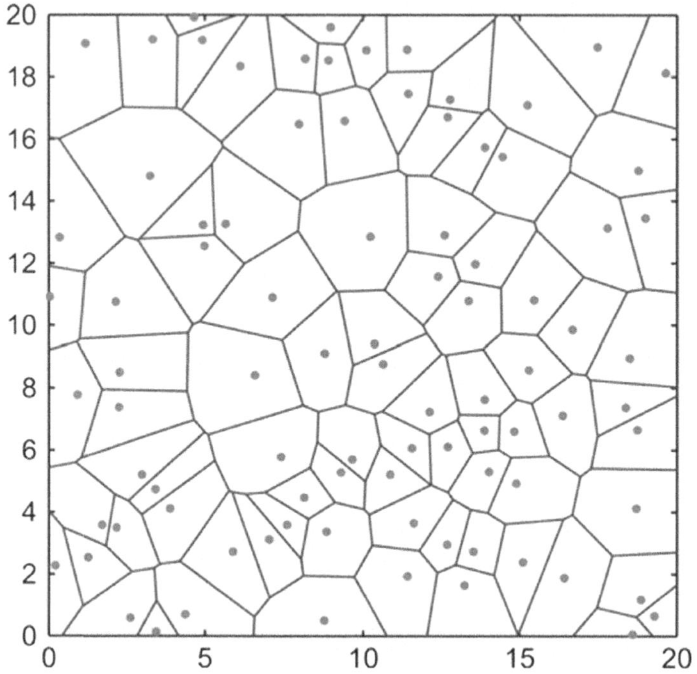

Fig. 2 FBSs are distribute as spatial PPP and femtocell boundaries conform to a Voronoi tessellation

2.1 Cell Edge Coverage Probability (CECP)

In this subsection, we obtain the CECP analytically using FFR scheme. The CECP has already been analyzed in [12]. However, we will show that our analytical expressions are computationally inexpensive as compared to that of [12]. Also, it converges to the simulation outcomes in [12]. The CECP of a mobile user at a separation d_0 from the MBS in FFR scheme can be expressed as $P_{ce}(d_0) = \left[\gamma_{ce}(d_0) > \gamma_{trg} | \gamma_{cc}(d_0) < \gamma_{th}\right]$. Thus we have

$$P_{ce}(d_0) = P\left[\frac{P_0 \bar{h}_0 ||d_0||^{-\alpha}}{\sum_{j\in m} P_j \tau_j \bar{h}_j ||D_j||^{-\alpha} + N_0} > \gamma_{trg} \,|\, \frac{P_0 h_0 ||d_0||^{-\alpha}}{\sum_{j\in m} P_j \tau_j h_j ||D_j||^{-\alpha} + N_0} < \gamma_{th}\right], \tag{3}$$

where, $\gamma_{ce}(d_0)$ and $\gamma_{cc}(d_0)$ indicate the SINR of a user located at outer cell area (OCA) and inner cell area (ICA), respectively, $\bar{h}_0$ and $\bar{h}_j$ indicate composite multipath fading experienced by the users located at OCA, m is the number of interfering MBS at cell edge, γ_{trg} and γ_{th} indicate SINR threshold and target SINR, respectively. Here, (s) signifies that the SINR of a user located at cell edge is less than γ_{th}, i.e., any user whose SINR is lower than γ_{th} is to be considered as a cell edge user. Then allocates a resource to them by FFR scheme. By considering that the channel fading power is independent and identically distributed random variables between the sub-bands, we get

$$P_{ce}(d_0) = P\left[\gamma_{ce}(d_0) > \gamma_{trg} \mid \gamma_{cc}(d_0) < \gamma_{th}\right] = P\left[\gamma_{ce}(d_0) > \gamma_{trg}\right]. \tag{4}$$

The CECP of a typical user can be expressed as:

$$P_{ce} = \frac{\int_0^\infty P\left[\gamma_{ce}(d_0) > \gamma_{trg} \mid \gamma_{cc}(d_0) < \gamma_{th}\right] P\left[\gamma_{cc}(d_0) < \gamma_{th}\right] p_\gamma(d_0) dd_0}{\int_0^\infty [\gamma_{cc}(d_0) < \gamma_{th}] p_\gamma(d_0) dd_0}, \tag{5}$$

where, $p_\gamma(d_0)$ indicates the probability density function (PDF) of distance, d_0, between cell user and nearest FBS, and it can be expressed as $p_\gamma(d_0) = 2\pi\rho\, d_0 e^{-\rho\pi d_0^2}, d_0 > 0$ [13]. Substituting (4) in (5), we get:

$$P_{ce} = \frac{\int_0^\infty P\left[\gamma_{ce}(d_0) > \gamma_{trg}\right] P\left[\gamma_{cc}(d_0) < \gamma_{th}\right] p_\gamma(d_0) dd_0}{\int_0^\infty [\gamma_{cc}(d_0) < \gamma_{th}] p_\gamma(d_0) dd_0}, \tag{6}$$

In [13], it has already been expressed:

$$P[\gamma_{cc}(d_0) > \gamma_{trg}] = \exp\left[-\pi\rho d_0^2 p(\gamma_{trg}, \alpha, \delta_{ICA}) - \gamma_{trg} N_0 d_0^\alpha\right], \tag{7}$$

$$P[\gamma_{cc}(d_0) > \gamma_{trg}] = \exp\left[-\pi\rho d_0^2 p(\gamma_{trg}, \alpha, \delta_{OCA}) - \gamma_{trg} N_0 d_0^\alpha\right], \tag{8}$$

where,

$$\begin{aligned} p(\gamma_{trg}, \alpha, \delta_{OCA}) &= \gamma_{trg}^{2} \alpha \int_{\gamma_{trg}^{-\frac{2}{\alpha}}}^{\infty} \frac{1}{u^{\frac{\alpha}{2}} + 1} du \\ &= \frac{2\gamma_{trg}}{\delta_{OCA}(\alpha - 2)} {}_2F_1(1, (\alpha-2)/\alpha, 2 - 2/\alpha, -\gamma_{trg}), \end{aligned} \tag{9}$$

where, ${}_2F_1(1, (\alpha-2)/\alpha, 2-2/\alpha, -\gamma_{trg})$ denotes the Gauss hypergeometric function [14]. Hence, P_{ce} can be further re-expressed as:

$$P_{ce} = \frac{\pi\rho \int_0^\infty e^{-\pi p d_0} e^{-\pi p d_0 p(\gamma_{trg}, \alpha, \delta_{OCA}) - \gamma_{trg} N_0 d_0^{\frac{\alpha}{2}}} \left(1 - e^{-\pi p d_0 p(\gamma_{th}, \alpha, \delta_{ICA}) - \gamma_{th} N_0 d_0^{\frac{\alpha}{2}}}\right) dd_0}{1 - \pi\rho \int_0^\infty e^{-\pi p d_0 (1 + p(\gamma_{th}, \alpha, \delta_{ICA})) - \gamma_{th} N_0 d_0^{\frac{\alpha}{2}}} dd_0}. \tag{10}$$

The respective derivation in [12] for CECP of a user in a FFR scheme can be expressed as:

$$P_{ce,[12]} = \frac{\pi\rho\int_0^{\infty} e^{-\pi p d_0(1+p(\gamma_{trg},\alpha,\delta_{OCA}))-\gamma_{trg}N_0 d_0^{\frac{\alpha}{2}}}dd_0}{1-\pi\rho\int_0^{\infty} e^{-\pi p d_0(1+p(\gamma_{th},\alpha,\delta_{ICA}))-\gamma_{th}N_0 d_0^{\frac{\alpha}{2}}}dd_0} - \frac{\pi\rho\int_0^{\infty} e^{-\pi p d_0} e^{-\pi p d_0 2\xi(\gamma_{trg},\gamma_{th},\alpha,\delta_{OCA})-\gamma_{trg}N_0 d_0^{\frac{\alpha}{2}}}dd_0}{1-\pi\rho\int_0^{\infty} e^{-\pi p d_0(1+p(\gamma_{th},\alpha,\delta_{ICA}))-\gamma_{th}N_0 d_0^{\frac{\alpha}{2}}}dd_0}, \quad (11)$$

where, $\xi(\gamma_{trg}, \gamma_{th}, \alpha, \delta_{OCA}) = \int_1^{\alpha}\left[1-\frac{1}{1+\gamma_{th}r^{-\alpha}}\left(1-\frac{1}{\delta_{OCA}}(1-\frac{1}{1+\gamma_{trg}r^{-\alpha}})\right)\right]dr$. In the noise free scenario, we can again re-write P_{ce} by solving integrals of (10) as:

$$P_{ce} = \frac{1+p(\gamma_{th},\alpha,\delta_{OCA})}{(1+p(\gamma_{trg},\alpha,\delta_{OCA}))(1+p(\gamma_{trg},\alpha,\delta_{OCA})+p(\gamma_{th},\alpha,\delta_{ICA}))}. \quad (12)$$

In the noise free scenario, the respective expressions in [12] can be re-written as:

$$P_{(ce,[12])} = \frac{1+p(\gamma_{th},\alpha,\delta_{ICA})}{p(\gamma_{th},\alpha,\delta_{ICA})}\left(\frac{1}{p(\gamma_{trg},\alpha,\delta_{OCA})} - \frac{1}{1+2\xi(\gamma_{trg},\gamma_{th},\alpha,\delta_{OCA})}\right) \quad (13)$$

At $\alpha = 4$, P_{ce} can be again expressed by:

$$P_{ce} = \frac{1+\sqrt{\gamma_{th}}arctan(\gamma_{th})}{\left(1+(\frac{1}{\delta})\sqrt{\gamma_{trg}}arctan(\gamma_{trg})\right)\left(1+(\frac{1}{\delta})\right)\sqrt{\gamma_{trg}}arctan(\gamma_{trg})+\sqrt{\gamma_{th}}arctan(\gamma_{th})} \quad (14)$$

Further, the respective expression in [12] can be written by:

$$P_{(ce,[12])} = \frac{1+\sqrt{\gamma_{th}}arctan(\gamma_{th})}{\sqrt{\gamma_{th}}arctan(\gamma_{th})}\times \left(\frac{1}{1+(\frac{1}{\delta})\sqrt{\gamma_{trg}}arctan(\gamma_{trg})} - \frac{1}{1+2\xi(\gamma_{trg},\gamma_{th},4,\delta_{OCA})}\right). \quad (15)$$

To find the coverage probability (CP) of the FFR scheme, we initially denote CP of a particular CU by P_{cp}. Thus,

$$P_{cp} = \int_{d_0>0}(P_{cc}P[\gamma_{ce}(d_0) > \gamma_{th}] + P_{ce}P[\gamma_{ce}(d_0) < \gamma_{th}])p_{\gamma}(d_0)dd_0, \quad (16)$$

where, the term $P_{cc}P[\gamma_{ce}(d_0) > \gamma_{trg}]$ stands for the CP due to ICUs and the another term $P_{ce}P[\gamma_{ce}(d_0) < \gamma_{trg}]$ stands for the CP due to OCUs. The CP at cell centre can be written as in [12, Theorem 2]:

$$P_{cc} = \frac{P[\gamma_{ce}(d_0) > max(\gamma_{trg},\gamma_{th})]}{P[\gamma_{ce}(d_0) > \gamma_{th}]}. \quad (17)$$

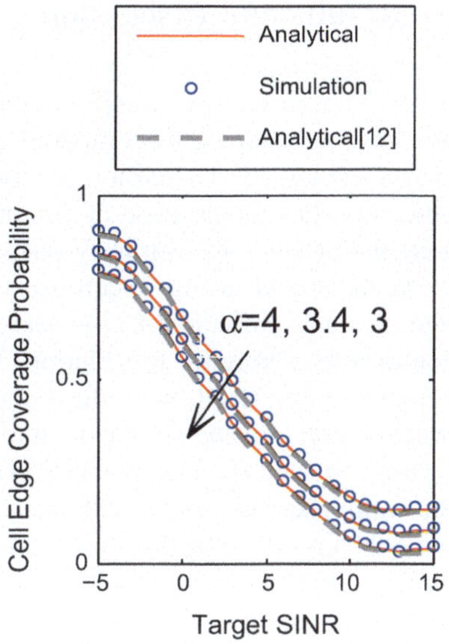

Fig. 3 Contrast of our analytical results, results provided in [12] and simulations for $\alpha = 3, 3.4, 4$, where, $\delta_{OCA} = 3$, $\gamma_{th} = 1\,\text{dB}$, $\rho = 0.3\,\text{Km}^2$

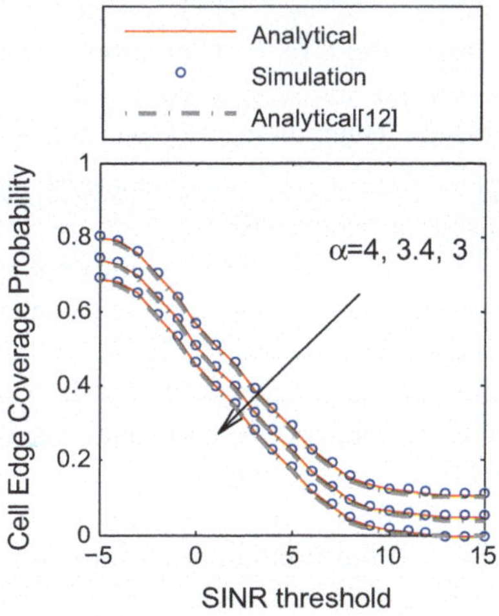

Fig. 4 Contrast of our analytical results, results provided in [12] and simulations for a range of γ_{th}, where, $\delta_{OCA} = 3$, $\gamma_{th} = \gamma_{trg}$, $\rho = 0.3\,\text{Km}^2$

Applying (12) and (17), P_{cp} can be further simplified as:

$$P_{cp} = \frac{1}{1 + p(max(\gamma_{trg}, \gamma_{th}), \alpha, \delta_{ICA})} + \frac{1}{1 + p(\gamma_{trg}, \alpha, \delta_{OCA})} - \frac{1}{p(\gamma_{th}, \alpha, \delta_{ICA}) + p(\gamma_{trg}, \alpha, \delta_{OCA})} \tag{18}$$

3 Results and Discussion

It can be seen that the CECP expressions in (10), (12) and (14) are computationally easy to compute as compared to corresponding expressions in (11), (13) and (15), successively. Thereafter, we contrast the analytical outcomes to the simulation outcomes. For the purpose of computations, we choose a 20 Km $\times$ 20 Km surface area and produce the arbitrary positions of MBSs applying spatial PPP of density ρ. As illustrated in Fig. 2, cell users (CUs) have scattered uniformly in this specified area and individual CU is served by the close FBS and femtocell boundaries conform to a Voronoi tessellation. We consider inner cell and for individual CU belongs to inner cell, we induce composite multipath fading belonging to its own transmission link and composite multipath fading belonging to the rest FBSs, which is unwanted interference power and compute SINR per CU. Further, we categorize the entire users into outer cell users (OCUs) and inner cell users (ICUs) depending upon the preset SINR threshold. Then, for the OCUs, we further induce composite multipath fading belonging to its own transmission link and composite multipath fading belonging to the outer region of the rest FBSs as multipath fading is independent between OCUs and ICUs. Then, we again compute the new SINR and compare the new SINR of each user with γ_{trg}, and iterate this 20,000 times to achieve the CECP. Thereafter, we take on spectrum assignment in FFR scheme, where the entire spectrum and spectrum assigned to the ICUs are denoted by f and f_0. Thus, as in [12], we get, $\frac{f_0}{f} = \int_{d_0=0}^{\infty} P\left[\gamma(d_0) > \gamma_{trg} p_\gamma(d_0) dd_0\right]$. Besides, $f_1 = f_2 = f_3 = \frac{(f - f_0)}{3}$, where f_1, f_2, f_3 are spectrum allocated to the OCUs for different macrocells.

Figure 3 differentiates our contributions obtained from the analysis, the results presented in the analysis of [12], and the simulations results for $\alpha = 3, 3.4, 4$, respectively. Furthermore, in Fig. 4, both the outcomes achieved from the analysis and the simulations are contrasted for the variation of γ_{th}. This can be found that our contributions obtained from the analysis, the outcomes obtained from the analysis in [12], and the outcomes from the simulations are entirely in so close match, in both Figs. 3 and 4. Moreover, our contributions obtained from the analysis are comparatively closer to the simulations results for all values of α.

4 Conclusion

We have presented the numerical expressions of the CECP metric for FFR scheme and then successfully demonstrated through the simulation to demonstrate that the accuracy of our analytical work. We conclude that both our contributions and the analytical results in [12] are exactly converged to the simulation results.

Acknowledgements This work was funded, in part, by the Hamad Bin Khalifa University, Qatar Foundation, Qatar, by the framework of Competitiveness Enhancement Program of the National

Research Tomsk Polytechnic University, Russia TPU CEP_IC_110/2017. Authors also acknowledged the contribution of the COST Action on Inclusive Radio Communications (IRACON) CA15104.

References

1. Yu FR, Zhang X, Leung VCM (2012) Green communications and networking. CRC Press, New York
2. Chen Y, Zhang S, Xu S, Li GY (2011) Fundamental trade-offs on green wireless networks. IEEE Commun Mag 49(6):3037
3. Chandrasekhar V, Andrews J, Gatherer A (2008) Femtocell networks: a survey. IEEE Commun Mag 46(9):5967
4. Ghosh J, Roy SD (2015) Qualitative analysis for coverage probability and energy efficiency in cognitive-femtocell networks under macrocell infrastructure. Electron Lett 51(17):1378–1380
5. Liang Y, Chung W, Ni G, Chen I, Zhang H, Kuo S (2012) Resource allocation with interference avoidance in OFDMA femtocell networks. IEEE Trans Veh Technol 61(5):22432255
6. Cheung W, Quek T, Kountouris M (2012) Throughput optimization, spectrum allocation, and access control in two-tier femtocell networks. IEEE J Sel Areas Commun 30(3):561574
7. Yang X, Fapojuwo AO (2015) Coverage probability analysis of heterogeneous cellular networks in Rician/Rayleigh fading environments. IEEE Commun Lett 19(7):1197–1200
8. Chinnadurai S, Selvaprabhu P, Jeong Y, Sarker AL, Hai H, Duan W, Lee MH (2017) User clustering and robust beamforming design in multicell MIMO-NOMA system for 5G communications. AEU Int J Electron Commun 78:181–191
9. Ghosh J, Jayakody DNK, Qaraqe, M (2017) Cognitive-Femtocell based resource allocation in Macrocell network. In: 28th annual IEEE international symposium on personal, indoor and mobile radio communications (IEEE PIMRC 2017)- workshop on communications for networked smart cities (CorNer), Montreal, Canada
10. Hansen F (1977) Mobile fading-Rayleigh and log normal superimposed. IEEE Trans Veh Technol VT–26:332–335
11. Simon MK, Alouini MS (2002) Fading channel characterization and modeling. Digital communication over fading channels. Wiley, New York
12. Novlan TD, Ganti RK, Ghosh A, Andrews JG (2011) Analytical evaluation of fractional frequency reuses for OFDMA cellular networks. IEEE Trans Wirel Commun 10(12):4294–4305
13. Andrews JG, Baccelli F, Ganti RK (2011) A tractable approach to coverage and rate in cellular networks. IEEE Trans Commun 59(11):31223134
14. Gradshteyn IS, Ryzhik IM (2007) Table of integrals, series, and products, 7th edn. Elsevier/Academic Press, Amsterdam

Smart Raspberry Based GSM Gateway

Belkacem Benadda, Karam Medjahdi and Bilal Beldjilali

Abstract Mobile telecommunications systems face nowadays-rapid increasing services based on the third and new generations. This emergence influences the offered services as well as related costs. It is interesting to have reduced communications charges within an institution. One solution consists to use Gateways ensuring an intelligent management for outgoing calls by selecting the appropriate operator and telecommunication services to deal with best price selection. In this work, we are building low-cost smart GSM gateway based on Raspberry Pi and Microsoft .NET technologies to allow institutional or companies intranet networks to communicate via one of the local Algerian national operators.

Keywords Dot NET · Embedded · Gateway GSM · Raspberry Pi

1 Introduction

The different telecommunications mobile generations are continuously influencing the everyday individual's lives. Currently, companies are seriously considering the budget allocated to telecommunications infrastructures and services. The idea of combining the company intranet network advantageously with the phone or mobile networks makes it possible to achieve significant communications gains in performance and costs. To achieve such combination the use of a gateway is required.

B. Benadda (✉) · K. Medjahdi
Faculty of Technology, Abou Bekr Belkaid University of Tlemcen, Tlemcen, Algeria
e-mail: benadda.belkacem@gmail.com

B. Beldjilali
Department of Space Geodesy Algerian Space Agency Arzew, Oran, Algeria
e-mail: bel.bilal@yahoo.fr

A. Boyaci et al. (eds.), *International Telecommunications Conference*,
Lecture Notes in Electrical Engineering 504,
https://doi.org/10.1007/978-981-13-0408-8_4

Our proposed Global System for Mobile communications GSM Gateway aims to provide multiple benefits: sending Short Messages SMS and Multi Media Messages MMS via local web portal, ensuring better quality of communication at low cost, taking advantage of the temporary promotions offered by telephony operators. However, investment in this equipment brand remains relatively expensive, a point not appreciated by small and medium-sized Algerian companies [1]. We also note that actually the use of GSM Gateways is occasional in Algeria [1].

This work presents hardware and software implementation of smart GSM Gateway, with reduced investment, adapted to Algerian operators and services context. The built GSM Gateway acts as an embedded system based on the Raspberry Pi board combined with intelligent management software developed through the .NET technology. This article is organized in six sections. We state in Sect. 2 the place and advantages of the GSM Gateways. Section 3 describe the hardware integration we have done based on Raspberry Pi. The management software implementation is state on Sect. 4. Section 5 conclude this article.

2 Place of the GSM Gateways

Mobile phone operators offers are increasing; simple survey done in Algeria [1] shows that the biggest services consumers are usually companies no matter their sizes. In addition to that, the use of computer networks is also intense, companies computers are permanently connected to each other and to the Internet. These reasons inspire telecommunications specialists to the Gateway concept that allows communications to switch between intranet computer and GSM network. When preparing our GSM Gateway design we have taken note of the following items.

2.1 The GSM Gateways Place on the Market

Large telecommunications equipment manufacturers as Cisco, DLINK and ALPHATECH have embarked on the GSM Gateway very lucrative industry [2, 3]. Different models on sizes and capacities are commercialized, making the choice of the equipment determined by the user needs. Indeed, different conditions influence gateway users, but the main usage is always linked in one way or another to financial aspect. The simplest model is promoted as an USB dongle integrating SIM card, to ensure simple interface between computers and the GSM network [2, 3].

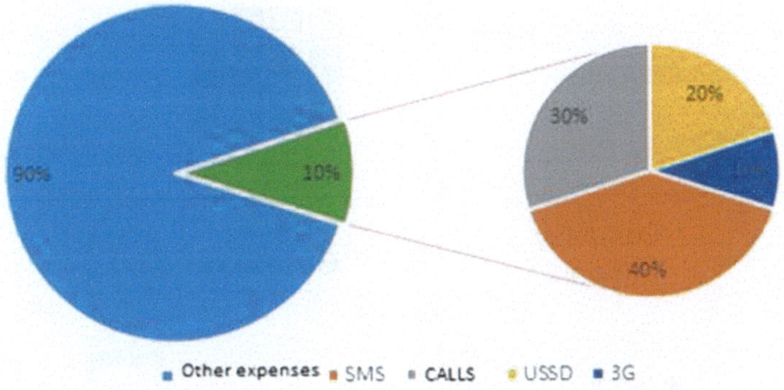

Fig. 1 Typical Algerian companies communications charges. Common communications charges are split into 3 parts, 40% used for short messages (SMS) then 30% used for voice and phone calls, followed by 20% for USSD and 10% for services related to wide area networks [1]

2.2 *The GSM Gateways Benefits*

In Algeria, companies attempt to reduce the maximum as possible the telephone services costs (see Fig. 1). However, they come across difficulties since the lack of tools. We have searched the best approach for Algerian companies to decrease telecommunications charges by doing price study in the local market. Services costs always remains favorable if the phone call use the same operator network and increase if other operators' network parts are involved. Although the three Algerian mobile operators are constantly placing on the markets attractive offers as well as facilities and packages available to reduce communications expenses. The use of GSM gateway will reduce mobile call costs by 40–80%, in addition of several advantages:

- Optimal sound quality [2, 3].
- No extra charge for international calls, by combining the Internet and voice over IP [2, 3].
- Dynamic SIM cards management by defining the duration use for each SIM cards
- and the authorized services. This means that it is possible to switch off automatically the SIM cards that have used all of their communications packages. This ensures that each SIM is used to its maximum capacity and that no additional charges are made.
- Development and deployment of new applications and services.

3 The Gateway Hardware Implementation

Traditionally Gateways are connected to a dedicated server that handles the communications exchanges as shown on Fig. 2. The Gateway placement in the company is an important condition related to optimal radio signal quality and secure connection

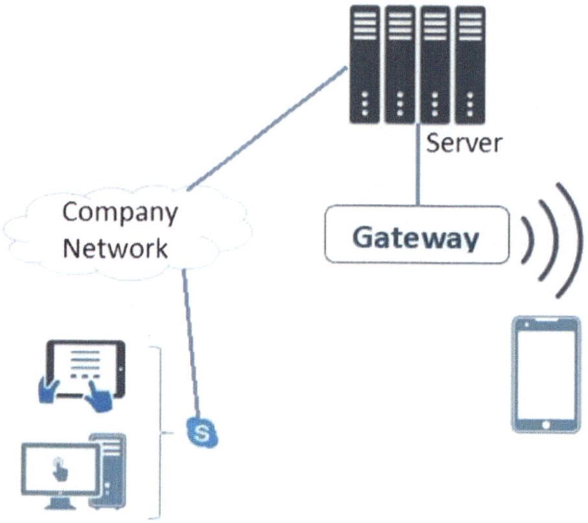

Fig. 2 Standard gateway position on the company's network

Fig. 3 The built Raspberry based Gateway with two connected USB GSM Modems

to company's local network. The GSM Gateway we are building, shown on Fig. 3, fulfil this configuration. Indeed, our built Gateway acts in addition as a server and bring the benefit to be connected to any point on the company network.

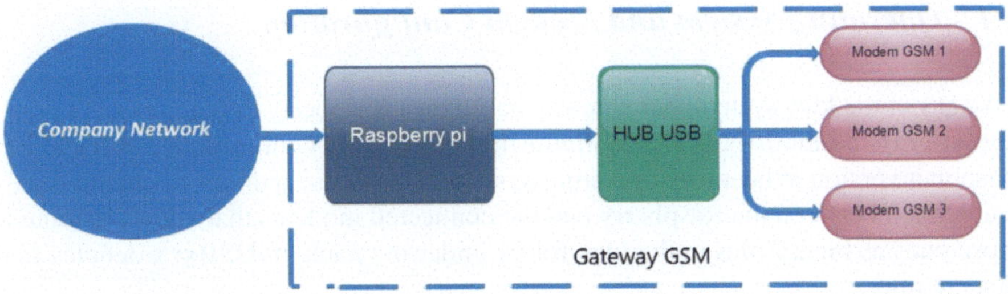

Fig. 4 Synoptic diagram of our built GSM gateway

The Raspberry Pi is a small computer, with the dimensions of a bankcard, which boots from a Secure Digital SD card [4, 5]. In this work, we propose to use the Raspberry for the gateway integration. The main reason is the easy low cost embedded development provided through the raspberry based solution. The used Raspberry is equipped with an ARM processor operating with a clock frequency of 700 MHz. Undeniably, ARM processors are used in many devices as Apple's: iPod, iPhone, and iPad, or Android based smartphones, home broadband routers and modems, as well as digital TV boxes. For connectivity the Raspberry Pi has a 10/100 Mbps Ethernet port.

In our solution, we use third-generation GSM USB modules. The modems are plug into the USB connector of the Raspberry through an USB HUB, with an external power supply, as shown by Fig. 3. Indeed, the Raspberry Pi 322 mA, 5.09 V power is insufficient to supply all the connected 3G USB Modems. In addition, the USB connectors of the HUB allows us to use easily several 3G GSM Modems. The diagram in Fig. 4 explains how our built Gateway works. Connected GSM modem can be used at the same time to communicate over the mobile network. To control the different modem communication tasks, the used modems support the extended AT command set.

4 The Gateway Software Integration

The software of our gateway is divided into three parts: operating system installation and configuration, server development, and client development. We have also assure security for all exchanged data associated with the gateway by developing encryption scheme. All the developments are made using .NET technology with the C # language.

The execution on Rasbian, the raspberry related operating system, is performed exhausting the functionalities of the Mono project [6].

4.1 Operating System and Modem Configuration

Like any embedded system, our gateway based on a Raspberry Pi requires an operating system to be able to boot and configure the hardware. For our solution, we installed Raspbian version Wheezy as operating system [7]. We use a driver to ensure communication between the Raspberry and the connected modems that allows to mount those ones as binary files in the /dev/folder under the name ttyUSBx: x denotes the modem rank.

4.2 The Gateway Server

The server installed within the Raspberry, performs several operations simultaneously as shown by Fig. 5:

- Handles hardware management and interfacing. This concerns the detection of connected modems using the ttyUSBx interface files of the Raspbian operating system.
- Supports configuration files that contain the information needed to run each modem, and allows routing algorithms to recognize the correct modem. Configuration files use XML classes and object collections. The configuration files also contain the path of the files that store the favorite numbers assigned to each modem. These files allow routing functions to prioritize the use of specific modem with class of numbers.
- Handles all the features that manage communication in the network, this mainly concerns sockets, encryption and management of connected clients.

Once the server application starts and configures the modems, other listening processes are executed simultaneously to be able to receive commands over the company's network. Indeed the implemented server socket listening on port 17000 to client requests, gives the possibility to receive commands from the overall company's network. If the server receives authorized client request, it randomly answers in a port greater than 1024. Each server client communication is started in separate thread. The server frees the port 17000 to allow other connections to take place. Accepted clients frames are reformulated into AT commands to be executed by the modems (Fig. 6).

Frames sent by clients have a unique format shown by Fig. 7. Once read and verified a first algorithm tests the number of separation characters. In our case we use pipeline char "| " to separate the different datagram parts. Valid frames are split into three strings: header, number or indicative and useful data. The number or indicative specify the selected modem. The header states the AT commands to be executed.

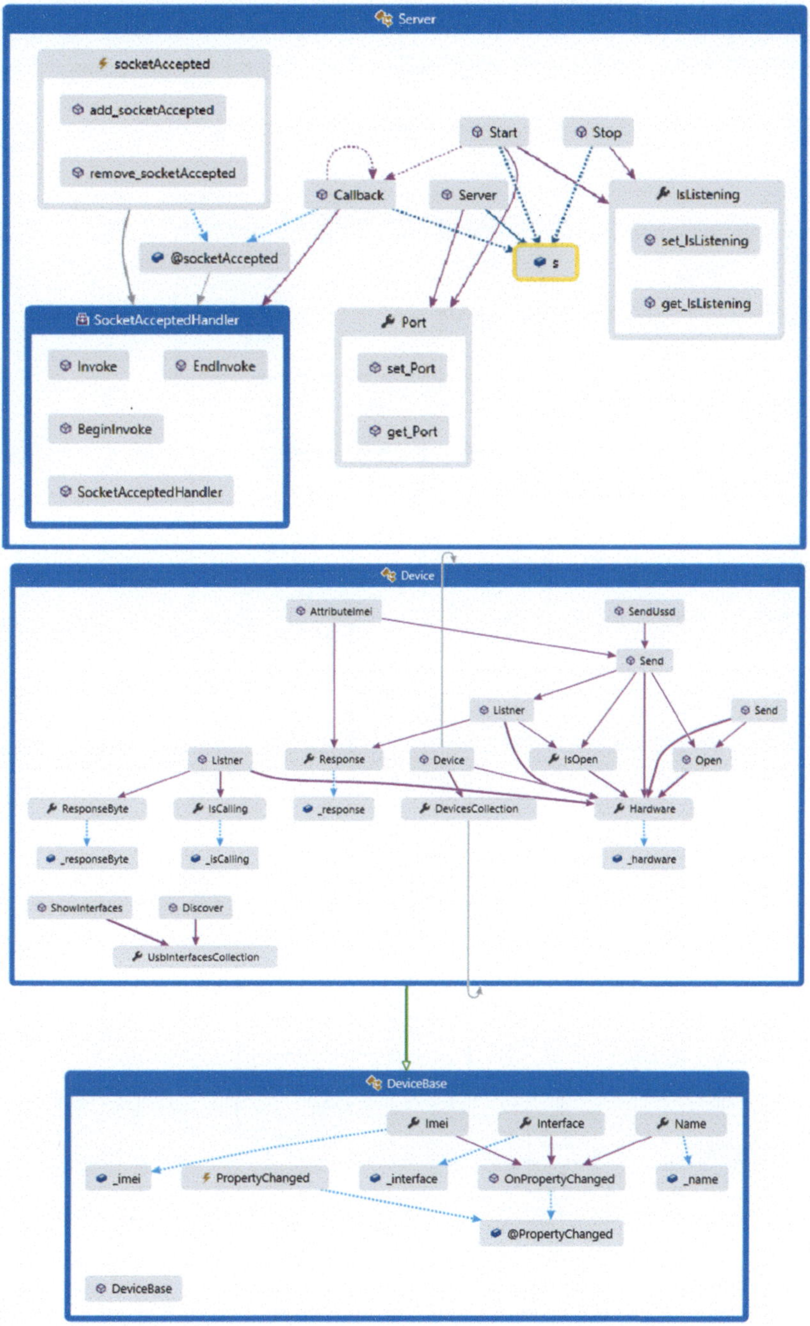

Fig. 5 Server code map. The server begins by reading the configuration files that contain the modem's system interface, the operator's name, the parameters for the local and international dialing codes, and the favorite number files path. The server runs based on recursive call

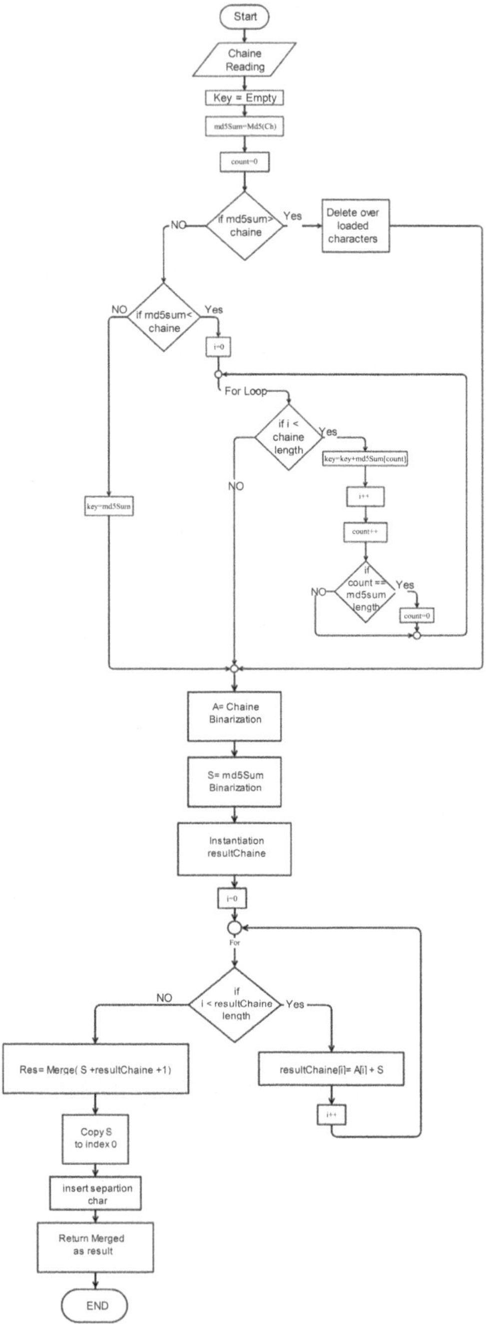

Fig. 6 Main Server management flow chart

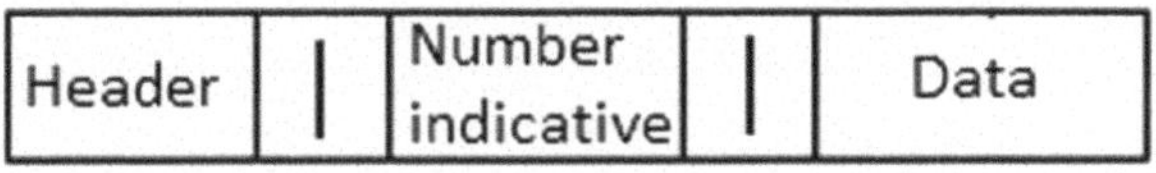

Fig. 7 Proposed exchanged data frame format. The header precise the commands to be executed by the modem. Number or Indicative the selected operator and service. Data as voice or SMS

4.3 The Gateway Client Application

The client, based on graphical interface, has been programmed to pass the messages in the given datagrams format. It requests connection and services via the IP address of the server. Each client application manages its own favorite numbers by sending them in a special frame interpreted and saved thru the server in configurations files. The developments is done based on Microsoft Model View View Model MVVM framework.

4.4 Security and Encryption

Since our solution uses a local network, the data sent can easily be spoofed. Indeed, if the messages exchanged in clear can be analyzed, hackers are able to inject similar frames and control our GSM gateway. We propose a customized encryption algorithm introduced in both sides applications client and server. Our algorithm uses the Message Digest 5 (MD5) [8] as a basis for the verification also for the encryption key construction. The flowchart of Fig. 8 detail our cryptographic approach.

5 Conclusion

During the development of the GSM Gateway, several problems were identified in respect to hardware and software resources. The segmentation of the GSM gateway problem with low investment costs makes it possible to first design the hardware aspect based on a Raspberry Pi board interfaced with several 3G generation GSM modems.

All the software development is done based on the.Net technology. The server is executed on Rasbian Linux taking advantage of the mono project services. The client application is distributed over the company network stations. Simple encryption approach based on the MD5 algorithm is used to protect the exchanged Data grams of our Gateway.

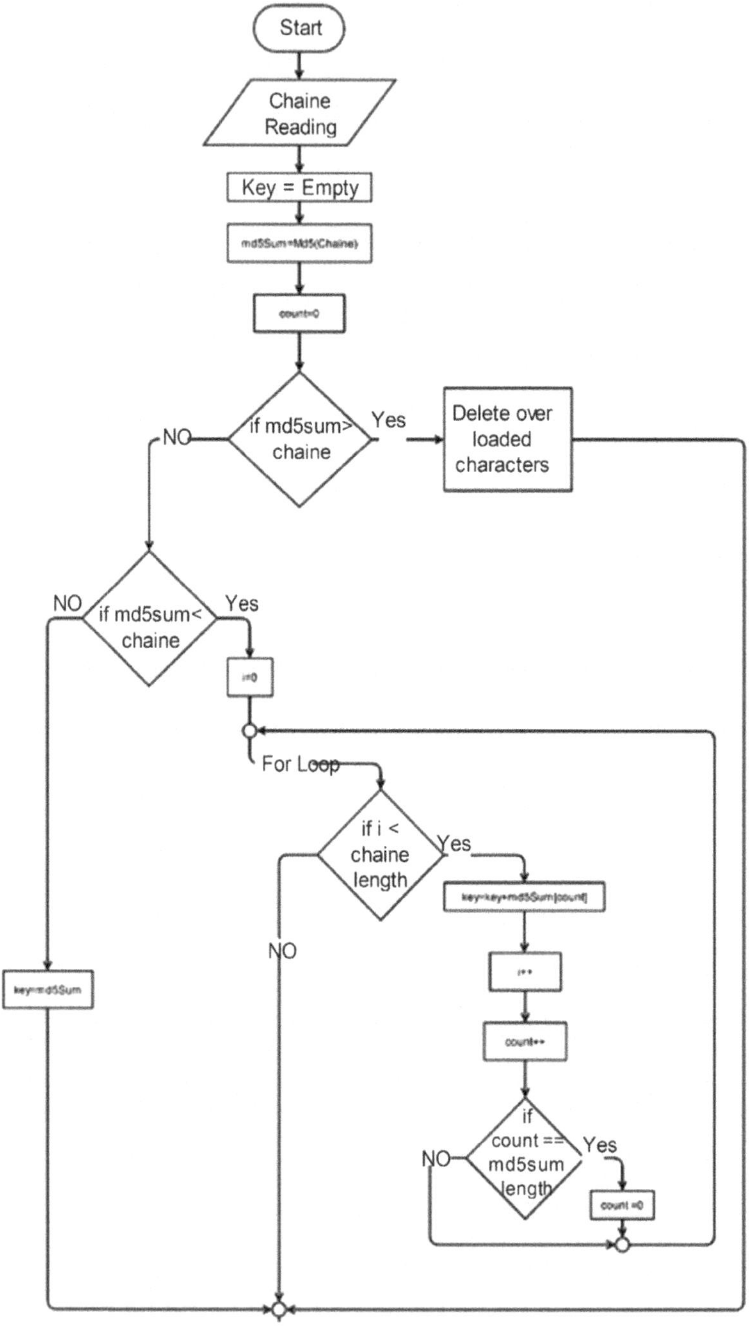

Fig. 8 Flowchart for our proposed encryption algorithm

References

1. Medjahdi MK (2015) La Gateway GSM, Master in Telecommunications. University Abou Bekr Belkaid Tlemcen (2015)
2. Olga Z (2016) VOIP for business users using internet Elephony to increase service levels and cut costs, 1rd edn, Level 7 Systems LTD
3. Doherty J, Anderson N (2006) Internet Phone services simplified: an illustrated guide to understanding, selecting, and implementing VolP-based Internet Phone services for your home. CISCO Press (2006)
4. Simon M (2016) Raspberry PI Cookbook, 2nd edn. O'Reilly (2016)
5. Benadda ME, Beldjilali B (2017) Embedded BeagleBone based Wi-Fi intrusions detector and vulnerabilities checker, 2017 Seminar on Detection Systems Architectures and Technologies (DAT), Algiers
6. Solin D (2003) Developing on Linux with C# and .Net: The Mono Project. Apress
7. William H (2015) Learning Rasbian. PACKT Publishing
8. Malik S (2003) Network security principles and practices. CISCO Press

IEEE 802.11s Mesh Network Analysis for Post Disaster Communication

Mehmet Ali Ertürk, Muhammed Ali Aydin, Luca Vollero and Roberto Setola

Abstract Wireless mesh networks (WMN) provide rapidly deployable, cost effective and flexible communication infrastructures. WMNs are particularly useful in the case of disaster, allowing rescuers to build communication infrastructures that may facilitate their work and help them in saving lives. IEEE 802.11s is today the reference standard for wireless mesh networks (WMNs). The aim of this paper is to evaluate the performance of IEEE 802.11s in disaster scenarios when a robust Wireless infrastructures has to be built quickly out of nothing. In such scenarios, parameters like the time elapsed from the distribution of nodes in the area and the successful transmission of the first data packet, or the network delay, may be essential in order to save lives. The aim of this paper is to evaluate such parameters.

Keywords IEEE 802.11s · Mesh network · Post disaster communication

1 Introduction

Recovering from a disaster is a process that may be roughly split into two phases, (i) emergency and (ii) reconstruction. An emergency following traumatic events like natural disasters or terrorist attacks poses serious challenges for rescue teams and victims. Especially the destruction of communication infrastructures makes extremely critical the exchange of information and the coordination of rescue teams. More-

M. A. Ertürk · M. A. Aydin (✉)
Department of Computer Engineering, Istanbul University, Istanbul, Turkey
e-mail: aydinmali@istanbul.edu.tr

M. A. Ertürk
e-mail: mehmetali.erturk@istanbul.edu.tr

L. Vollero · R. Setola
Department of Computer Engineering, Università Campus Bio-Medico di Roma, Rome, Italy
e-mail: vollero@ieee.org

R. Setola
e-mail: r.setola@unicampus.it

A. Boyaci et al. (eds.), *International Telecommunications Conference*,
Lecture Notes in Electrical Engineering 504,
https://doi.org/10.1007/978-981-13-0408-8_5

over, the progressive extinction of old fashion broadcast technologies, such as FM and AM radios, makes even more complex the distribution of information to the people present in the area of the disaster, being almost impossible in the absence of today more common communication technologies such as 4G and WiFi. In such scenarios, the availability of rapidly deployable network infrastructures is critical in order to implement more effective recovery actios in the emergency phase and better communication with victims.

Wireless mesh networks (WMN) provide rapidly deployable, cost effective and flexible communication infrastructures. WMN technology can be used as a range extender or structure enabler for public safety. In this paper we want to evaluate these features and we want to examine the applicability of standard WMN in post-disaster scenarios. Among the numerous technologies proposed until now for WMN, we focus on IEEE 802.11s which is probably the most important standard for distributed access mesh networking [1]. Moreover, the native integration with the most diffuse standard for WiFi communications (IEEE 802.11) is a key aspect that makes such a technology of particular interest in the application scenario under analysis.

The aim of this paper is to evaluate the performance of IEEE 802.11s in disaster scenarios when a robust Wireless infrastructures has to be built quickly. In such scenarios, parameters like the time elapsed from the distribution of nodes in the area and the successful transmission of the first data packet, or the minimum available throughput among nodes, may be essential in order to save lives.

Paper is organized as follows: Sect. 2 discusses the application of mesh network solution in the post disaster communications, whereas Sect. 3 describes the IEEE 802.11s standard. Section 4 introduces simulation environment and results, whereas Sect. 5 concludes the paper.

2 Related Work

Wireless mesh networks (WMNs) has gained today significant attention as a promising technology for providing access services in different application scenarios. WMNs are especially promising in settings where a communication infrastructure is required without or with limited support of existing systems, such as the scenarios of natural or human induced disaster.

In [2] Baryun proposes a new protocol for a cost effective and easy to deploy hybrid communication system. The protocol aims at achieving the requirements of rescuers and cooperation in scenarios of disasters. In [3], Xu et al. propose a planning scheme for the construction of a common emergency packet transport network, whereas in [4] Jia et al. consider the problem of power consumption in the case of Unmanned aerial vehicles (UAVs) used as flying base stations and in [5] Albuquerque et al. investigate and evaluate the performance of most popular DTN (Delay Tolerant Network) routing protocols in the same scenario. Reference [6] considers the On-the-fly establishment of multihop wireless access networks (OEMAN) approach. The authors of the paper propose a linear program for overload-aware routing, and consider wireless interference and integrate it into the proposed routing optimization model. In [7] Tanha et al. propose an optimized routing scheme maximizing the end-to-end throughput for mission critical applications. The proposed scheme outperforms the schemes based on shortest path routing in terms of total throughput and fairness.

All the considered works are based on some abstract modeling of the communication technologies and make assumptions on expected performance. The approach of this paper is different, we start with a specific technology, the IEEE 802.11s standard, and we aim at evaluating what we can expect in terms of performance and set-up time. Results of our work represent the baseline for the application of such a technology in disaster scenarios and may represent the starting point for improvements on the standard.

3 IEEE 802.11s

IEEE 802.11s is today the reference standard for wireless mesh networking. IEEE 802.11s is defined in [1] and it is based on the PHY and MAC layers of the IEEE 802.11 standard [8]. IEEE 802.11s defines two types of nodes: mesh stations (MSs) and mesh points (MPs). MSs are the client nodes, whereas MPs can be of two types: (i) mesh access points (MAPs) and (ii) mesh portal points (MPPs). MPPs work as gateways and connect the mesh network with the Internet. Conversely, MAPs work as backbone nodes and forward traffic between a STA and a MPP or between STAs.

The IEEE 802.11s MAC layer defines a mesh coordination function (MCF), which allows contention-based or scheduled channel access [9, 10]. The MCF is based on functionality of both the enhanced distributed channel access (EDCA) and the mesh coordinated function controlled channel access (MCCA). EDCA is a QoS CSMA/CA protocol under which devices share statistically the wireless channel while exchanging data, whereas MCCA is a reservation based protocol. MCCA performs better than EDCA when the network backbone is known and in case of highly loaded network scenarios. However, EDCA match better the requirements of non stable networks such as those used in the areas of disasters.

In this paper we focus on IEEE 802.11s based on EDCA and we assume that the reader has a basic knowledge of EDCA's functionalities (see [11] for details on EDCA).

4 Simulation Results

In the study, we analyzed different distributions of Mesh Nodes with different configurations to evaluate the performance obtained under saturation condition. The goal is to have clues on critical parameters like configuration times and communication performance.

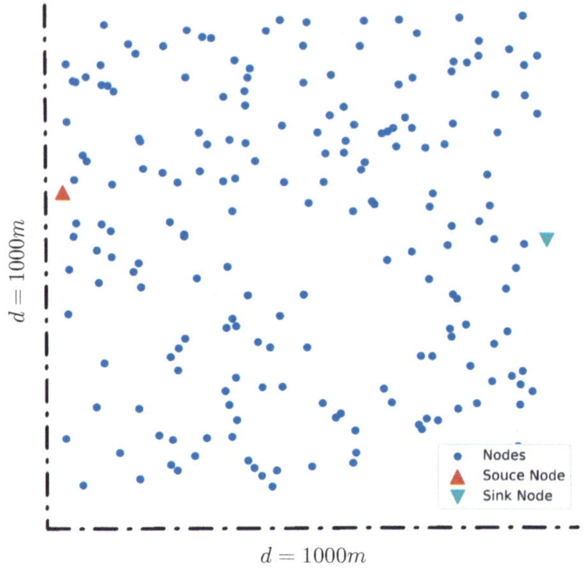

Fig. 1 Simulation scenario

In particular, we consider a number of mesh nodes N, ranging into the set $\{150, 175, 200\}$. Each node is placed in order to have at least an adjunct node to which it establishes a wireless link over a non-interfering channel. We assume to have the control over this condition during the distribution of the nodes in the area. Eventually, we assume that the area that has to be managed by the IEEE 802.11s nodes is a d by d plane territory. d is 1000 m in our study.

Two STA nodes are placed in the edges of the areas to work as source n_s and sink/destination n_d of a new established communication. n_s is configured to transmit to n_d a CBR flow composed of UDP packets 1000 bytes long with a fixed constant inter-transmission time $t_c = 0.1$ s.

Figure 1 shows the network setup. Network performance are evaluated in terms of:

- Network setup time; i.e. the time from node distribution and the first successful packed received by n_d.
- Network delay; i.e. the difference between the packet transmission time and the reception time of the same packet.
- Packet loss ratio; i.e. the average fraction of packets lost during the transmission.

The general purpose discrete-event network simulator ns-3 is used to simulate the scenarios [12]. For each configuration, 10 different simulations are performed, each simulation run 180 s and obtained results are summarized in Figs. 2, 3 and 4.

Figure 2 shows the network status as measured the first received packet by n_d. Network delay is obtained from all successfully delivered packets and presented in Fig. 3. Network performance of the mesh network can be determined by packet delays

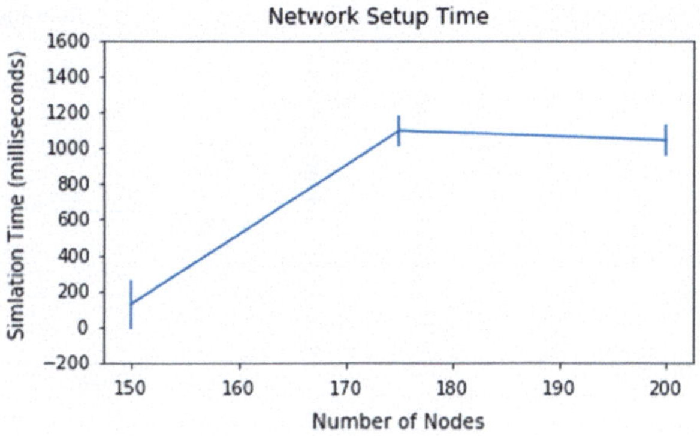

Fig. 2 Network startup time

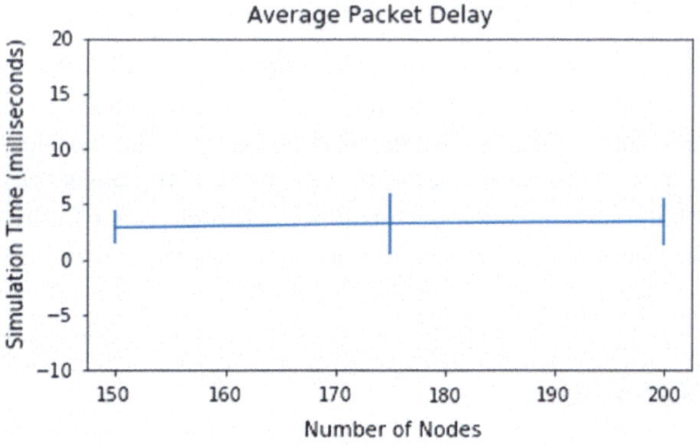

Fig. 3 Network delay time

and drops. Figure 4 presents ratio of the UDP packets dropped during simulation. All the results are reported with confidential interval of 95%.

From the results we observe that, network setup depends on number of nodes placed in the area. The higher the number the slower is the setup due to the complexity of the configuration phase that increases with the number of nodes. In all the considered scenarios, however, the configuration time is under control (less than any dangerous threshold). Delay also and packet drops are reasonable for all scenarios. Based on the proposed results we evaluated the performance for a reduced number of nodes (125) and we observed an abrupt reduction of performance which indicates the presence of a density threshold under which setup is hard to obtain (results are not reported being the communication between the source and the destination unavailable in the majority of the simulations).

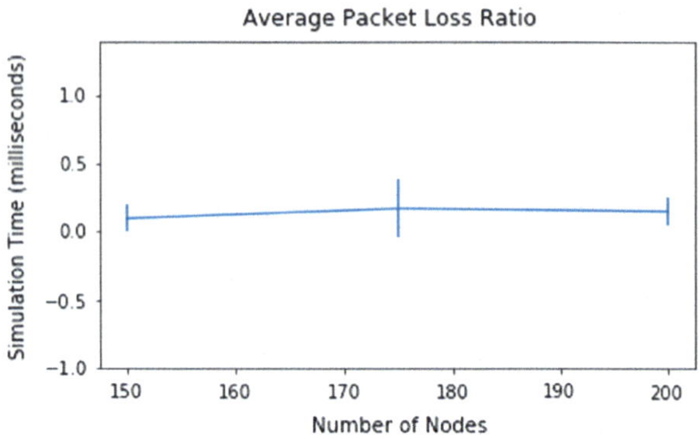

Fig. 4 Network packet loss ratio

5 Conclusion

In this paper we evaluate the performance of IEEE 802.11s in disaster scenarios. Assuming a plane area and a random distribution of IEEE 802.11s nodes, we evaluate how quick is the network setup and the average performance that users may expect from such a network. Our work is the basis for future studies aiming at better characterizing such performance, evaluate other figure of merit related, for instance, to power consumption and proposing optimization of the protocol.

References

1. IEEE Standard for Information Technology–Telecommunications and information exchange between systems–Local and metropolitan area networks–Specific requirements Part 11: Wireless LAN Medium Access Control (MAC) and Physical Layer (PHY) specifications Amendment 10: Mesh Networking, IEEE Std 802.11s-2011 (Amendment to IEEE Std 802.11-2007 as amended by IEEE 802.11k-2008, IEEE 802.11r-2008, IEEE 802.11y-2008, IEEE 802.11w-2009, IEEE 802.11n-2009, IEEE 802.11p-2010, IEEE 802.11z-2010, IEEE 802.11v-2011, and IEEE 802.11u-2011), pp. 1–372
2. Baryun A, Al-Begain K, Villa D (2011) A hybrid network protocol for disaster scenarios. In: 2011 5th International Conference on Next Generation Mobile Applications, Services and Technologies (NGMAST). IEEE, pp 129–136
3. Xu S, Yoshikane N, Shiraiwa M, Tsuritani T, Harai H, Awaji Y, Wada N (2017) Multi-carrier interconnection-based emergency packet transport network planning in disaster recovery. In: Proceedings of the 13th International Conference on DRCN 2017-Design of Reliable Communication Networks. VDE, pp 1–8
4. Jia S, Fadlullah ZM, Kato N, Zhang L (2016) ECO-UDC: an energy efficient data collection method for disaster area networks
5. de Albuquerque JC, de Lucena SC, Campos CA (2016) Evaluating data communications in natural disaster scenarios using opportunistic networks with Unmanned Aerial Vehicles. In: 2016 IEEE 19th International Conference on Intelligent Transportation Systems (ITSC). IEEE, pp 1452–1457

6. Phung CV, Minh QT, Toulouse M (2016) Routing optimization model in multihop wireless access networks for disaster recovery. In: 2016 International Conference on Advanced Computing and Applications (ACOMP). IEEE, pp 135–140
7. Tanha M, Sajjadi D, Tong F, Pan J (2016) Disaster management and response for modern cellular networks using flow-based Multi-hop Device-to-Device communications. In: 2016 IEEE 84th Vehicular Technology Conference (VTC-Fall). IEEE, pp 1–7
8. Association IS et al (2012) 802.11-2012-IEEE Standard for information Technology–Telecommunications and information exchange between systems Local and metropolitan area networks–Specific requirements Part 11: Wireless LAN Medium Access Control (MAC) and Physical Layer (PHY) Specifications. Accessed http://standards.ieee.org/about/get/802/802.11.html
9. Carrano RC, Magalhães LC, Saade DCM, Albuquerque CV (2011) IEEE 802.11s multihop MAC: a tutorial. IEEE Commun Surv Tutor 13(1):52–67
10. Chakraborty S, Nandi S (2013) IEEE 802.11s mesh backbone for vehicular communication: Fairness and throughput. IEEE Trans Veh Technol 62(5):2193–2203
11. Ertürk MA, Vollero L, Aydin MA, Turna OC, Bernaschi M (2014) A framework for modeling and implementing QoS-aware load balancing solutions in WiFi hotspots. In: 2014 11th International Symposium on Wireless Communications Systems (ISWCS). IEEE, pp 33–38
12. The ns-3 network simulator (2019). https://www.nsnam.org. Accessed 1 Oct 2017

Fuzzy Logic Approach for Layered Architecture Cognitive Radio Systems

Ali Riza Ekti

Abstract Cross layer design seems to be one of the most fundamental characteristics of next generation wireless networks and cognitive networks. Traditional layered architecture still exists and seems to be present in different forms in future as will. One should keep in mind that traditional architecture solves many problems even though they might be suboptimal solutions however crucial shortcomings still exist and hinders the evolution of wireless communications from being completely adaptive and cognitive. Therefore, it is wiser to maintain the fundamental structure of layers in different forms especially modifying the algorithmic order. In addition, an extra block that operates whenever necessary by carrying out cross layer tasks. This way cross layer can be achieved through the use if this extra block in order to fully utilize both traditional architecture and the use of this extra block. A decision mechanism are required sitting conceptually between these two. This mechanism switches accordingly depending on the complexity of the problem considered and solution required in the light of constraints present such as relative power, battery consumption, computational complexity and so on.

Keywords Cognitive radio · Fuzzy logic · Layered architecture

1 Introduction

In wireless communication history as in most of the systems, there are two major distinct characteristics observed in terms of its evolution; the first characteristic is to extend and to insert new features into the existing structure or design, the second one is to improve the already existing capabilities. When we first consider the first characteristics, we see that insertion of new features arises from the fact that wireless systems develop very fast. Therefore, they always require new services and application evolving through the time. One of the most striking examples for such situations

A. R. Ekti (✉)
Electrical and Electronics Engineering Department, Balıkesir University, Balıkesir, Turkey
e-mail: arekti@balikesir.edu.tr

A. Boyaci et al. (eds.), *International Telecommunications Conference*,
Lecture Notes in Electrical Engineering 504,
https://doi.org/10.1007/978-981-13-0408-8_6

is cell phones. Earlier cell phones were used only for voice transmission along with limited text messaging applications however contemporary cell phones are capable of transmitting multimedia along with an operating system running on. When we consider the second characteristics, it is clear that improvement of already existing capabilities is a must because every physical concept is finite. Therefore, either with the same resources the new developments achieved better performances or the same tasks should be performed with lower use of system resources compared to earlier designs. In addition to this improvement, adaptation and optimization should always be there as driving forces behind the development under dynamically changing conditions [1].

In this aspect, the propose of this study is to provide a conceptual step in wireless system design in the light of recently emerging technology called Cognitive radio (CR). The idea of CR is first presented by Joseph Mitola III. and Gerald Q. Maguire, Jr. "The point in which wireless personal digital assistants and the related networks are sufficiently and computationally intelligent about radio resources and related computer-to-computer communication to detect user communications needs as a function of use context, and to provide radio resources and wireless services most appropriate to those needs" (Joseph Mitola III, Cognitive Radio: An Integrated Agent Architecture for Software Defined Radio, PhD dissertation, Royal Institute of Technology (KTH) Stockholm, Sweden, 8 May, 2000). This is the definition of CR by Joseph Mitola. There are many advantages of cognitive radios in wireless communications. CR is basically an intelligent wireless device which is aware of the environment, spectrum, location, radio frequency (RF) environment, power, user, spectrum etc. and can easily adapt/optimize itself to the necessity of the system [2].

In this paper, a new design for next generation wireless networks and cognitive radios is proposed based on earlier designs along with currently available methods and approaches to better adapt and optimize the overall system performance. The rest of the paper is organized as follows. The rest of the paper is structured as follows. In Sect. 2, statement of the problem is presented. Contemporarily used layered architecture of cognitive radio systems is provided in Sect. 3. Proposed approach is investigated in Sect. 4. Finally, concluding remarks are drawn in Sect. 5.

2 Statement of the Problem

Basically the scenario considered in this study deals with the connection of two nodes where both of the nodes which try to convey messages to each other. Note that in order to convey the messages, two nodes need to be connected to each other. The connection can be either wired or wireless depending on the application or system design specifically to the work. The focus of this work will be on wireless connection which exploits radio transmission. However the approach and concepts mentioned in this paper are also valid for wired connections.

One might wonder why we have two different types of connections even if the proposed approach and the concepts are the same for both wired and wireless systems, the answer lies in the heart of transmission over the connection type which is called propagation channel characteristics. It is known that wireless propagation channel differs significantly from wired propagation channel. This difference leads different designs for both systems based on the same approach and the concept. The first step to establish a connection between two nodes is to provide a physical connectivity between the nodes. Hence we need a specific set of functionalities that can take care of every single issue related to physical connectivity of the two nodes. Physical connectivity means signaling over the air for radio transmission since electromagnetic waves are used in RF-systems, at this point the physical connectivity here refers to every single item related to physical signaling beginning from the use of electromagnetic waves.

Since radio propagation employs electromagnetic waves one of the biggest concerns for wireless connectivity occurs in the use of RF-spectrum. Depending on the design, communication systems use the RF-spectrum in different ways. Wireless systems should control the use of resources by employing spectral set off functionalities since RF-spectrum is finite and might not be available for use when it is desired. Similar concerns exist when managing network is considered for example in a mobile network transferring an ongoing transmission from one access point to another one must be handled appropriately. When this reasoning is taken further steps ahead, we see that we end up having a layered architecture. Note that this layered architecture is based on units which can carry out fundamental tasks in an algorithmic order.

3 Contemporarily Used Layered Architecture of Cognitive Radio Systems

In contemporarily used layered architecture the fundamental design relies on following seven layers which as known as open system interconnection (OSI). Layers of this architecture are defined as follows [3]:

- Physical Layer: Defines the physical characteristics of the network.
- Data Link Layer: The appropriate physical protocol is assigned to the data. Also, the type of network and the packet sequencing is defined.
- Network Layer: Path determination and logical addressing. It provides switching, routing technologies and creating logical paths.
- Transport Layer: This layer maintains flow control of data and provides for error checking and recovery of data between the devices.
- Session Layer: This layer maintains and ends communication with the receiving device.
- Presentation Layer: This layer takes the data provided by the Application layer and converts it into a standard format that the other layers can understand.
- Application Layer: This layer supports application and end-user processes.

Back in time, one of the popularly used instant messaging services, MSN, is given as example. In MSN applications assumed that a person wants to voice chat with another person. The first thing is the MSN application on the computer which constitutes the application layer. When one of the parties begins to talk through MSN, The microphone captures the voice and converts it to voltage levels. Presentation layer converts voltage levels into bits in an appropriate format. Next the online availability of the user is checked and put into the message by session layer. Next the information is ordered accordingly by the transport layer. Next the packages delivered into network by following different paths which are determined by network layer for each individual user the packet ordering and resource use are checked at MAC layer. Finally the information is carried on the transmission medium by physical layer.

As we see the layered architecture seems to be easiest way of solving the problem considered here in terms of the proposed approach this comes with several advantages however as will be discussed subsequently such an approach comes at the expense of several very critical shortcomings in the design and the development in below, disadvantages and advantages summarized [4–7].
Advantages:

- Modularity: Each Layer can be designed independent of others which enable simplicity.
- Standardization: Design only requires having the knowledge of explicit definitions and abstractions which enables interoperability.
- Expandability: Layers can be updated, altered, or expanded independently which enables individual flexibility.

Disadvantages:

- Ordering: Execution of any process in any layer has to be after the execution of previous processes in former layers which causes inefficiency and latency.
- Interaction: Due to strict isolation, information cannot cross other layers which cause unawareness, redundant processes, sub-optimal performance.
- Adaptation: In wireless communications, rapid channel variations cannot be responded immediately which causes decrease in capacity, sub-optimal performance.
- Topologies: Some of the network topologies need flexible layer architecture which causes inefficiency.

When disadvantages and advantages are investigated the shortcomings of layered architecture are very critical, despite these advantages layered architecture can not immediately be abolished because of the contemporary and vastly deployed standards such as global system for mobile communications (GSM), worldwide interoperability for microwave access (WIMAX), code division multiple access (CDMA) networks are based on the traditional architecture. Also backward compatibility should always be in the system design process for next generation wireless networks [8–10].

4 Proposed Approach

In this section, two items are going to be investigated;

- Reasoning behind proposed approach
- Detail explanation of the proposed approach

Before developing a new model it is mandatory to contemplate the drawbacks of the traditional architecture. Next step is trying to overcome the major problems encountered in the traditional architecture with minimum amount of intervention and modification. Finally appending a novel set of units will complete the task. Recall that there are four major disadvantages of the traditional layered architecture;

- Algorithmic ordering
- Strictly isolated layers
- Slow or no adaptation to rapidly changing conditions
- Lack of flexibility in the architecture

Since the technology is highlighted explicitly, the easiest way to solve the problems started here is to attack the shortcomings listed above. When we look at the shortcomings, we can group them into two broad classes. First class includes ordering, adaptation topologies which can all be labeled with structural design. Strictly isolated layered architecture constitutes the second group on its own in this manner. In this aspect it is better to begin with the second class first.

In order to overcome interaction problem in the traditional layer architecture, the basic approach would be to increase the amount of information crossed between the layers. This way the information provided by lowest layer has more chance to reach the uppermost layer and vice versa. When we consider the second group, we see that slight modifications will not be sufficient to overcome the problems itemized in this group. This class requires a drastic change or modification in the design. Therefore many of the problems listed as disadvantages falling in this group [11–14].

4.1 Outline of the Proposed Approach

Conceptually the approach proposed should still include the traditional layered architecture to some extent in a slightly different form. First and foremost, the algorithmic ordering must be broken into pieces in which layers remain the same however the ordering is totally removed. Second, at the expense of removing ordering, an additional structure should be included into the design in order to recognize the disordered layers and not to lead to chaos. Furthermore, keeping in mind that traditional layered architecture with ordering might be the optimum solution for specific cases, it is better to have the flexibility of putting the layers into their original algorithmic order for some occasions. Hence a unit that allows the intervention of the extra modification to interrupt or not, is a must in the proposed approach [15, 16].

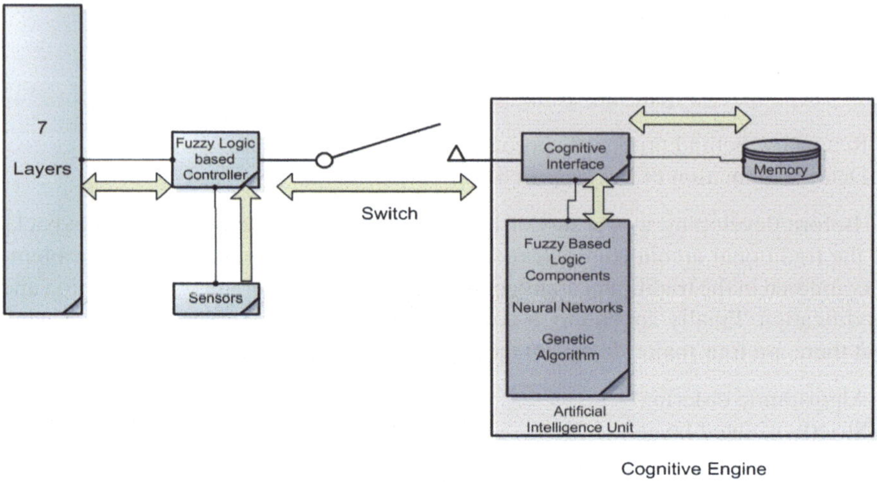

Fig. 1 Proposed approach fuzzy logic approach

When we think about the traditional architecture, we came up with a solution which we need to use traditional layer in more efficient way and we wanted to add a new layer which changes by the USER. What do we mean by USER layer, lets say that we get a message while driving and it is dangerous to read this message while driving because, we can lose our focus on driving and have an accident. Instead of having text message, our phone can sense that we are driving and converting text to voice or one more example cognitive radio can increase the volume if we are in a stadium or crowded place. When we add this 8th layer, we saw that the best example to describe this is to imagine an octopus which can use its arms individually and without depending on other arms.

Having said that the conceptual model proposed is still had a traditional layered architecture it is desirable to tie this up with the specific blocks that from the original proposed approach. In the light of the discussion carried up until now. I developed the following model. The model consists of four main blocks, including seven layer stacks as one of the block. The main idea behind the proposed approach is to have both seven layer stacks and cognitive engine simultaneously. However, the most important characteristic is to have another unit which decides on whether the cognitive engine will take the control or not during the process. This way a hybrid approach which takes the advantages of each layer is established. In other words, cognitive engine gets into the picture when it necessary that can be seen in Fig. 1.

4.2 Details of the Units Employed in the Proposed Approach

Sensors: In next generation wireless networks many of the decisions can be fortified by the presence of additional information collected from the environment.

Information collection can be done in two ways. The cheapest way to extract information in wireless systems is to use no additional hardware unit in the design. However, this comes at the expense of unreliable information and increased computational complexity which are two important characteristics to be considered in the system design stage. There is another way of collecting information and processing it in especially recently emerging wireless systems. Sensors, which are the additional add-on units, can provide very valuable information at no cost at the expense of extra cost of inserting these units. For instance, in most of the contemporary cell phones are equipped with global positioning system (GPS) hardware which provides precise location information when it is available. There might be some other sensors for different purposes in the same context such as temperature sensors.

Artificial Intelligence: In the proposed approach one of the crucial tasks is to fuse of the bids of information and to find the best way to adopt the system simultaneously to environment changes and/or system requirements. In order to do this, the unit that is responsible for these tasks needs to run different algorithms which are mainly categorized in artificial intelligence area. Some of them are neural networks, genetic algorithms, Hidden Markov Models (HMMs), some linear and some nonlinear optimization tools. Note that parallelism and stochasticity are two very important characteristics of these tools. Another important aspect is to find the optimum solution for generally multivariate problems. It is worth mentioning that such tools might not be able to provide the global optimization due to the following these reasons. First such a global optimization might not exist in the first place. The second, the tools might reach a local optimization rather than a global one and cannot proceed further and third one must remember that these tools are fed by finite support therefore instant periods of time reaching even a solution might not be possible.

Memory: One of the most desired properties of the proposed approach is to be able to learn about actions and experiences past. The fundamental tool for such capability is to employ a storage mechanism or memory combined with Artificial Intelligence Unit. Memory can provide very valuable, critical, and timely pieces of information for the proposed model.

Cognitive Interface: When sensors, memory, artificial intelligence tools are considered individually the tasks related to each unit are well defined. However, the proposed approach somehow needs to fuse and merge the input and output of all of the units and makes a decision based on the feedback. Since each individual unit performs a different task by a different mechanism it is expected to have different ways of interpreting the outcomes of each individual unit. Hence mechanism or tool that can understand, translate all these feedback provided by each individual is required. This task is established by the unit called cognitive interface in the approach proposed. Cognitive interface takes the advantage of semantic language used in describing different contexts especially in digital domain to exchange valuable information from one unit to another. A very well-known language for cognitive radio is known as radio knowledge representation language (RKRL), which is used in cognitive interface for this purpose [17].

Fuzzy Logic Unit: Recall that the traditional architecture achieves very good performances in many scenarios despite its critical shortcomings. Simultaneously, there are very crucial disadvantages of the traditional architecture which cannot be alleviated in any form of modification. Therefore, it is wiser to have two different architectures attend simultaneously in which only one of them is employed at a time when it is desired. For instance, layered architecture can still be used for very straightforward tasks. In these tasks the intervention of the additional architecture is redundant furthermore it might even be very costly. On the other hand especially when the cross layered optimization is considered the traditional architecture will yield nothing but suboptimal results and leads to inefficiencies and poor performances. In such cases the intervention of the additional unit is a must at every cost. Hence, a mechanism that distinguish the cases in which either the traditional or the extra block will be employed is required.

Considering human analogy the additional block corresponds to the source of conscious which is known as brain. However, same very basic at the same time very crucial actions are carried out without the intervention of the human brain in the human body. Two of the examples for such tasks would be involuntary (stereotype) reflex actions and digestive systems. Note that human digestive system operates on an algorithmic order for instance in testiness cannot perform unless stomach finishes its task. Furthermore brain does not take full control in digestion operation which corresponds to the case where the traditional architecture is selected by the fuzzy logic. The second example which does not require an algorithmic order but still excludes the information of extra block is the involuntary reflex actions. Again in human body when a sharp needle is jobbed into human feet the body immediately pulls of the leg without intervention of the human brain. Note that this example corresponds to decision mechanism which resembles the sensory information provided by the sensors in the proposed approach. The common point of the both examples is not to have the intervention of the extra units, the brain in human analogy [18].

One might wonder why fuzzy logic based unit is employed in switching mechanism. First it is appropriate to consider as to the options that can be employed in such a mechanism. Technically almost all of the artificial intelligence unit tools can be employed in switching mechanism, however there are several concerns in employing these tools for instance neural networks can be a promising approach for the same task. However it requires parallel processing and an extensive training stage. Similarly genetic algorithms can be considered as another option but they are very prone to reaching local optimization and might stay there as if they reach local optimization. On the other hand fuzzy logic is based on well-defined rules all of which are carefully identified previously, it does not require parallel processing and learning stage and they can be extended very easily by adding or modifying the existing rules. Hence, in the approach proposed, fuzzy logic is preferred in the switching mechanism.

In CR domain most of the problems fall into the category of multi-objective optimizations. However, there are efforts in the literature that do not apply directly multi-objective optimization problems (MOPs). They rather apply single-objective optimization with multi-objective function. The basic idea behind this approach is to gather all the relevant parameters [7]. In this aspect there are three main approaches to find solution to these problems [19, 20]:

- Bottom-Up Approach: The lower layers strive to save the upper-layer from possible loses note that this approach cannot establish an overall optimization since multimedia applications are prone to loses.
- Application-Centric Approach: In contrast to bottom up approach, application layer has higher priority in terms of controlling the optimization process. However, this approach fails in overall optimization as well since the response time for the application layer is huge compare to those of on lower layers.
- MAC-Centric Approach: The main idea behind this approach is to find a part in between and to keep the control of the optimization process around center of traditional layer. This way both upper and lower layers feed the center and center decides upon the pieces of information fed.

Note that even in these approaches somehow center tried to be located over the traditional protocol layer however the shortcoming of the inherent algorithmic ordering prevent these methods to reach global optimization. However, the proposed approach technically might achieve global optimization since inherit ordering is abolished and additional information crossing between layers is allowed. Reader might refer to [21] for further details.

4.3 Challenges for Cross Layer Design

The merging of the proposed cognitive engine and other units requires a complex hand shaking protocols in the aspect of traditional protocol stack. The hand shaking must be so comprehensive that one single protocol must include all the necessary operations required by the flow between traditional layers, sensors and cognitive engine. Proposed approach requires cross layer design and optimization. Some of the parameters and constraints on different layers might conflict with each other. Especially for wireless communications, the solution speed (convergence rate) is very important. In other words, the solution must be carried out in very short duration of time. The solution must be performed considering limited amount of resources such as battery and power consumption. All the pieces of information necessary for optimization process must be present right before the optimization step is initiated. This requires synchronization and control between different parts of the units of the proposed method (sensors, fuzzy logic, traditional layers, etc.) Note that even though all the items listed above are solved without any further problems, my proposed method is based on individual optimization. In a multi access network optimization of an individual node does not mean a global network wise optimization [22].

5 Concluding Remarks

Cross layer design seems to be one of the most fundamental characteristics of next generation wireless networks and cognitive networks. Traditional layered architecture still exists and seems to be present in different forms in future as will. One should keep in mind that traditional architecture solves many problems even though they might be suboptimal solutions however crucial shortcomings still exist and hinders the evolution of wireless communications from being completely adaptive and cognitive. Therefore, it is wiser to maintain the fundamental structure of layers in different forms especially modifying the algorithmic order. In addition an extra block that operates whenever necessary by carrying out cross layer tasks. This way cross layer can be achieved through the use if this extra block in order to fully utilize both traditional architecture and the use of this extra block. A decision mechanism are required sitting conceptually between these two. This mechanism switches accordingly depending on the complexity of the problem considered and solution required in the light of constraints present such as relative power, battery consumption, computational complexity and so on.

The proposed approach addresses all of these considerations and issues appropriately. However, global network wise optimization has not been addressed yet, since the current form of proposed approach focuses only on individual optimization of nodes in the network.

In individual optimization, each cognitive radio follows a greedy approach trying to maximize its own system performance and throughput without considering the other devices in the network. There is no fairness involved. Each radio tries to optimize its own system performance to adapt the environment. When one radio tries to maximize its own throughput, this might hurt another radios (in the same network) throughput. There can be conflicts for resources, as specified in [18]. For e.g., if the network layer of one CR decides on using the most optimum network route to route its packets, this CR device starts dominating that route by using the same route for all its packets. This might lead to congestion on that route (as discussed in the presentation). This makes the route useless for the other cognitive radios in the network and hence their performance suffers. Since each CR doesnt consult other devices in the network but decides on its own, it is not aware of the overall situation in the network. For e.g., sometimes other CR devices may have higher priority applications running on them like real-time video transmission and might need more network resources while some other CR devices can afford to wait as they just carry non-real time data or text messages. In individual optimization, this factor is not taken into account. The CR with text messages tries to find the fastest and best route for its own transmission and reception though it can afford to wait. But the CR with real-time video is the one that needs to use the fastest route as it cannot wait. If the CR with the text message gets access to that optimum fastest route, the CR with real-time video cannot transmit and must wait. This decreases the overall network performance and throughput as delays and interruptions in real-time video will affect the QoS of the network. Hence, when a single radio performs optimization independent of other devices in the network, it is unaware of such issues in the network which need to be addressed to achieve overall global network optimization. So though each device is using the optimum technique, when it comes to the overall picture, this might hurt other radios performance in the network as every radio is being greedy and selfish in trying to achieve its own optimum performance without considering the overall network performance. Hence individual optimization may not always be the best solution for the complete network and hence doesnt lead to global network optimization.

References

1. Rappaport TS et al (1996) Wireless communications: principles and practice. Prentice Hall PTR, New Jersey
2. Mitola J, Maguire GQ (1999) Cognitive radio: making software radios more personal. IEEE Pers Commun 6(4):13–18
3. Day JD, Zimmermann H (1983) The OSI reference model. Proc IEEE 71(12):1334–1340
4. Cooper GH (1983) An argument for soft layering of protocols. Technical report, Massachusetts Inst of Tech Cambridge Lab for Computer Science
5. Clark DD, Tennenhouse DL (1990) Architectural considerations for a new generation of protocols, In: ACM SIGCOMM computer communication review, vol 20. no 4. ACM, pp 200–208
6. Haas Z (1991) A protocol structure for high-speed communication over broadband isdn. IEEE Netw 5(1):64–70
7. Arslan H (2009) Adaptation techniques and enabling parameter estimation algorithms for wireless communications systems. CRC Press, USA
8. Srivastava V, Motani M (2005) Cross-layer design: a survey and the road ahead. IEEE Commun Mag 43(12):112–119
9. Shakkottai S, Rappaport TS, Karlsson PC (2003) Cross-layer design for wireless networks. IEEE Commun Mag 41(10):74–80
10. Lin X, Shroff NB, Srikant R (2006) A tutorial on cross-layer optimization in wireless networks. IEEE J Sel Areas Commun 24(8):1452–1463
11. Balakrishnan H, Padmanabhan VN, Seshan S, Katz RH (1997) A comparison of mechanisms for improving tcp performance over wireless links. IEEE/ACM Trans Netw 5(6):756–769
12. Raisinghani VT, Iyer S (2004) Cross-layer design optimizations in wireless protocol stacks. Comput Commun 27(8):720–724
13. Xylomenos G, Polyzos GC (1999) Internet protocol performance over networks with wireless links. IEEE Netw 13(4):55–63
14. Haas ZJ (2001) Design methodologies for adaptive and multimedia networks. IEEE Commun Mag 39(11):106–107
15. Rappaport TS, Annamalai A, Buehrer R, Tranter WH (2002) Wireless communications: past events and a future perspective. IEEE Commun Mag 40(5):148–161
16. Mitola J (2000) Cognitive radio—an integrated agent architecture for software defined radio
17. Fourier J (1822) Theorie analytique de la chaleur, par M. Fourier. Chez Firmin Didot, père et fils
18. Arslan H, Yarkan S (2007) Cross-layer adaptation and optimization for cognitive radio. Cognitive radio, software defined radio, and adaptive wireless systems (Signals and communication technology). pp 421–452
19. van Der Schaar M et al (2005) Cross-layer wireless multimedia transmission: challenges, principles, and new paradigms. IEEE wirel Commun 12(4):50–58
20. van Der Schaar M, Tekalp M (2005) Network and content-adaptive cross-layer optimization for wireless multimedia communication by learning. IEEE international symposium on circuits and systems
21. Dantzig GB (1957) Discrete-variable extremum problems. Oper Res 5(2):266–288
22. Kawadia V, Kumar PR (2005) A cautionary perspective on cross-layer design. IEEE Wirel Commun 12(1):3–11

Part II
Antennae and Propagation

A Model Selection Strategy of Gaussian Process Regression for Modeling Inset-Fed Microstrip Patch Antenna

Karim Ferroudji, Abdelmaled Reddaf, Islem Bouchachi and Boudjerda Mounir

Abstract This paper presents a modeling of inset-fed microstrip patch antenna using Gaussian Process Regression (GPR) technique. The vast majority of the studies employ a readily existing model, using a fixed mean and covariance functions without further investigation. In this paper we propose a strategy to choose the most appropriate parameters of Gaussian process regression technique for modeling inset-fed microstrip patch antenna. We evaluate the influence of the choice of mean and covariance functions on the performance of the GPR models. Moreover, the dependency of the antenna resonant frequencies on the physical and geometrical properties of the materials involved, dimensions of the patch, and the feed location is investigated. In order to validate the performance of the proposed GPR model, we evaluate different algorithms with main focus on Radial Basis Function Neural Networks, and Multilayer Perceptron Neural Network. The obtained results show that the proposed method outperforms the neural network models in terms of mean square error and determination coefficient. The results give a good agreement with the results obtained using HFSS software, which ensures the validity of our proposed model in the evaluation of the resonant frequency over a spectrum range of 1–10 GHz.

Keywords Gaussian process regression · Model selection · Microstrip patch antenna · Resonant frequency

K. Ferroudji (✉) · A. Reddaf · I. Bouchachi · B. Mounir
Research Center in Industrial Technologies CRTI, P.O. Box 64, 16014 Cheraga, Algiers, Algeria
e-mail: k.ferroudji@crti.dz

K. Ferroudji
Laboratoire d'Automatique Avancée et d'Analyse des Systèmes (LAAAS), Université de Batna-2, Fesdis, Algeria

A. Boyaci et al. (eds.), *International Telecommunications Conference*, Lecture Notes in Electrical Engineering 504,
https://doi.org/10.1007/978-981-13-0408-8_7

1 Introduction

Microstrip patch antennas were the foci of a great deal of research in a wide range of applications from modern communication systems, to life saving biomedical applications, due to its compactness ability, light weight, and low fabrication cost [1–3]. Microstrip patch Antenna consists of a dielectric substrate with a fixed dielectric constant [4]. Radiating patch is present on one side of a dielectric substrate and the ground plane on other side of a substrate. The Radiating patch is designed with several shapes like circular, rectangular, triangular, elliptical etc. The resonant frequency of these antennas depends on their physical properties of the materials involved, dimensions, geometrical shape, and the feed location and type. Consequently, these antenna parameters can be modified in order to acquire the desired resonant frequency.

Artificial neural networks (ANNs) [5–7] has been widely employed in antenna modeling in the past two decades [5–7]. It has been demonstrated that various types of ANN such as Radial Basis Function Networks (RBF) and Multilayer Perceptron (MLP) are universal function approximators. However ANNs are not always easy to implement due to the ambiguity for making certain decisions regarding the type of activation function, the number of hidden units that should be chosen, or the learning rate to select [8–10].

In this paper, we propose Gaussian process regression (GPR) as an alternative to neural networks for the modeling of Inset-Fed Microstrip Patch Antenna [11]. Moreover, we describe a strategy to choose the best GPR model for this application by evaluating several mean and covariance functions [12]. We analyze the dependency of the antenna resonant frequencies as a function of physical and geometrical properties, materials involved, dimensions of the patch, and feed location using GPR algorithm. Various values of patch width, patch length, inset feed point, substrate thickness as well as dielectric constant are used in the training process. Moreover a comparison study between GPR method and Artificial Neural Networks namely MLP, and RBF is performed.

The results give a good agreement with the results obtained using HFSS software, which ensures the validity of our proposed model in the evaluation of the resonant frequency. Therefore, the GPR model can accurately estimate the resonant frequency over a spectrum range of 1–10 GHz.

The rest of this manuscript is organized in four sections. Section 2 presents the microstrip patch antenna used in this study. Section 3 describes the general block diagram as well as the three methods employed in this paper. Results and discussion are given in Sect. 4. Section 5 draws the conclusion of this study.

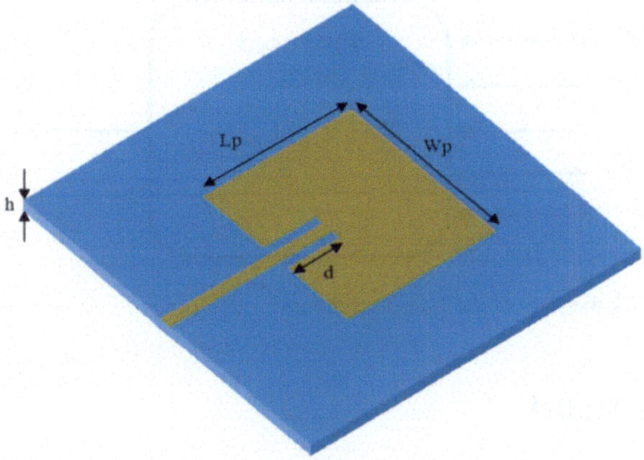

Fig. 1 Geometry of inset-fed microstrip patch antenna

Table 1 Parameter ranges of inset-fed microstrip patch antenna

Parameter	Dimension
Operating frequency f_r	1–10 GHz
Dielectric constant of the substrate ε_r	[2.1, 2.2, 4.4, 5.5, 10]
Length of the patch Lp	0.65–1.5 mm
Width of the patch Wp	0.6–12 cm
Position of inset feed point d	0.4–10 cm
substrate thickness h	0.1–3.5 cm

2 Microstrip Patch Antenna

The patch geometry of inset-fed rectangular antenna is illustrated in Fig. 1. The notch width is placed in a symmetrical way in the patch width. The parameter's dimensions have been changed in order to create a dataset for the proposed models. The thickness of the substrate must satisfy the equation below [11]:

$$h \leq \frac{c}{4f_r\sqrt{\varepsilon_r - 1}} \tag{1}$$

where, ε_r is the dielectric constant of the substrate, h is the height of the substrate, c is the air velocity, andf_r is the operating frequency. The antenna is designed in HFSS Simulator for different values of operating frequency, length of the patch, width of the patch, position of inset feed point, and substrate thickness. Moreover, several dielectric constant of the substrate are be used for the design of microstrip patch antennas and their dielectric constants are generally in the range of $2.1 \leq \varepsilon_r \leq 10$. The dataset is created using HFSS software. The Physical dimensions of inset-fed microstrip patch antenna are illustrated in Table 1.

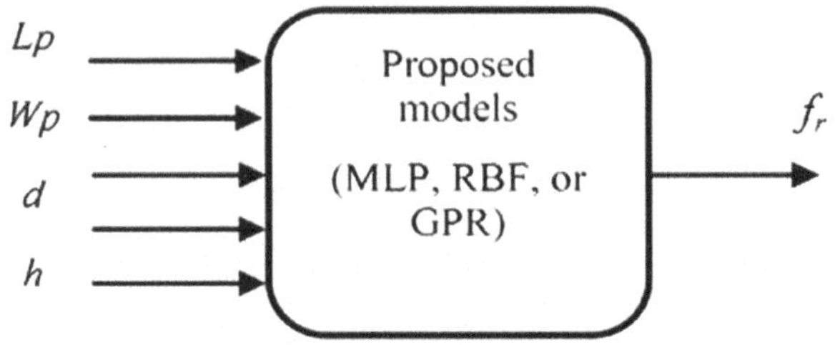

Fig. 2 General block diagram of the proposed model

3 Proposed Model

The general block diagram of the proposed model is illustrated in Fig. 2. We evaluate different algorithms with main focus on Radial Basis Function Neural Networks, Multilayer Perceptron Neural Network, and Gaussian Process Regression. The input data are first collected using HFSS simulator. The dataset is arranged as (*Lp*, *Wp*, *d*, *h*, ε_r, f_r), where *Lp* is the length of the patch, *Wp* is the width of the patch, *d* is the position of inset feed point, *h* is the substrate thickness, εr is the dielectric constant of the substrate, and *fr* is the operating frequency. Then, the dataset is randomly divided into two subsets, test set and training set. After selecting the best model using cross-validation technique, the test set is used to assess the chosen model. The training set is used to adjust the models parameters. Therefore, the algorithms have only access to the training set, the test set is kept unseen. The testing precision is obtained by applying the selected model on the testing data.

3.1 Radial Basis Function Neural Networks

Radial-basis function neural networks can be employed in a large range of applications generally because of the fact they can approximate any regular function [13]. The training of RBF networks can be divided into a linear supervised part and an unsupervised part. RBF consists of an input layer, a hidden layer (radial basis function layer), and an output layer [14]. The input nodes in the input layer are equal to the dimension of the input vector. The optimal number of neurons in the hidden layer as well as the spread of the RBF Gaussian are determined experimentally using cross validation technique.

3.2 Multilayer Perceptron Neural Networks

Multi-layer perceptron neural network is a feedforward artificial neural network model. MLP consists of a number of nodes organized into multiple layers, with each layer fully connected to the next one. MPL involve three or more layers of processing nodes: an input layer in which the input data are feed, one or more hidden layers, and an output layer. Each node is a neuron with a nonlinear activation function except for the input nodes. The MLP neural network is trained using a supervised learning by the iterative back-propagation method. The complexity of the Multi-layer perceptron network can be modified by changing the number of nodes in each layer and the number of layers [14].

3.3 Gaussian Process Regression

Gaussian process regression attract more and more attention as it becomes a powerful statistical tool for data modeling. GPR algorithms are Bayesian non parametric method. It can be exploited to solve regression and classification supervised machine learning problems. It has been applied in a wide range of applications such as: image processing problems [15], recognition of motions and activities of objects in videos [16], medical imaging [17], system identification [18], and modeling of coplanar waveguide [19].

A Gaussian process represents a collection of arbitrary variables $\{f(\mathrm{p})|\mathrm{p} \in P\}$ in which any finite subset of realizations of process $\mathrm{f} = \{f(\mathrm{p_i})\}_{i=1}^{n}$ is jointly Gaussian distributed [12]. This process is a generalization of a normal multivariate distribution into infinite dimension: it depicts a distribution over functions. A Gaussian process is characterized by its mean function:

$$\mu(\mathrm{x}) = \mathbb{E}[f(\mathrm{p})] \tag{2}$$

and covariance function:

$$(f(\mathrm{p}), f(\mathrm{p}')) = \mathbb{E}[(f(\mathrm{p}) - \mu(\mathrm{p}))(f(\mathrm{p}') - \mu(\mathrm{p}'))] \tag{3}$$

where p is the input data.

In our case we employ Gaussian process to investigate the relationship between the physical and geometrical properties of the inset-fed microstrip patch antenna $\mathrm{p} \in \mathbb{R}^2$, and the Operating frequency (s); this process is denoted by:

$$s_i = f(\mathrm{p_i}) + \in_i \tag{4}$$

The regression process is conducted by interfering the non-linear function $f(\cdot)$ from noisy observations that occurred in identical independent distribution with the variance σ_{ϵ}^{2}.

The training dataset is organized as $\{p_i, s_i\}_{i=1}^{n}$ with n observation, then we create a new 2xn matrix containing input data P (the physical and geometrical properties of the inset-fed microstrip patch antenna) and the outputs data s (operating frequency). Estimating the posterior distribution corresponds to the interference of $f(\mathrm{p})$ from a Bayesian point of view:

$$p(f(\mathrm{p})|s, P) = \frac{p(s|f(\mathrm{p}), P)p(f(\mathrm{p}))}{p(s|P)} \tag{5}$$

$p(s|f(\mathrm{p}), P)$ is the likelihood function, indicates the probability of obtaining the operating frequency s, given the function $f(\mathrm{p})$, where $p(f(\mathrm{p}))$ is the prior distribution of the latent function, which is a Gaussian process in itself.

In order to estimate operating frequency, GPR employs the covariance of neighboring (s). Gaussian process' covariance functions rely on kernel functions. The choice of a covariance function depends to the problem at hand; the most common covariance function (kernel) is the squared exponential kernel:

$$cov\left(f(\mathrm{p_i}), f(\mathrm{p_j})\right) = k\left(\mathrm{p_i}, \mathrm{p_j}\right) = \exp\left(-\frac{1}{2}\|\mathrm{p_i} - \mathrm{p_j}\|^2\right) \tag{6}$$

where K(P, P) is the covariance matrix for all pairs of training outputs.

If the error term ϵ_i in (4) is independent and identically Gaussian distributed then:

$$cov(s) = K(P, P) + \sigma_{\epsilon}^{2} I \tag{7}$$

where I and σ_{ϵ}^{2} are the identity matrix and variance of model error respectively. In order to predict new data, one can write:

$$s^* \equiv f(P^*) \tag{8}$$

where P* is the new input data (test data) and s* is the new corresponding outputs.

By assuming a zero mean prior distribution, we might depict the joint Gaussian distribution of output data as:

$$\begin{bmatrix} \mathrm{S} \\ \mathrm{S}^* \end{bmatrix} \sim \mathcal{N}\left(0, \begin{bmatrix} K(P, P) + \sigma_{\epsilon}^{2} I & K(P, P^*) \\ K(P^*, P) & K(P^*, P^*) \end{bmatrix}\right) \tag{9}$$

where $K(P, P^*)$ is a $n \times n^*$ matrix containing the covariances between all pairs of s and s*, and n^* is the size of the test data.

Conditioning this distribution on the training data gives again a Gaussian process with mean Function (10) and covariance Function (11):

$$\bar{s}^* = K(P, P)[K(P, P) + \sigma_\in^2 I]^{-1} s \tag{10}$$

$$\mathrm{cov}(s^*) = K(P^*, P^*) - K(P^*, P)[K(P, P) + \sigma_\in^2 I]^{-1} K(P, P^*) \tag{11}$$

This is the predictive distribution (posterior distribution).

The covariance and mean functions construct the principle configuration of the GPR algorithm, however the optimal hyperparameters of these functions need to be optimized in order predict the new observations correctly. This is generally performed by maximizing the marginal likelihood function, the probability of the data given the algorithm, with respect to the hyperparameters. For Gaussian distributed noise, its logarithmic form is:

$$\log p(s|X, \theta) = -\frac{1}{2} s^T [K(P, P) + \sigma_\in^2 I]^{-1} s - \frac{1}{2} \log\left|K(P, P) + \sigma_\in^2 I\right| - \frac{n}{2} \log 2\pi \tag{12}$$

where θ presents the model's hyperparameters and T is the transpose matrix. This function balances between data fit and model complexity, and it is essential to GPR model.

Furthermore, GPR framework can deal with non-Gaussian datasets. This can be done by changing the noise term of the covariance function ($\sigma_\in^2 I$) by a new covariance matrix. In addition, we choose different mean functions by selecting a set of fix basis functions and inferring their parameters from the data [12]. For more details about Gaussian process regression, please refer to [12].

3.4 Model Validation

The performance of the three proposed models is assessed by comparing the measured values with estimated values the using different statistical methods such as; Mean Square Error (MSE) and Determination Coefficient (R^2).

The MSE is the difference between the measured values and the predicted values. It is calculated by:

$$\mathrm{MSE} = \frac{1}{n} \sum_{i=1}^{n} (s - s')^2 \tag{13}$$

The determination coefficient (R^2) indicates the strength of the fitting model that illustrates the relationship between the predicted and measured values. If the regression is perfect: $R^2 = 1$, if there is no linear relationship between the predicted and the actual values, then R^2 is equal to 0:

$$R^2 = 1 - \frac{\sum_{i=1}^{n} (s - s')^2}{\sum_{i=1}^{n} (s - \bar{s}')^2} \tag{14}$$

where,

s: The targets of test data.

s': Predicted model values.

$\overline{s}$: mean of targets of test data

4 Results and Discussion

In this paper, the physical and geometrical properties of the antenna are modified to resonate the inset-fed microstrip patch antenna in a range of 1–10 GHz. The proposed models are adapted for the calculation of the resonant frequency of the antenna when the values of patch width, patch length, inset feed point, substrate thickness as well as dielectric constant are given. Thus the input data are (Lp, Wp, d, h, ε_r, f_r) and the output is the resonant frequency f_r. This dataset is used to train the proposed models. The ranges of dataset are among $2.1 \leq \varepsilon_r \leq 10$, $0.65 \leq Lp \leq 1.5$ mm, $0.6 \leq Wp \leq 12$ mm, $0.4 \leq d \leq 10$ mm, $0.1 \leq h \leq 3.5$ mm, and $1 \leq f_r \leq 10$ GHz. The dataset used in this paper is obtained from HFSS simulator, it consists of 1850 samples, this dataset is divided into two subsets; training set (1110 samples) and test set (740 samples).

In order to select best Gaussian process prior to predict the operating frequency of the antenna, we evaluated a variety of mean and covariance functions. The squared exponential (*SEard*) and the Rational Quadratic covariance function (*RQard*) are the kernels we employed to build the different covariance functions. These basis covariance functions are exploited directly or they are exploited in combination with each other, for example as multiplicand (prod of … and …) or summand (sum of …and …), to form a new covariance functions [20].

Furthermore, we employed four different mean functions; (i) the *Zero* (zero mean function); (ii) *Const* (the constant mean function), (iii) *One* (the one mean function), and (iv) the *Linear* (linear mean function). Moreover, we combined Linear and Const functions to produce another mean function using a sum function.

The evaluation of GPR models is conducted using MES and R^2 of the test. The test error is obtained from the average of tenfold cross-validation. The combination (mean and covariance functions) that results in a model with lowest test error is selected as the best GPR model.

The results for the GPR models with each of mean and covariance functions used in the evaluation algorithm are summarized in Table 2. Since no other parameters than the mean and covariance functions changed, the results give a good indication on the suitability of the mean and covariance functions for the prediction of the operation frequency of the antenna. The best results appear in bold.

Table 2 indicates that the performance of the GPR model greatly depends on the selection of the of mean and covariance functions.

The best performances corresponding to the lowest test error are obtained using zero mean function and squared exponential covariance function which are the most appropriate mean and covariance functions for this application.

Table 2 Test error of GPR models for different mean and covariance functions

Covariance functions				
Mean functions	SEard	RQard	Sum(SEard, RQard)	Prod(SEard, RQard)
Zero	$\mathbf{1.45 \times 10^{-4}}$	19.90×10^{-4}	13.65×10^{-4}	203.67×10^{-4}
Const	5.79×10^{-4}	13.47×10^{-4}	12.14×10^{-4}	14.45×10^{-4}
One	1.46×10^{-4}	58.23×10^{-4}	26.36×10^{-4}	4.68×10^{-4}
Linear	2.58×10^{-4}	25.28×10^{-4}	11.57×10^{-4}	172.32×10^{-4}
Sum (Linear and Const)	2.76×10^{-4}	16.82×10^{-4}	19.28×10^{-4}	278.42×10^{-4}

Table 3 Comparison results of the three proposed techniques

	MSE	R^2
GPR	$\mathbf{1.45 \times 10^{-4}}$	0.9997
MLP	1.54×10^{-4}	0.9979
RBF	6.94×10^{-4}	0.9904

In order to evaluate the performances of the proposed GPR model, we compared the performance of the obtained GPR model with two neural networks methods (MLP and RBF), the results are illustrated in Table 3. The test error is calculated by comparing the data obtained using HFSS software and those estimated using the proposed models.

Table 3 indicates that the best results (lowest test error) are obtained using GPR algorithm based on the selected mean and covariance function. It should be noted that, for the present work, we used Multilayer Perceptron Neural Networks with two hidden layers (12 and 11 neurons in the first and the second hidden layer respectively), and radial basis function with 12 neurons in the hidden layer. The choice of the MLP and RBF parameters is conducted using tenfold cross-validation.

Another measure of how well the model has recognize the data samples is the regression plot (refer to Figs. 3, 4, and 5). Here the regression is plotted using test values. The regression plot shows the predicted outputs of the model plotted in terms of the associated true values (targets). If the model recognize the data well, the linear fit to this output-target relationship should cross the top-right corners and bottom-left of the plot. R^2 coefficient is an indication of the relationship between the targets and predicted values. If $R^2 = 1$, this means that there is an exact linear relationship linking targets and outputs and. If R^2 is zero or close to zero, then there is no linear relationship between targets and outputs. It should be noted that the best R^2 is obtained using the GPR model.

The evaluation of the three models (MLP, RBF, and GPR) is performed using 10- fold cross-validation to guarantee statistical validity of the results and avoid overfitting [21].

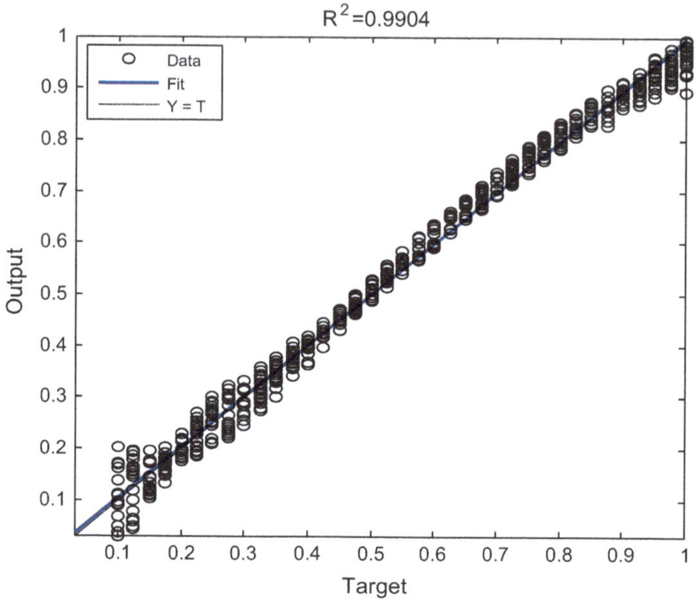

Fig. 3 RBF outputs against target values

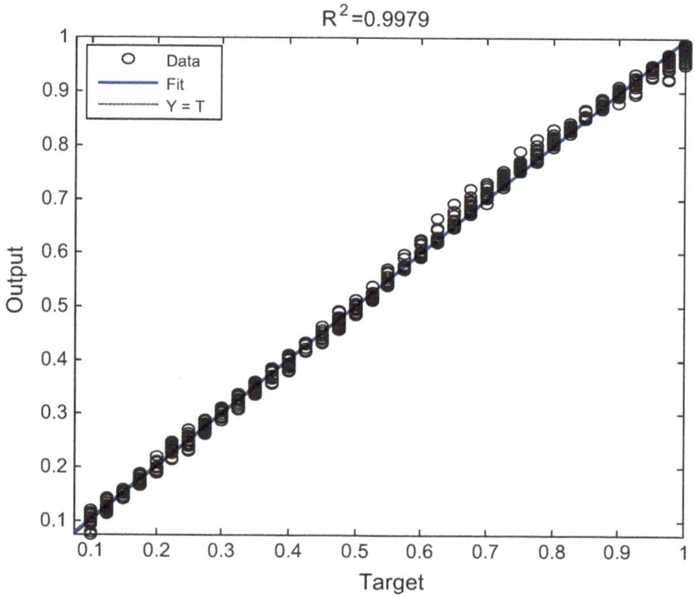

Fig. 4 MLP outputs against target values

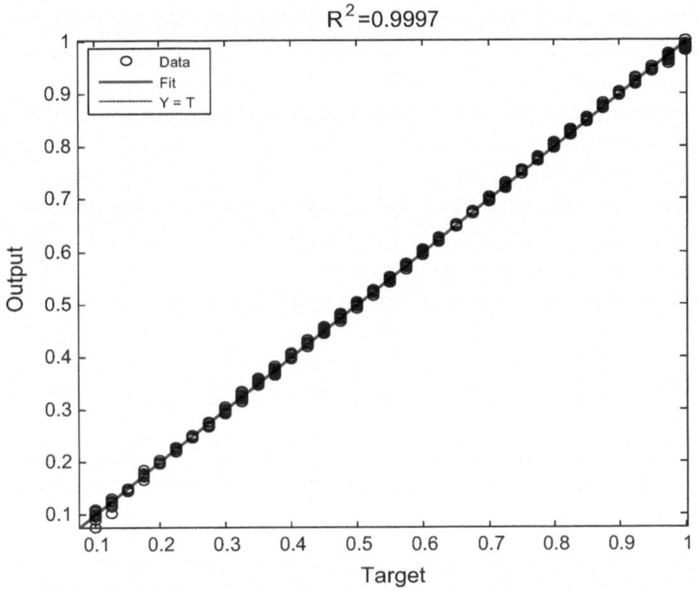

Fig. 5 GPR outputs against target values

The main objective of this study is to assess the performances of model selection in GPR approach in predicting the operating frequency of the antenna. Therefore, we did not focus on the processing time of the algorithms. However, it should be noted that the training stage for the MLP and RBF neural networks takes few minutes while the training stage for GPR model lasts for less than one minute. The processing time for the test stage is less 1 s.

5 Conclusion

This paper presented a modeling of an inset-fed microstrip patch antenna using Gaussian Process Regression technique. We investigated the dependency of the antenna resonant frequencies on the physical and geometrical properties of the materials involved, dimensions of the patch, and the feed location. Since the performances of the GPR model highly depends on the choice of mean and covariance functions, we proposed in this paper a strategy to select the most suitable mean and covariance functions for predicting the operating frequency of the antenna. Considering the test error, it has been demonstrated that the GPR model based on squared exponential covariance function and zero mean function provides better performances (lowest test error) compared to the other proposed GPR-models. The results indicated that GPR technique is indeed appropriate for predicting the operating frequency of the antenna in a range of 1–10 GHz. Furthermore, we compared the obtained results based on the selected mean and covariance functions with two Neural Networks methods (MLP, and RBF). The results demonstrated that GPR model achieved better performances compared to the Neural Networks approaches. The superiority of GPR technique over Neural Networks methods can be explained by the fact that: (i) GPR models are less sensitive to relatively small training data sets, (ii) The performance of GPR models relies on the choice of mean and kernel functions, but this dependence is less influential. (iii) The advantage of the GPR model is related mainly to its simplicity of implementation, and fast processing time.

References

1. Ali Z, Singh VK, Singh AK, Ayub S (2013) Wide band inset feed microstrip patch antenna for mobile communication. In: International conference on communication systems and network technologies (CSNT), pp 51–54
2. Kalyan M, Partha Pratim S (2016) Half hexagonal broadband high gain microstrip patch antenna for mobile and radar applications. Microwave Opt Technol Lett 58(5):1028–1032
3. Çalışkan R, Gültekin SS, Uzer D, Dündar Ö (2015) A microstrip patch antenna design for breast cancer detection. Proc-Soc Behav Sci 195:2905–2911
4. Balanis CA (2016) Antenna theory: analysis and design. Wiley, New York
5. Karaboga D, Guney K, Sagiroglu S, Erler M (1999) Neural computation of resonant frequency of electrically thin and thick rectangular microstrip antennas. IEE Proce-Microwaves Antennas Propag 146(2):155–159
6. Gupta VR, Gupta N (2005) An artificial neural network model for feed position of the microstrip antenna. Elektronika Ir Elektrotechnika 60(4):82–89
7. Sarkar BD, Shankar S, Thakur A, Chaurasiya H (2015) Resonant frequency determination of rectangular patch antenna using neural network. In: 1st International conference on next generation computing technologies (NGCT). IEEE, pp 915–917
8. Rasmussen CE, Williams CK (2006) Gaussian processes for machine learning, vol 1. MIT Press, Cambridge
9. Zhang QJ, Gupta KC, Devabhaktuni VK (2003) Artificial neural networks for RF and microwave design-from theory to practice. IEEE Trans Microw Theory Tech 51(4):1339–1350
10. Angiulli G, Cacciola M, Versaci M (2007) Microwave devices and antennas modelling by support vector regression machines. IEEE Trans Magn 43(4):1589–1592
11. Hansen RC, Burke M (2000) Antennas with magneto-dielectrics. Microwave Opt Technol Lett 26(2):75–78
12. Rasmussen CE, Williams CK (2004) Gaussian processes in machine learning. Lect Notes Comput Sci 3176:63–71
13. Gupta M, Jin L, Homma N (2004) Static and dynamic neural networks: from fundamentals to advanced theory. Wiley, New York
14. Benoudjit N, Ferroudji K, Bahaz M, Bouakaz A (2011) In vitro microemboli classification using neural network models and RF signals. Ultrasonics 51(3):247–252
15. He H, Siu WC (2011) Single image super-resolution using Gaussian process regression. In: IEEE conference on computer vision and pattern recognition (CVPR). IEEE, pp 449–456
16. Kim K., Lee, D., Essa, I.: Gaussian process regression flow for analysis of motion trajectories. In IEEE international conference on Computer vision (ICCV), pp. 1164–1171. IEEE. (2011)
17. Zhu F, Carpenter T, Gonzalez DR, Atkinson M, Wardlaw J (2012) Computed tomography perfusion imaging denoising using Gaussian process regression. Phys Med Biol 57(12):N183

18. Chan LLT, Liu Y, Chen J (2013) Nonlinear system identification with selective recursive Gaussian process models. Ind Eng Chem Res 52(51):18276–18286
19. Jacobs JP, De Villiers JP (2010) Gaussian-process-regression-based design of ultrawideband and dual-band CPW-fed slot antennas. J Electromagn Waves Appl 24(13):1763–1772
20. Duvenaud DK, Nickisch H, Rasmussen CE (2011) Additive gaussian processes. In Advances in neural information processing systems, pp 226–234
21. Arlot S, Lerasle M (2016) Choice of V for V-fold cross-validation in least-squares density estimation. J Mach Learn Res 17(208):1–50

A Comparison of Genetic Algorithm and Practical Swarm Optimization for the Design of Waveguide Filters

I. Bouchachi, K. Hamdi-Cherif, K. Ferroudji, M. Boudjreda, A. Reddaf and M. L. Riabi

Abstract When high specifications are required in the design of microwave structures, and there is no analytical method or it is very complicated with high range of variables, the use of optimization method becomes inevitable. In this paper a comparison of two population based optimization methods, inspired from nature is presented. We employed genetic algorithm and practical swarm optimization techniques. To carry out this comparison, two filters are synthesized. The comparison is based on the rate of success to solve problems, the number of iterations, and the calculation time.

Keywords Genetic algorithm · Practical swarm optimization · Filter synthesis

1 Introduction

Several research papers deal with Direct filter synthesis [1, 2]. These papers concentrate on the synthesis of the coupling matrix, or the equivalent electrical circuit of the filter. However, they didn't address the direct waveguide filter synthesis. In the aforementioned papers, the authors stated that they used software or optimization methods to convert the coupling matrix or equivalent circuit to a waveguide filter. This inevitable use of optimization method is due to the complexity of the problem that emerges from the electromagnetic fields behavior, and the number of variables (dimension of the filter, the huge number of high order modes appearing at each waveguide discontinuities).

In the last two decades, genetic algorithm (GA) attract more and more attention in a wide range of applications such as economy, medicine, electronic....etc. due

I. Bouchachi (✉) · K. Hamdi-Cherif · K. Ferroudji · M. Boudjreda · A. Reddaf
Research Center, Idustrial Technologies CRTI, P.O. Box 64, 16014 Cheraga, Algiers, Algeria
e-mail: i.bouchachi@crti.dz

I. Bouchachi · M. L. Riabi
Laboratory of Electromagnetism and Telecommunications, University of Constantine 1, Constantine, Algeria

A. Boyaci et al. (eds.), *International Telecommunications Conference*, Lecture Notes in Electrical Engineering 504,
https://doi.org/10.1007/978-981-13-0408-8_8

to its ability of making a global search and explore the search space using different kind of crossover. On the other hand, practical swarm optimization PSO, proposed by Kennedy and Eberhart in the mid 1990s, is one of the latest optimization method. PSO has become increasingly a powerful optimization tool because of its simplicity in programming and its speed of execution.

This paper presents a synthesization of two filters: dual-mode filters of second and fourth order respectively that have been proposed in [3]. We chose these two filters because we are treating numerous similar microwave structures using the mode-matching method [4]. Note that this is one of the most powerful and suitable method for analyzing this kind of structures.

2 Practical Swarm Optimization

Particle Swarm Optimization (PSO) is a relatively new technique of optimization used to solve continuous non-linear problems. It was first presented in 1995 by Jim Kennedy who discovered the method through simulation of a simplified social model [5]. PSO is a very simple concept, the paradigms are implemented in a few lines of computer code. It requires only primitive mathematical operators. Therefore, it is computationally inexpensive in terms of both memory requirements and speed. PSO has been recognized as an evolutionary computation technique; it is basically developed through simulation of bird flocking in two dimensional spaces. The position of each agent is represented by the XY axis position, whereas the velocity is expressed by vx (the velocity of X axis) and vy (the velocity of Y axis.)

Each individual i in the swarm contains parameters for the position x_i and the velocity y_i. The position of each particle represents a potential solution to the optimization problem.

The swarm behavior in conventional particle swarm optimization is influenced by the following: number of particles, neighborhood size, inertia weight, maximum velocity, and acceleration calculation to modify the velocity.

2.1 PSO Notation

For each particle i:
x_i is a vector denoting its position and y_i denotes its objective function value.

- y_i is the vector denoting its velocity.
- x_i is a vector denoting its position and y_i denotes its objective function value.
- y_i is the vector denoting its velocity.
- p_i is the best position that has found so far and $Pbest_i$ denotes its objective function score.

- g_i is the best position that has been found so far in its neighborhood and $Gbest_i$ denotes the objective function value of g_i.
- $\vec{U}(0, \varphi_i)$ is a random vector uniformly distributed in $[0, \varphi_i]$ generated at each generation for each particle.
- φ_1 and φ_2 are the acceleration coefficients determining the scale of the forces in the direction of p_i and g_i
- $\otimes$ denotes the element-wise multiplication operator.

2.2 PSO Algorithm

- Randomly initialize particle positions and velocities.

  For each particle i:
  While not terminate:

- Evaluate fitness y_i at current position x_i.
- If y_i is better than $Pbest_i$ then update $Pbest_i$ and p_i.
- If y_i is better than $Gbest_i$ then update $Gbest_i$ and g_i.

For each particle

- Update velocity y_i and position x_i using:

$$\vec{v}_{i+1} \leftarrow \vec{v}_i + \vec{U}(0, \varphi_1) \otimes (\vec{p}_i - \vec{x}_i) + \vec{U}(0, \varphi_2) \otimes (\vec{g}_i - \vec{x}_i) \quad (1)$$

$$\vec{x}_{i+1} \leftarrow \vec{x}_i + \vec{v}_i \quad (2)$$

3 Genetic Algorithm

Genetic algorithms are stochastic global-search algorithms. This procedure begins by creating an initial M individuals random population of chromosomes [6, 7]. Each chromosome represents the coding of a data set X (potential solution of the problem). After the evaluation of every chromosome, they will be sorted from the best to the worst. Only the healthiest half of the population is allowed to survive to the next generation. To choose the parents, each chromosome is assigned a probability of selection on the basis of its rank in the sorted population (refer to Fig. 1).

The reproduction by crossover and mutation is then performed on the selected parents to generate the offspring. The evolution optimization process is reiterated until satisfaction of halting criterion or if a total number of generations is reached [8].

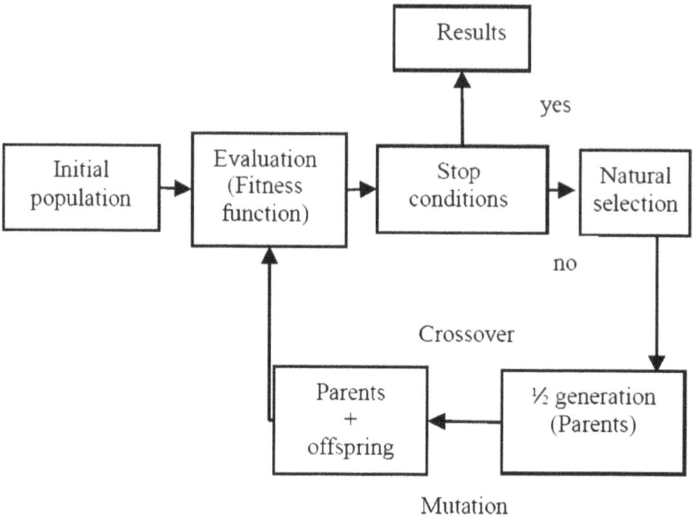

Fig. 1 Flowchart for the Genetic algorithm

4 Fitness Function

The fitness function is the most important part of any optimization method. It should correspond to an evaluation of how good the candidate solution is. Subsequently, the optimal solution is the one which minimizes the fitness function [9]. Moreover, it has an important influence on the computing time, because this step includes the analysis of the structure to optimize (mode matching method in this work) to get a better idea of this influence. For example, if we have 20 individuals in our population and we need 20 generations to solve the problem, by taking into consideration that every individual (potential solution) must be analyzed in 1/3 s, then this step of the program alone, will take 20*20/3 s. That is more than 2 min, which is almost the duration of all program execution time.

4.1 Discrete Target Approximation

Is a simple evaluation of the specifications fulfillment, it consist of assigning to each one of the N frequency points a value of 0 if the response function does not satisfy the corresponding specification and a value of 1 if it satisfies the specification. The final fitness is the sum of the {0, 1} values for all the response points, divided by the number of points [9].

5 Numerical Results

As mentioned above, we resynthesize two known filters proposed by Jarry et al. [3] to compare these two optimization methods. We used the first filter (dual-mode filter of second order) to compare the convergence speed, the calculation time, and the success rate. The second filter (dual-mode of forth order) is employed to study the influence of variable range on the calculating time (number of iteration). It should be noted that the initial filter response before optimization is shown in Fig. 2. The response of the same filter after optimization using GA and PSO is illustrated in Figs. 3 and 4 respectively.

With regard to the effectiveness recall that, both methods have a stop criterion, which is either by satisfying the filter specification or by reaching the maximum iteration number. If the filter specification is not satisfied we consider that the program failed to solve the problem. Then, by 50 trials we notice that the GA is slightly more effective than the PSO with success rate of 93% for the first and 85% for the second (Fig. 4).

For the same 50 trials we calculate and draw the average of the convergence curve for the two methods. We notice that the tested evaluation functions work well and converge at the almost same iteration number. From the curves in Fig. 5 we can observe that the solution is reached approximately at the 12th iteration for the PSO method, and at the 20nd iteration for the GA from the curve in Fig. 6, which is a considerable difference that allows us to conclude that PSO is twice faster than GA.

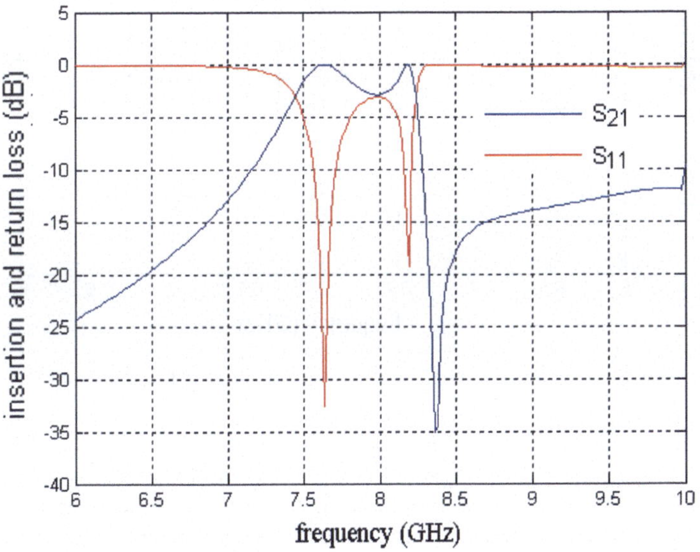

Fig. 2 Initial filter response before optimization

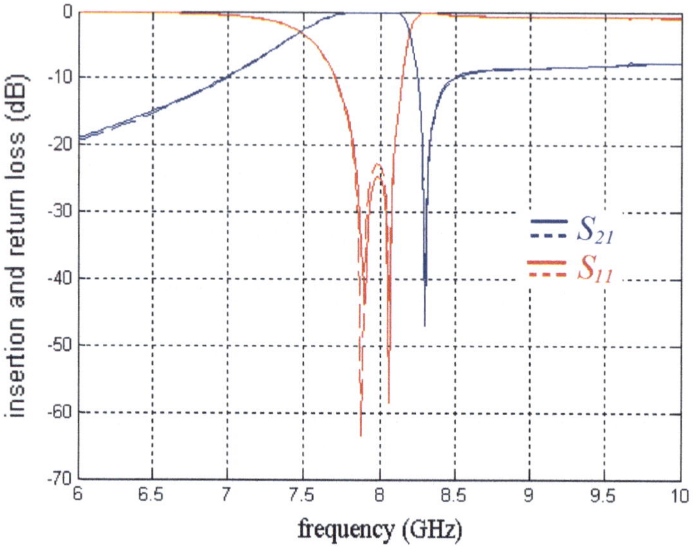

Fig. 3 Superposition of optimized filter using GA (broken line) and response in [3]

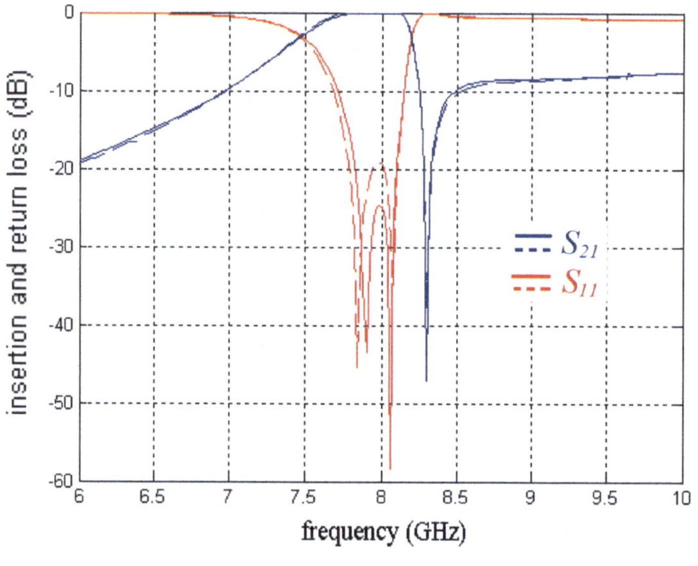

Fig. 4 Superposition of optimized filter using PSO (broken line) and response in [3]

In the second part of this work, we use a simple technique to analyze the influence of variable number on the iteration number. It consists of synthesizing the 4th order filter Fig. 7. It consists of finding the ten dimensions of the waveguide filter (Fig. 8).

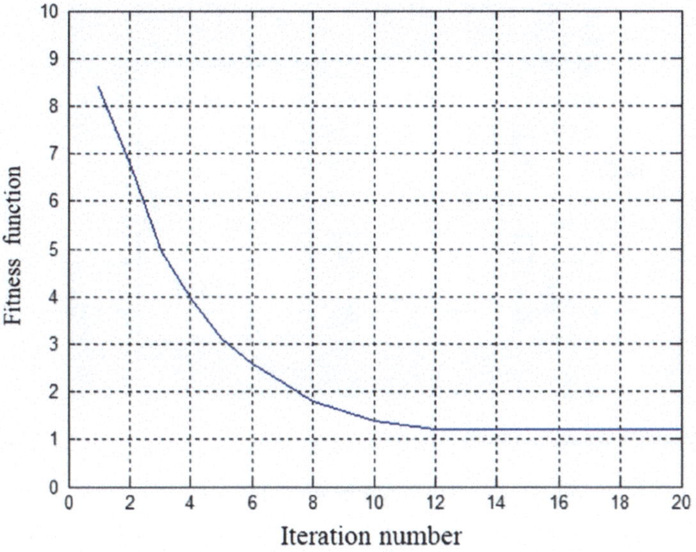

Fig. 5 PSO solution convergence according to Iteration number

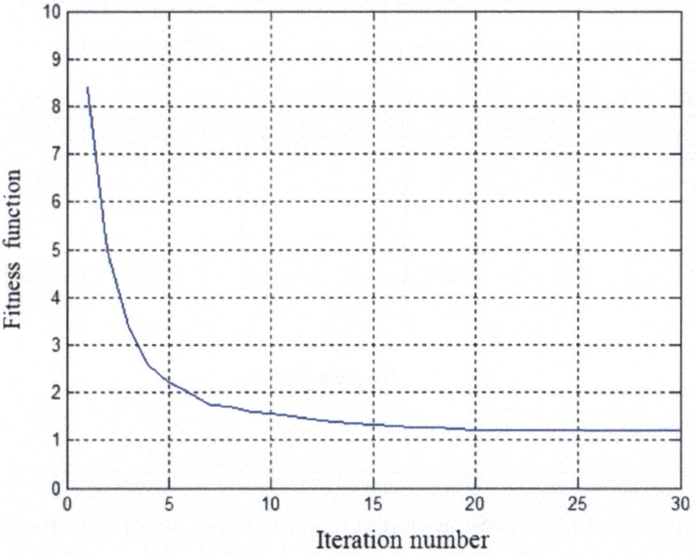

Fig. 6 GA solution convergence according to iteration number

After saving the results, we resynthesize the same filter by inserting nine of the ten dimensions which means reducing the problem to find only one dimension. Then we repeat the maneuver for two, three …until nine variables. We did 20 trials for

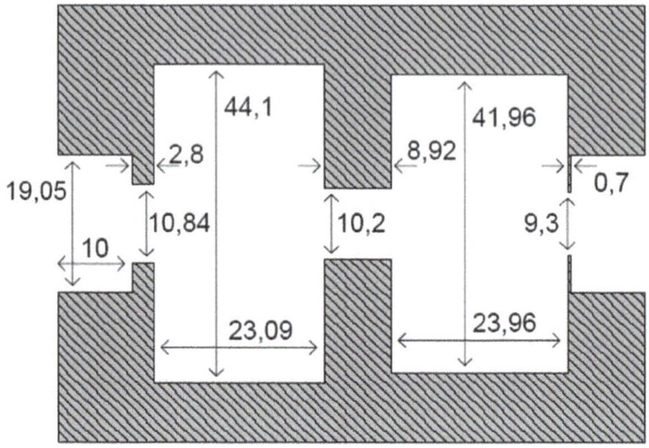

Fig. 7 Waveguide filter section from [3]

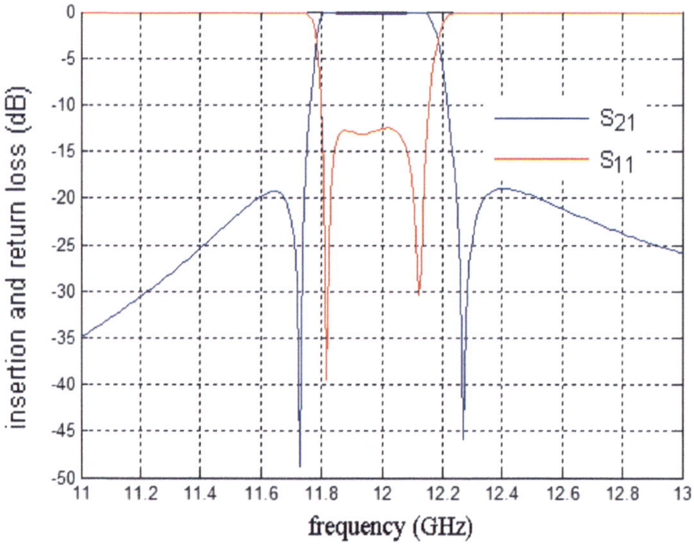

Fig. 8 Fourth order waveguide response

each maneuver in order to plot the iteration number variation according to variable number (blue curve in Fig. 9) Note that the red curve in Fig. 9 is a mathematical interpolation of our results by ($y = x^2 + 1$).

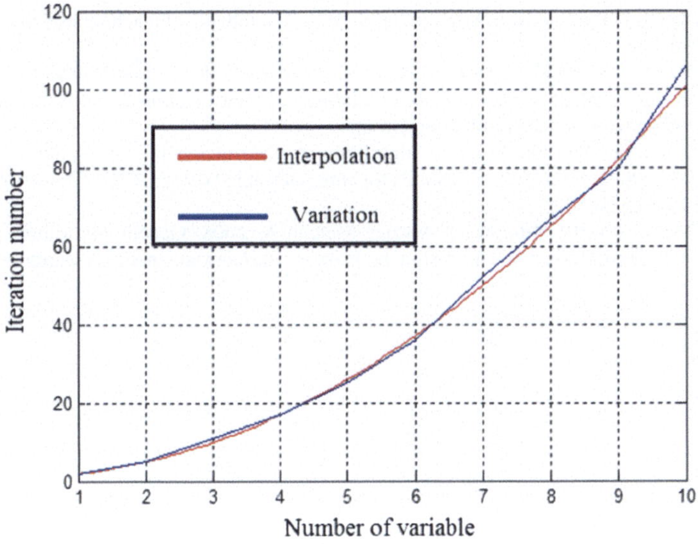

Fig. 9 Iteration number variation according to variable number compared with mathematical interpolation

6 Conclusion

PSO is similar to the GA in the sense that they are both population-based search approaches and that they both depend on information sharing among their population members to enhance their search. The two methods are favorable to perform with high success rate the tasks of synthesizing microwave structures, with a slight advantage for the GA.

The PSO makes the difference in the simplicity of programming (few line of program code) and its speed of executing time, as compared to the GA with its relative complicated program, numerous operations (natural selection, cross over, mutation, binary conversion) and the need of more iterations to solve problems.

References

1. Amari S, Rosenberg U (2004) Direct synthesis of a new class of bandstop filters. IEEE Trans Microw Theory Tech 52(2):607–616
2. Wu KL, Meng W (2007) A direct synthesis approach for microwave filters with a complex load and its application to direct diplexer design. IEEE Trans Microw Theory Tech 55(5):1010–1017
3. Jarry P, Kerherve E, Pham JM, Roquebrun O, Guglielmi M (2004) Synthesis and realizations of a new class of dual-mode microwave rectangular filters. J Microw Optoelectron Electromag Appl (JMOe) 3(5):41–46
4. Bornemann J, Vahldieck R (1990) Characterization of a class of waveguide discontinuities using a modified TE_{mn}^{x} mode approach. IEEE Trans Microw Theory Tech 38(12):1816–1822

5. Kennedy J (2011) Particle swarm optimization. In: Encyclopedia of machine learning. Springer, US, pp 760–766
6. Holland JH (1992) Genetic algorithms: Computer programs that "evolve" in ways that resemble natural selection can solve complex problems ever their creators do not fully understand. Scientific American, Nature Publishing Group
7. Haupt RL, Werner DH (2007) Genetic algorithms in electromagnetics. Wiley, New Jersey
8. Bandler JW (1969) Optimization methods for computer-aided design. IEEE Trans Microw Theory Tech 17(8):533–552
9. Nogales MJ, Garcıa JP, Hinojosa J, Alvarez-Melcón A (2008) Genetic algorithms applied to microwave filters optimization and design. In: Progress in electromagnetics research symposium, pp 99–103

Part III
Localization

Deep Learning in Indoor Localization Using WiFi

Zeynep Turgut, Serpil Üstebay, Gülsüm Zeynep Gürkaş Aydın and Ahmet Sertbaş

Abstract In this study, the indoor localization was performed on indoor networks. WiFi technology is located in almost every building. For this reason, WiFi technology has been selected to perform positioning, and RSSI values from WiFi technology access points have been examined. For this purpose, RFKON_MB_WIFI dataset in RFKON database which is a sample database is used and data of 18480 signal strength are analyzed. The Fingerprinting method of Scene Analysis methods is used to perform the localization process. As a first step, the signal strengths in the data set are normalized by preprocessing. In the second step, positioning was performed using SVM, PCA, LDA, KNN, N3, BNN, Naive Bayes Classification, and Deep Learning methods. When the results obtained are compared, the most successful result is obtained from deep learning that is known to have a high accuracy on big data with an accuracy of 95.95%.

Keywords Indoor · Localization · Deep learning · Stacked autoencoder Fingerprinting

---

Z. Turgut
Halic University, Istanbul, Turkey
e-mail: zeynepturgut@halic.edu.tr

S. Üstebay · G. Zeynep Gürkaş Aydın (✉) · A. Sertbaş
Istanbul University, Istanbul, Turkey
e-mail: zeynepg@istanbul.edu.tr

S. Üstebay
e-mail: serpil.ustebay@istanbul.edu.tr

A. Sertbaş
e-mail: asertbas@istanbul.edu.tr

A. Boyaci et al. (eds.), *International Telecommunications Conference*,
Lecture Notes in Electrical Engineering 504,
https://doi.org/10.1007/978-981-13-0408-8_9

1 Introduction

Indoor localization is one of the most frequently used topics in recent years. Thanks to the use of GPS systems, location information of persons or objects can be easily determined with high accuracy in outdoor. However, due to the difficulties of indoor networks, GPS systems cannot achieve the same success in closed areas. Communication challenges such as obstacles, fading, reflections, multipath encountered in indoor localizations are at the top of these difficulties. WiFi technology is the most studied technology because almost every building has a WiFi system inside and a result of that WiFi is one of the key technologies to design a positioning system that can work effectively in indoor areas. X. Zhang, J. Wang, Q. Gao, X. Ma, and H. Wang [1] suggest a WIFI positioning based on deep learning in their studies. In order to cope with various and unpredictable wireless signals, a deep artificial neural network has been proposed to perform the indoor localization process. This method is very useful in applications where it is desired to learn reliable features and where the working space is composed of large noisy samples. In the study KNN (K- Nearest Neighbor), LLE (Locally Linear Embedding) and DNN (Deep Neural Network) methods were compared using fingerprinting technique. Thus, it has been proved that the proposed method is not only able to cope with the difficulty of signal diversity, but also is successful with large data clusters. For this reason, WiFi networks are used in this study in order to determine the location in indoor localization. Deep Learning method which is more successful on large data is compared performed with Support Vector Machine (SVM), Principal Component Analysis (PCA), Linear Discriminant Analysis (LDA), KNN, N Nearest Neighbors Algorithms (N3) and Naïve Bayes methods by using RSSI values in RFKON database (an open source database includes Received Signal Strengths gathered from Eskişehir Osmangazi University).

2 Indoor Localization Using WiFi

In indoor areas which have got WiFi technology, the methods of localization are divided into two as deterministic and probabilistic methods [2]. Deterministic methods can determine position more quickly using conventional calculation methods such as Trilateration, Triangulation, Proximity, and Dead Reckoning. The hierarchy of indoor localization techniques is shown in Fig. 1 [3]. The distance from the mobile stations to the stations is measured by the trilateration method. In order for the signal propagation models to work correctly, at least 3 access point positions must be known. Trilateration methods measurement techniques such as Received Signal Strength (RSS), Time of Arrival (ToA), and Time Difference of Arrival (TDoA) for measuring distance.

It is appropriate to use triangulation technique when area has a direct field of view between the mobile user and the reference point. They can use methods such as Angle of Arrival (AoA), Angle of Departure (AoD) and so on. Proximity technique

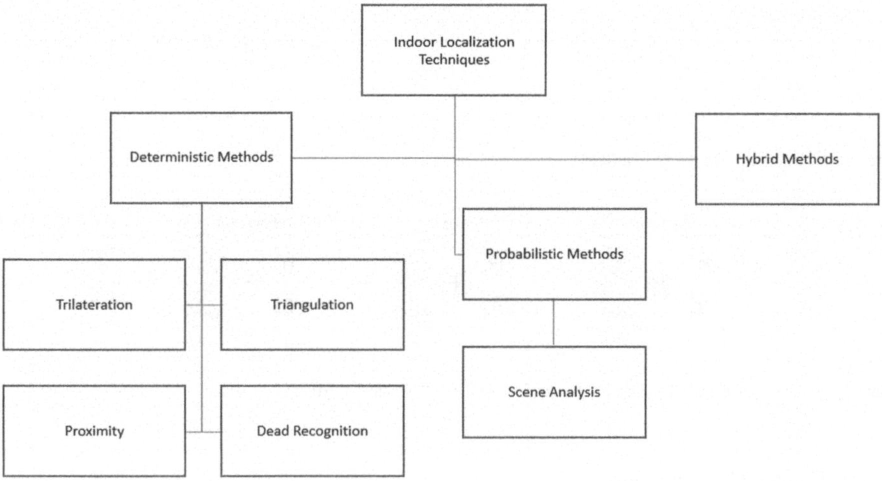

Fig. 1 The hierarchy of indoor localization techniques

checks whether two devices are connected or within a specific zone [2]. It is the positioning technique which is measured by measuring the values such as direction, speed, acceleration through the sensors. Dead Reckoning method is a process for estimating the position of the end point with known or predicted speed information according to the latest known position. Deterministic methods, as a difference from probabilistic methods, use simple calculation techniques. Probabilistic methods predict the position of the objects in the next step, taking into account the fact that they are not exactly in the current measurements. For this reason, probabilistic methods perform more accurate localization than deterministic methods [4, 5]. Therefore, in this study fingerprinting method which has a region based approach is used from Scene Analysis methods. The fingerprinting method consists of two phases; training and positioning. During the training phase, RSSI values of a number of access points is calculated. RSSI values (usually offline) of a number of position coordinates are obtained before the positioning operation starts. Location detection is estimated by comparing the information obtained in the offline phase with the signal parameter (RSSI) obtained in the online phase [4].

3 Methodology

In the second phase, which is called as localization (online stage), the location is determined using SVM, PCA, LDA, KNN, N3, Naive Bayes and Deep Learning methods and the results are compared.

A. RFKON_MB_WiFi dataset

ID	Time	DeviceID	x	y	Floor	BatteryLevel	MAC 1	MAC 2	…	MAC n
1	11-05-2015 09:52:30	1	1.2	1.2	1	97	-67	-100	…	-100
2	11-05-2015 09:52:40	2	1.2	1.2	1	97	-100	-100	…	-100
3	11-05-2015 09:52:50	3	1.2	1.2	1	97	-65	-100	…	-100
4	11-05-2015 09:53:00	4	1.2	1.2	1	97	-64	-100	…	-100

Fig. 2 RFKON_MB_WIFI dataset

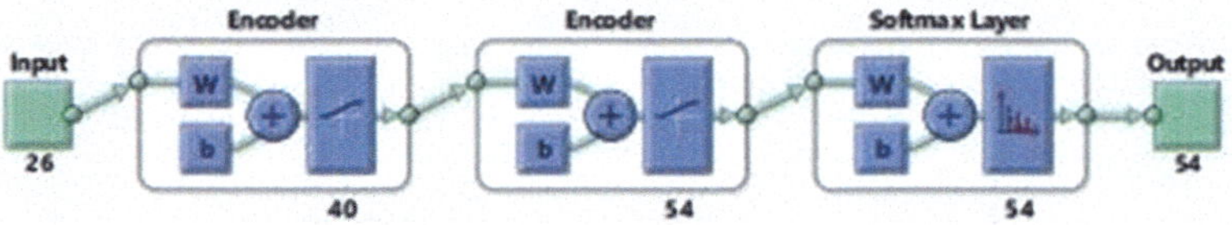

Fig. 3 Structure of autoencoders. https://www.mathworks.com/

3.1 RFKON Database

In this study, WiFi based positioning using the fingerprinting method was performed on a sample data set. In offline phase of the fingerprinting method RFKON_MB_WIFI data set is used which RFKON database contains. RFKON database is an open source database that has got data collected from Eskişehir Osmangazi University Technopark area. The data in the RFKON database is taken from the 800 m^2 area covered by a 2 floored building. The total area is divided into 86 square areas of each (1.2 m × 1.2 m), and from 26 different wireless access points located in the center of these areas the severity levels of the WIFI signals (RSSI) were collected. The RFKON_MB_WIFI dataset contains a total of 18480 measurements from 26 wireless access points in 54 different coordinates. Figure 2 shows the first 4 records of the RFKON_MB_WIFI data set [6].

3.2 Signal Preprocessing

The measurements in the RFKON_MB_WIFI dataset contain negative values. However, preprocessing has been carried out on the signals in order to compare deep learning with other methods used in this study. The data used in some classifiers should contain positive values. For this reason, the values in the data set are first converted to non-negative numbers using Eq. 1 and preprocessed on the signal [5]. After conversion, high values indicate strong signal intensity, while low values indicate weak signal intensity (Fig. 3).

$$Positive_i = \begin{cases} (RSS_i - min/100), If\ signal\ strength(RSS_i) \geq \tau \\ 0, otherwise \end{cases} \quad (1)$$

Where the RSSi used in the conversion process represents value of the ith measurement, min represents the minimum RSS value in the data set. The minimum value for the RFKON_MB_WIFI dataset is –100 dBm. The RSSI signal values vary between 100 and –100 dBm. Here –100 dBm indicates no signal strength. For this reason, the threshold value, τ, is assumed to be –100 dBm. Equation 2 shows the first phase. In the first phase called offline, RSS values of M access points are collected from known l (Locations) reference points and these datas are processed and stored for later use [7].

$$r_i(\tau) = [r_{i1}, r_{i2}, \ldots, r_{iM}]^T, \tau = 1, 2, \ldots t\, t > 1 \tag{2}$$

In the fingerprinting method, a previously recorded 4- tuple information on the RFKON database is used in the offline phase for an RSS value which can be seen in Eq. 3.

$$(r_i(\tau), l_i) \tag{3}$$

where l_i represents the known position taken from the x and y coordinates of the reference point in Eq. 4.

$$l_i = [x_i, y_i]^T, i = 1, \ldots, L \tag{4}$$

3.3 Compared Localization Methods

- Support Vector Machine (SVM): One of the recommended approaches for the linear classification problem is marginal classifiers, which is the most popular Support Vector Machines (SVM) classifiers of these classifiers [8]. In the SVM approach, an equation separating the positive and negative class samples from a two-dimensional S space is calculated. According to this equation, the data are classified.
- Principal Component Analysis (PCA): Basic Component Analysis is a statistical method used to monitor the distribution of two- or three- dimensional space data depending on the size of the (feature) [9]. The main purpose of the method is to find a new data space with base vectors (basic components) that can better distinguish the ones in the different group from the ones that make up the data set, and to move the data to the new distance by means of these vectors, in other words transferring the data. It is targeted that the amount of change in each basic component of the new space which the transferring is reduced is the largest.
- Linear Discriminant Analysis (LDA): Linear Discriminant Analysis method [10] groups the samples by class labels and aims to find the basic components of the new space that are to be moved in a similar manner to the TBA method for two main purposes. These two purposes for the samples in the new data space are:

  - To maximize the exchange of centers of different groups,
  - To minimize the variation within each group.

- K-Nearest Neighbors Algorithm (KNN): The data to be categorized depending on the closest K neighbors to itself. The multiple class of the K neighbors includes the class of the new data. Uses Euclidean distance as the distance measurement. The equation for distance measurement is shown in Eq. 5. Although different distance distances have been used in the literature [5], the most common use is the Euclidean.

$$Distance = \sqrt[2]{(X_1 - X_2)^2 + (Y_1 - Y_2)^2} \tag{5}$$

- N-Nearest Neighbors Algorithm (N3): It is an algorithm based on KNN. All neighbors are used when the classification is done. The similarity value, which is the data to be classified for each neighbor, is calculated and these values are ordered from highest to lowest. These data are used as a weight vector. The class that most closely resembles itself is identified and labeled with the weight vector [9].
- Binned Nearest Neighbors Algorithm: BNN works similarly to the NN algorithm. The similarity values between all neighbors and the data which is desired to be classified are calculated. A circle is formed around the data to be classified by a structure called a defined bin. Categorization is performed considering the neighbors with the highest similarity values within this circle [11].
- Naïve Bayes Classification: Naive Bayes makes statistical estimation according to Bayes Theorem [12]. For calculating similarity value probability is used. By using probability values which are obtained from teaching data the class of test data are determined. As the number of teaching data increase, the success rate of classifying test data increase.

3.4 Localization with Deep Learning

In this study autoencoders are used. An autoencoder is a type of artificial feedforward neural network which aims to reconstruct its own inputs. The parameters contained in the autoencoder are trained to get their inputs on output. For this reason, the number of inputs of an autoencoder is always equal to the number of outputs. An autoencoder can be used for dimensionality reduction if the number of hidden units are less than the number of input units. The autoencoder is responsible for preprocessing and extracting the features of the incoming layers. An autoencoder originally has got two different parts for feature extraction: encoder and decoder. In a deep neural network structure, one autoencoder also can be used as Stacked Autoencoders. Extracted features from autoencoders are stored and evaluated in a stack [13].

In the applied deep learning method; a deep neural network structure with two autoencoders (AE) and a softmax layer was created. The autoencoder is responsible for preprocessing and extracting the features of the incoming layers. In a deep neural

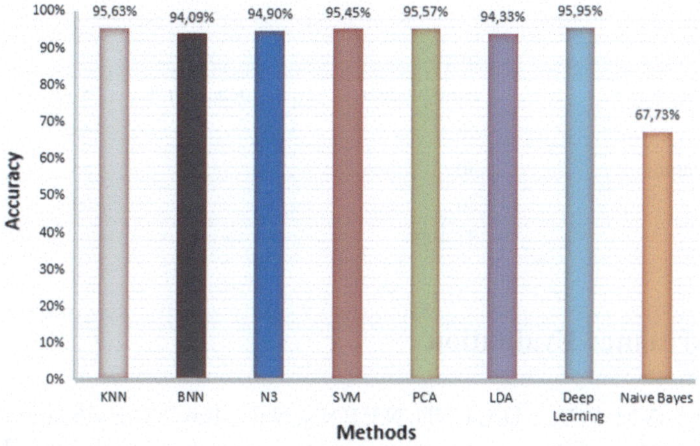

Fig. 4 Accuracy results

network structure, one autoencoder can be used as well as a stacked autoencoder (SAE) can be created using multiple autoencoders as in this work.

The dataset contains 26 RSSI values for 26 wireless access points as well as 26 features. A total of 54 different coordinate signals are collected. For this reason, it is necessary to increase the input size by going to the output result of 26 features from the samples taken from the features. 268 features are given as inputs to the first autoencoder (AE1) resulting in 40 output units. 40 features were extracted from the 26 inputs and then reduced to 26 outputs. (X) input data, (x) the output data size M; The hidden layer output of the autoencoder with the number of neurons in the hidden layer N is given in Eq. 6. c is the hidden layer output, w is the weighting parameter between the input and hidden layer, b is the bias vector, and f is the neuron activation function.

$$c = f(b + W^T X) \tag{6}$$

Second autoencoder (AE2) takes 40 features from the AE1and extracts 54 new features. As shown in Fig. 4, it is desirable to create 54 classes from 26 features so the hidden layers of the autoencoders are incrementally growing. The sigmoid transfer function (7) is used as the activation function in the encoding phase of the autoencoders.

$$f(z) = \frac{1}{1 + e^{-z}} \tag{7}$$

Linear transfer function (8) is used in decode phase.

$$f(z) = z \tag{8}$$

Table 1 Case I

Value type	Steps		
	Autoencoder 1	Autoencoder 2	Softmax layer
Inputs	26	30	54
Extracted feature	30	54	54
Output	26	30	54
Accuracy	87.40%		

4 Performance Evaluation

In this study, SVM, PCA, LDA, KNN, N3, BNN, Naive Bayes Classification methods are run 100 times and the average accuracy value is calculated as shown in Fig. 4. At random, 70% of the data set in each iteration is divided into two parts for train and 30% slice test. For KNN algorithm, Euclidean distance was used for k-value distance measurement and 95.63% accuracy is obtained. For SVM method, the linear core function is used and the penalty value (C) is chosen as 1, with an accuracy rate of 95.45%. In the LDA method, the pseudo-linear analysis type is used. A multinomial distribution is used in Naïve Bayes method with 67.73% accuracy. Different alpha values are tested in N3 and BNN algorithms. The 0.75 value, which gives the highest accuracy in the test results, is preferred. 94.89% accuracy was obtained with N3 method and 94.88% accuracy with BNN method. In the PCA method, the data set has been moved to a new space so that differences between the data have been clarified. The KNN approach has been preferred for classification of the new space in the test phase. As a result of the PCA method, 95.57% accuracy is determined.

By using parameters which are defined in Localization in Deep Learning section Deep Learning method obtained 95.95% accuracy which is the highest value as compared to the other methods used in this study. When the parameters are changed between autoencoders, the accuracy of the Deep Learning method also changes. Tables 1, 2, and 3 show three different cases as to how accuracy changes when the parameters are changed. In Table 1 autoencoder 1 is run with 30 inputs instead of 40 and accuracy decreased by 8.55% in this case. In Table 2 autoencoder is run with 35 inputs and accuracy of 93.60% has been reached. This results in 2.30% lower accuracy than the accuracy which is obtained when the first autoencoder is run with 40 inputs.

As different from the cases in Tables 1 and 2 in Table 3 the number of the extracted features in autoencoder 1 is risen to 45. But still, the accuracy is higher when the number of inputs is 40.

These three case improves that in stacked autoencoders feature extraction numbers between autoencoders must be chosen carefully. If one of the autoencoders in the stack extracts more feature than the others accuracy may decreased.

Table 2 Case II

Value type	Steps		
	Autoencoder 1	Autoencoder 2	Softmax layer
Inputs	26	35	54
Extracted feature	35	54	54
Output	26	35	54
Accuracy	93.60%		

Table 3 Case III

Value Type	Steps		
	Autoencoder 1	Autoencoder 2	Softmax layer
Inputs	26	45	54
Extracted feature	45	54	54
Output	26	45	54
Accuracy	92.53%		

5 Conclusion

This study compares positioning methods for an indoor network that has WiFi technology and RSSI values. SVM, PCA, LDA, KNN, N3, BNN methods were compared with the Deep Learning method which can obtain high accuracy on large data. Deep Learning method is found to be more successful than other methods with 95.95% accuracy. Figure 4 shows the accuracy comparison of used methods. In the future studies, Deep Learning method is planned to be used for person or object tracking by using online data and supported by various estimation methods.

Acknowledgements This work is also a part of the Ph.D. thesis titled "Mobility Management for Internet of Things" at Istanbul University, Institute of Physical Sciences.

References

1. Zhang X, Wang J, Gao Q, Ma X, Wang H (2016) Device-free wireless localization and activity recognition with deep learning. In: 2016 IEEE international conference on pervasive computing and communications work PerCom workshop 2016
2. Bridal P, Diiha J, Krasnovsky M (2007) On the accuracy of weighted proximity based localization in wireless sensor networks 2 mobile localization in wireless networks In: IFIP—the international federation for information processing, vol 245, pp 423–432
3. Turgut Z, Aydin GZG, Sertbas A (2016) Indoor localization techniques for smart building environment. In: Procedia computer science, 2016, vol. 83, no. Ant, pp 1176–1181
4. Hern N (2014) A WiFi-based software for indoor localization

5. Torres-Sospedra J, Montoliu R, Trilles S, Belmonte Ó, Huerta J (2015) Comprehensive analysis of distance and similarity measures for Wi-Fi fingerprinting indoor positioning systems. Expert Syst Appl 42(23):9263–9278
6. Keser SB, Yayan U, Yazici A, Gunal S (2016) A priori verification and validation study of RFKON database. 5(1): 20–27
7. Siller M (2016) A fingerprinting indoor localization algorithm based deep learning, pp 1006–1011
8. Alpaydin E (2013) Yapay Ogrenme. Boğaziçi Üniversitesi Yayınevi
9. Hotelling H (1933) Analysis of a complex of statistical variables into principal components. J Educ Psychol 24(6):417–441
10. McLachlan GJ (2004) Discriminant analysis and statistical pattern recognition. Wiley Interscience, New Jersey
11. Todeschini R, Ballabio D, Cassotti M, Consonni V (2015) N3 and BNN: Two new similarity based classification methods in comparison with other classifiers. J Chem Inf Model 55(11):2365–2374
12. Dogrusoz A (2007) Makine Ögrenmesi Teknikleri ile Metinlerin Otomatik Olarak Sınıflandırılması. Yıldız Teknik Universitesi
13. Karabulut EM (2016) Investigation of deep learning approaches for biomedical data classification. Cukurova universitesi

LocSwayamwar: Finding a Suitable ML Algorithm for Wi-Fi Fingerprinting Based Indoor Positioning System

Beenish Ayesha Akram, Ali Hammad Akbar, Bilal Wajid, Omair Shafiq and Amna Zafar

Abstract Indoor localization has been a challenging problem for over a decennium. Wi-Fi Fingerprinting based solutions stand out in comparison with Angle-Of-Arrival (AOA), Time-Of-Arrival (TOA), Time-Difference-Of-Arrival (TDOA) approaches as they inherently incorporate radio propagation models in fingerprints (FP) which provide more realistic information than radio signal propagation models as well as do not need extra hardware. Diverse Location Based Services (LBS) heavily rely on the performance of localization algorithms used for pattern matching with the collected FP database. This work investigates the performance of several machine learning algorithms as a multiclass classifier for room-level indoor localization including K*, k-NN, Random Forest, FURIA, Multi-Layer Perceptron, and J48. We report results of top five algorithms along with five algorithms selected from various algorithmic categories obtaining an accuracy greater than 95%. Data was generated by collecting 14,080 fingerprints from 20 Access Points at 180 reference points in 1209 m^2 area of Software Engineering Center, University of Engineering and Technology (UET), Lahore to construct real-world FP dataset. The results obtained indicate that the best performance is achieved by K* followed by k-NN, Random Forest, FURIA, Multilayer Perceptron, J48 with accuracies 99.52, 99.06, 98.76, 97.26, 97.05, and 95.91% respectively.

Swayamwar: in ancient India, was a practice of choosing a husband, from among a list of suitors, by a girl of marriageable age

B. A. Akram · A. H. Akbar · A. Zafar (✉)
Department of Computer Science and Engineering, University of Engineering and Technology, Lahore, Pakistan
e-mail: amnazafar@uet.edu.pk

B. A. Akram
e-mail: beenish.ayesha.akram@uet.edu.pk

B. Wajid
Department of Electrical Engineering, University of Engineering and Technology, Lahore, Pakistan

O. Shafiq
Carleton School of Information Technology, Carleton University, Ottawa, ON, Canada

A. Boyaci et al. (eds.), *International Telecommunications Conference*,
Lecture Notes in Electrical Engineering 504,
https://doi.org/10.1007/978-981-13-0408-8_10

Keywords Wi-Fi · Fingerprinting · Indoor localization · Indoor positioning system · Indoor positioning algorithms

1 Introduction

The problem of localization including both outdoor and indoor localization has seen massive industrial and academic research efforts due to their high market potential. Outdoor localization has been successfully commercialized in form of GPS, GLONASS and COMPASS. For indoor location estimation, satellite based techniques do not work well due to No-Line-Of-Sight (NLOS) and occlusion. Measurement based localization systems typically include Time of Arrival (TOA), Time Difference of Arrival (TDOA), and Angle of Arrival (AOA) [1]. All these aforementioned approaches share a common drawback of additional hardware requirements. On the other hand, FP based approaches do not need additional hardware support and use Received Signal Strength Indicator (RSSI) values as a notion of distance estimation from different Access Points (AP). Various sensory inputs have been explored other than Wi-Fi such as Bluetooth [2], Bluetooth Low energy [3], Magnetic readings [4, 5], Ultra Wide Band signals [6], ultrasound, images, inertial sensors etc. but they all usually require additional sensors for data acquisition. Nowadays, Wi-Fi AP constellation is the de facto standard and capturing Wi-Fi RSSI FP is possible with almost every common hand-held device such as smart phones, tablets, smart watches and laptops making it highly feasible for indoor localization without incurring any extra equipment cost.

In a typical Wi-Fi FP based Indoor Positioning System (IPS), the existing AP infrastructure creates a RF map (a.k.a. FP dataset) of the building which is used for preparing the localization system during the training phase. In the prediction phase, unseen FP is captured and using the pre-trained localization system, room number/label is estimated.

The same approach is used in this paper for evaluating performance of different machine learning (ML) algorithms used as multiclass classifier for room level location prediction of a user carrying an Android smart phone. Real world dataset was generated at ground floor of Software Engineering Center, UET, Lahore, Pakistan. The area covered was 1209 m^2, which was divided into 1.5×1.5 m square grid making a total of 180 Reference Points (RP). On average 78 samples were collected at each RP resulting in total of 14,080 samples collected.

2 Related Work

Comparison of various ML algorithms is a common preliminary phase of experimentation in order to find the most appropriate algorithm for a particular problem. This section discusses prior works on comparison of ML algorithms for indoor localization.

Esrafil Jedari et al. [7] evaluated performance of k-NN, JRip and Random Forest for indoor localization presenting accuracy of 95% (Random Forest), followed by 77.5% (k-NN) and 72.2% (JRip) with focus on k-NN parameter tuning. Li Zhang et al. [8] presented experimental results of k-NN, Neural Networks and SVM for Bluetooth based FP approach for indoor localization. They compared the algorithms on basis of accuracy, precision and required training time. They found k-NN, SVM and ANN to be suitable in descending order based on accuracy and precision noting k-NN needs no training, ANN (2–8 min) and SVM (3 h). Sinem Bozkurt et al. [9] evaluated Decision Tree C4.5, Naïve Bayes, Bayesian Network, k-NN, sequential minimal optimization, AdaBoost and Bagging for IPS based on accuracy and computation time, using publically available Wi-Fi FP dataset UJIIndoorLoc. Their results indicated k-NN achieved maximum accuracy. Tree classifier combined with bagging and AdaBoost was also found to be a good candidate.

3 Dataset Generation and Predictor Importance Estimation

3.1 Data Set Acquisition

We built an Android app for recording Wi-Fi APs RSSI vectors. Android smart phone Samsung Galaxy J5 (2016) was used to scan all observable APs at a RP of 2.4 GHz and 5 GHz bands. At each RP, samples were collected moving phone from 0° to 90° angle with floor, facing N, E, S, W, NE, SE, NW and SW directions in an attempt to avoid occlusion caused by human body [10, 11]. The acquired FPs were stored in the database along with marked room labels.

3.2 Pre-processing

The resulting dataset is sparse due to both visibility and invisibility of different APs at different locations of the building. The RSSI values from each AP varied from -15 dBm to -98 dBm (from strong to weak and near to far AP). The missing values were replaced, following the common practice [12, 13] with -100 dBm.

3.3 Predictor Importance Estimation

The total number of observed APs was 54 which determines the number of predictors for our classifiers. Random Forest was employed as multiclass classifier to determine the predictor importance and found 20 out of 54 APs to be important or more distinguishing for location prediction. Hence the dataset was filtered to contain only these 20 APs/predictors.

4 Machine Learning Algorithms

Diverse classification algorithms for IPS candidate were evaluated. This section briefly describes the top four performing algorithms out of the ten algorithms which scored higher than 95% on accuracy, as reported in Table 1, followed by the detailed experimentation on these very algorithms.

*4.1 K**

K* [14] is an instance based classifier where every new instance is compared with all previous ones in the dataset using distance or similarity measure where the closest one is used to allocate a class to the new instance. The distinctive feature of K* is entropy based similarity measure. The probability of transformation cost of one instance into another is ensured to be in random walk away manner. The sum of probabilities from the new instance to all other instances belonging to each class is computed and then class with the highest probability summation is selected as a final output.

4.2 K-NN

k-Nearest Neighbors is one of the simplest ML algorithms which has gained popularity in indoor localization since its utilization in RADAR [15] (the pioneer paper by Microsoft using Wi-Fi FPs with trilateration) to date. It does not require a training phase rather all the samples in dataset are stored along with class labels which leads to increasing response time with an increase in dataset. A sample whose class needs to be estimated during prediction phase is processed in the following manner; based on some similarity measure, distance is computed with every sample in the dataset. k is the user defined number of instances/samples to be considered as nearest neighbors determined by the aforementioned similarity measure. The majority vote is used to assign a class label to the sample at hand.

Table 1 Performance comparison of classification algorithms for Wi-Fi FP based indoor localization

Algorithm	Time to build model (S)	Accuracy	Kappa statistic	RMSE	Precision	Recall	F1	MCC	ROC area
K*	**0**	**99.52**	0.98	**0.02**	**0.99**	**0.99**	**0.99**	**0.99**	**1**
k-NN	**0**	99.06	**0.99**	0.03	**0.99**	**0.99**	**0.99**	0.98	0.99
Random forest	1.11	98.76	0.98	0.04	0.98	0.98	0.98	0.98	**1**
FURIA	5.92	97.26	0.96	0.05	0.97	0.97	0.97	0.96	0.99
Multilayer perceptron	25.84	97.05	0.96	0.06	0.97	0.97	0.97	0.96	0.99
J48	0.1	95.91	0.95	0.08	0.95	0.95	0.95	0.95	0.98
Dl4jMlpClassifier	26.31	94	0.93	0.08	0.94	0.94	0.94	0.93	0.99
SVM	5.82	90.6	0.89	0.12	0.93	0.90	0.90	0.9	0.94
NaiveBayes	**0.02**	89.79	0.88	0.12	0.91	0.89	0.89	0.89	0.99
AdaBoost with decisionstump	0.1	36.81	0.22	0.25	0.15	0.36	0.21	0.18	0.76

4.3 Random Forest

Random Forest [16] itself is an ensemble classifier that combines many tree classifiers. Three hyper parameters including number of trees, random number of predictors as the basis of split, and the maximum number of splits (depth of trees) are needed to be tuned for training a Random Forest ensemble. It also has the capability to provide predictor importance estimate, when trained. Random Forest uses bootstrapped data subset with replacement for each tree, choosing a small number of random features out of a total number of features/predictors for deciding best split using Gini index for classification. This approach makes it easy to train owing to mainly 2 and maximum 3 tunable hyper parameters, insensitive to noise and capable to avoid overfitting. Random Forest is also suitable for small dataset due to its built-in error estimation in terms of Out-Of-Bag (OOB) error calculation.

4.4 FURIA (Fuzzy Unordered Rule Induction Algorithm)

FURIA [17] is based on state-of-the-art rule learning algorithm named RIPPER. It inherits RIPPER's benefits of simplicity and easy comprehension with rules learning for each class in One-versus-All notion instead of following the least frequent to most frequent class order for rule learning, making class order irrelevant. Furthermore, it adds advanced features such as definition of fuzzy instead of crisp rules with provision of class' association certainty degree by each fuzzy rule. Support strength of each class is determined by the product of a fuzzy rule's certainty factor and firing degree, summed over all fuzzy rules for the respective class, final output is determined by maximum class support strength. Another feature is the unordered rule sets usage instead of ordered list which removes any default class bias. It also provides rule stretching method for uncovered examples where the available fuzzy rules are generalized by removing some of their ancestors.

5 Results and Discussion

Table 1 shows that time taken to train the classifier for one out of ten folds for cross validation is least by both K* and k-NN as there is practically no training involved. However, samples storage with class labels is needed. Thereafter, comes NaiveBayes, J48, AdaBoost with Decision Stumps, and Random Forest in ascending order with 0.02, 0.1, 0.1, and 1.11 s. FURIA with 5.92 s stands in the middle with maximum time required by neural network based approaches namely Multilayer Perceptron and Dl4jMlpClassifier (Deeplearning4 J based classifier). Training as well as testing times play critical role in IPS algorithm choice as real time system response requires testing time to be trivial. Furthermore, FP based solutions need periodic training of the

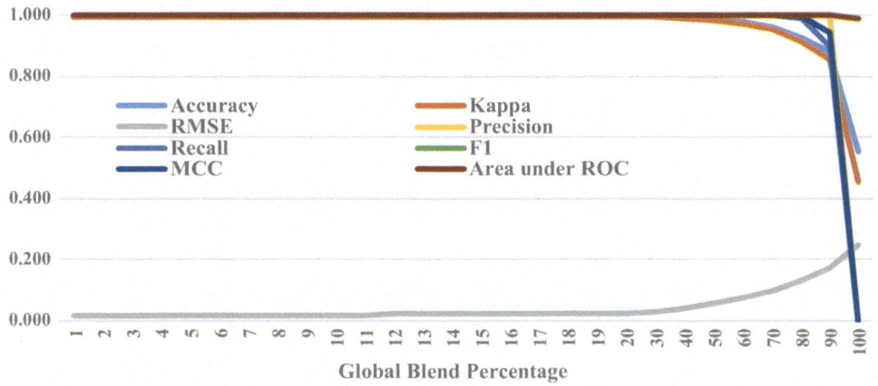

Fig. 1 Impact of global blend percentage

system to be practically useful due to changing nature of Wi-Fi FPs. Considering other performance measures from accuracy to ROC Area, best performance is achieved in descending order by K*, k-NN, Random Forest, FURIA, Multilayer perceptron, Dl4jMlpClassifier, SVM, NaiveBayes and AdaBoost with Decision Stumps. Further experiments were performed on top four performing algorithms in terms of hyper parameter tuning which are presented in Sect. 5.1 to 5.4 as follows, averaged over 10-fold cross-validation.

5.1 K Experiments*

For K*, the major tunable parameter is either 'auto entropic blend' or 'manual'. Enabling auto entropic blend did not produce good results in comparison with manual setting, as on average 80.4% accuracy was achieved over 10 folds. Figure 1 depicts the performance measures trend with entropic blend varied from 1 to 100%. Root Mean Squared Error (RMSE) started to grow after global blend increased over 30% with rapid accuracy reduction by crossing 70%. 90% drops accuracy further to 87%, and 100% global blend drags accuracy to 55%. Figure 2 represents training and testing times with growing global blend percentage which remains almost consistent except for decreased testing time at 100%.

5.2 K-NN Experiments

Figure 3 depicts performance measures for k varied from 1 to 10 along with distance weight for $k > 1$ kept as Nil, inverse of the distance and 1—distance. $k = 1$ was

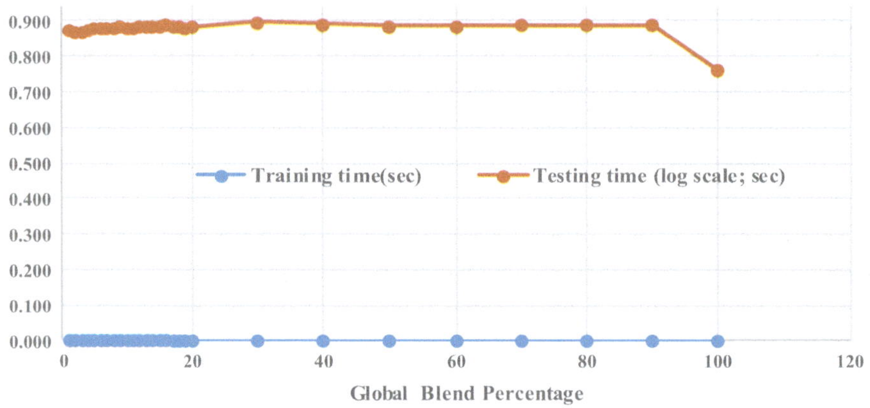

Fig. 2 Impact of global blend percentage on training and testing time

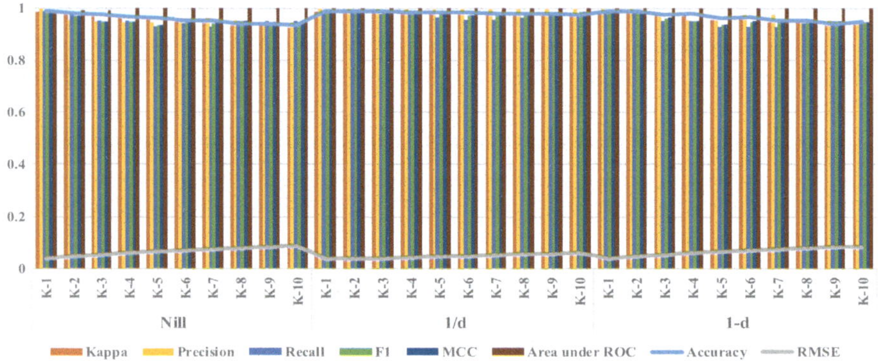

Fig. 3 Impact of increasing k and distance weight with Euclidean distance

optimal for our dataset as the accuracy starts to decrease except for the inverse of the distance weight case. Figure 4 represents the same measures w.r.t. growing neighbors, k along with Euclidean, Chebyshev, Dilca, Manhattan and Minkowski used as similarity measure indicating Manhattan to be the best choice followed by Minkowski, Euclidean, Dilca and Chebyshev in descending order. Figure 5 sheds light on respective training and testing times for k-NN.

5.3 *Random Forest Experiments*

There are mainly two tunable hyper parameters for Random Forest namely number of trees (numTree) in ensemble and number of random features (f) for tree split decision which is much less than the total number of predictors. Performance measures trend

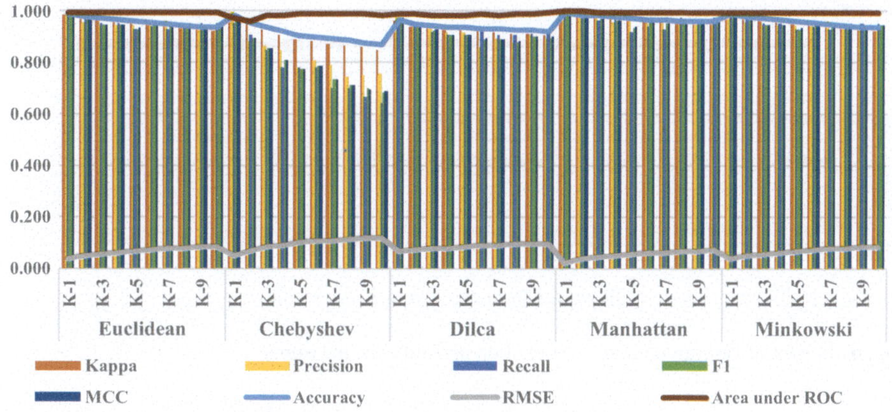

Fig. 4 Impact on different distance measures with increasing k

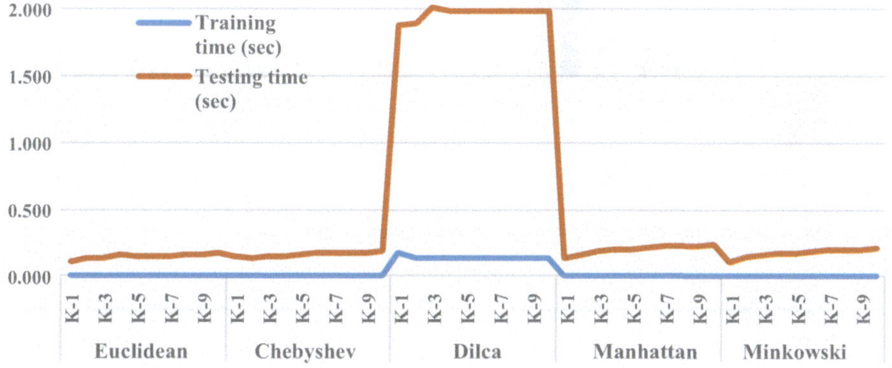

Fig. 5 Impact of distances with growing k on training and testing times

with increasing f is shown in Fig. 6 with just 1% accuracy change over the entire range of f from 1 to 15 with best performance measures obtained at f = 2. Training (shown at log10 scale) and testing time trends for f varied from 1 to 15 are shown in Fig. 7 with training time linearly increasing and testing time linearly decreasing with increasing f.

Performance measures' values for increasing numTree from 1 to 300 are represented in Fig. 8 indicating major performance improvement with 13 + trees with best performance with numTrees from 26 to 43. Training ($\log_{10}$) and testing time trends are shown in Fig. 9, both gradually increasing with growing numTrees with testing time axis marked on RHS and training time axis marked on LHS. It must be noted that difference between maximum and minimum of testing time for numTrees 1–300 is merely 7.18E-02 s.

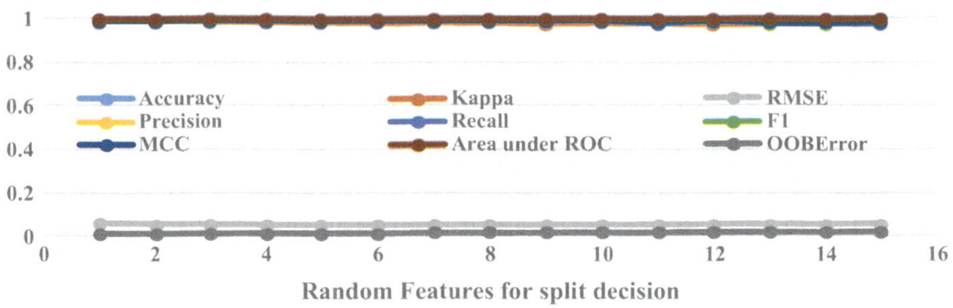

Fig. 6 Impact of varying random features f on performance measures

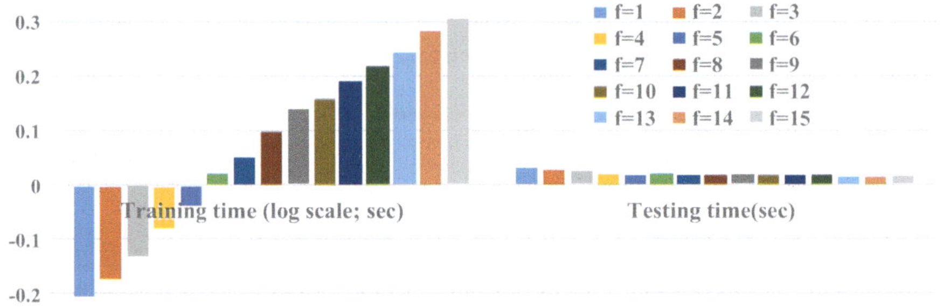

Fig. 7 Impact of varying random features f on training and testing time

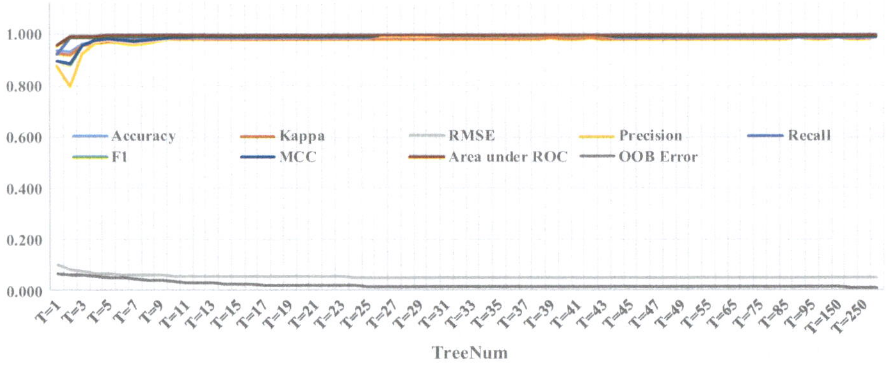

Fig. 8 Impact of increasing number of trees

5.4 *FURIA Experiments*

Experiments were conducted for FURIA with two perspectives. First, number of folds for fuzzy rule growth and pruning were varied from 2 to 10 whose results are presented in Fig. 10 indicating negligible variation in performance measures.

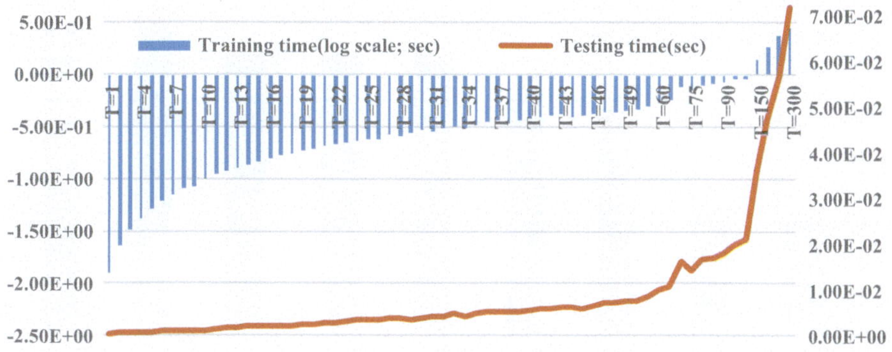

Fig. 9 Impact of number of trees on training and testing time

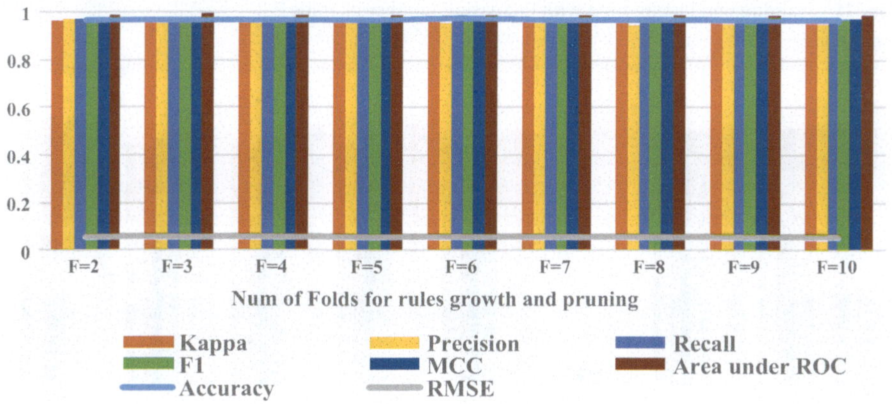

Fig. 10 Performance measures as a function of number of folds

A gradual increase in training time ($\log_{10}$ scale) with growing folds was observed with almost consistent testing time of 3.60E-03 s with exception of F = 3 where it became 5.90E-03 s as shown in Fig. 11. Second, minimum weight of instances within a split (MW) was varied from 1 to 10. The results from Fig. 12 depict that improvement of performance measures is visible with MW increased from 1 to 2, but with further increase the performance starts to degrade. Reduction of up to 3% accuracy is observed with MW = 10. MW = 2 was found to be optimal value for our data.

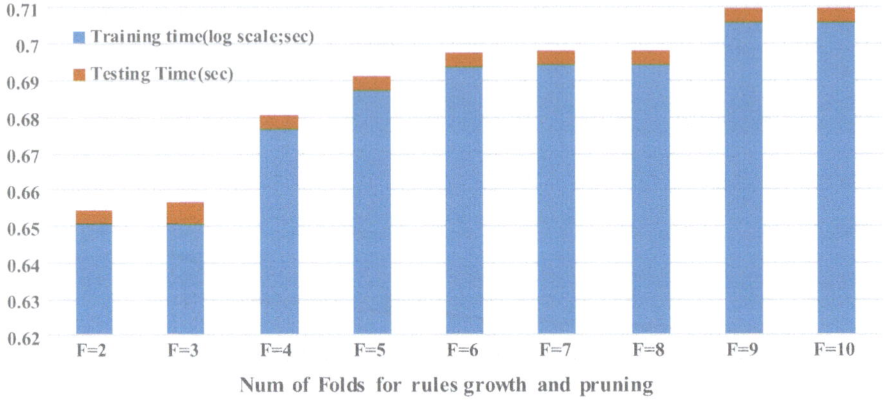

Fig. 11 Training and testing time with increasing number of folds

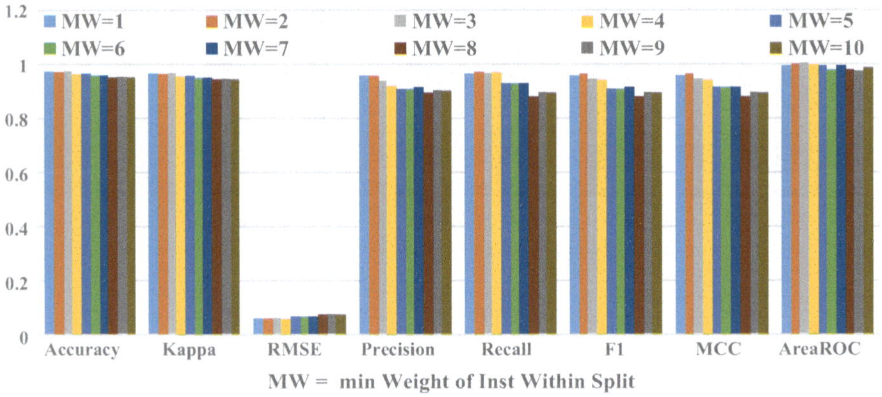

Fig. 12 Performance measures with increasing min.weight of instances within split

6 Conclusion

We evaluated various classifiers as a suitable algorithm to be used in indoor positioning system (IPS). In this paper, we presented results of 10 algorithms along with further detailed experimentation on top four best performing algorithms. It can be deduced that lazy and instance based approaches such as K* and k-NN are suitable for Wi-Fi FP based IPS as evident by the computed performance measures followed by Random Forest and FURIA with 98.7 and 97.2% accuracies respectively. However, this lazy approach also impacts the training and testing time discrepancy in which most of the algorithms follow the trend of large training time in contrast with negligible testing time whereas K* and k-NN need negligible training time with large testing time which grows with number of neighbors, number of samples/instances in the dataset. This trait makes them undesirable in situations where real time response is required, making Random Forest in particular and FURIA in general more suitable with negligible average testing time of 8.58E-03 and 3.66E-03 s respectively.

References

1. Zhang W, Liu K, Zhang W, Zhang Y, Gu J (2016) Deep neural networks for wireless localization in indoor and outdoor environments. Neurocomputing 194:279–287. https://doi.org/10.1016/j.neucom.2016.02.055
2. Gu Y, Ren F (2015) Energy-efficient indoor localization of smart hand-held devices using Bluetooth. IEEE Access 3:1450–1461. https://doi.org/10.1109/ACCESS.2015.2441694
3. Cooper M, Biehl J, Filby G, Kratz S (2016) LoCo: boosting for indoor location classification combining Wi-Fi and BLE. Pers Ubiquitous Comput 20:1–14. https://doi.org/10.1007/s00779-015-0899-z
4. Nagpal PS, Rashidzadeh R (2013) Indoor positioning using magnetic compass and accelerometer of smartphones. In: Proceedings of the international conference on selected topics in mobile and wireless networking, pp 140–145. https://doi.org/10.1109/mownet.2013.6613810
5. Zhang M, Shen W, Zhu J (2016) WIFI and magnetic fingerprint positioning algorithm based on KDA-KNN. In: Proceedings of the 28th Chinese control and decision conference CCDC 2016, pp 5409–5415. https://doi.org/10.1109/ccdc.2016.7531964
6. Luo J, Gao H (2016) Deep belief networks for fingerprinting indoor localization using ultrawideband technology. Int J Distrib Sens Netw. https://doi.org/10.1155/2016/5840916
7. Jedari E, Wu Z, Rashidzadeh R, Saif M (2015) Wi-Fi based indoor location positioning employing random forest classifier. In: IPIN 2015: international conference on indoor positioning and indoor navigation, pp 13–16. https://doi.org/10.1109/ipin.2015.7346754
8. Zhang L, Liu X, Song J, Gurrin C, Zhu Z (2013) A comprehensive study of bluetooth fingerprinting-based algorithms for localization. In: Proceedings of the 27th international conference on advanced information networking and applications workshops WAINA 2013, pp 300–305. https://doi.org/10.1109/waina.2013.205
9. Bozkurt S, Elibol G (2015) A comparative study on machine learning algorithms for indoor positioning
10. Yang S, Dessai P, Verma M, Gerla M (2013) FreeLoc: calibration-free crowdsourced indoor localization. In: Proceedings of the IEEE INFOCOM, pp 2481–2489. https://doi.org/10.1109/infcom.2013.6567054
11. Wu C, Yang Z, Liu Y, Xi W (2013) WILL: Wireless indoor localization without site survey. IEEE Trans Parallel Distrib Syst 24:839–848. https://doi.org/10.1109/TPDS.2012.179
12. Li N, Chen J, Yuan Y, Tian X, Han Y, Xia M (2016) A Wi-Fi indoor localization strategy using particle swarm optimization based artificial neural networks. Int J Distrib Sens Netw. https://doi.org/10.1155/2016/4583147
13. Calderoni L, Ferrara M, Franco A, Maio D (2015) Indoor localization in a hospital environment using Random Forest classifiers. Expert Syst Appl 42:125–134. https://doi.org/10.1016/j.eswa.2014.07.042
14. Cleary JG, Trigg LE: K*: an instance-based learner using an entropic distance measure. In: In Proceedings of the 12th international conference on machine learning, vol 5, pp 1–14 (1995). https://doi.org/10.1151.4098
15. Bahl P, Padmanabhan V (2000) RADAR: an in-building RF based user location and tracking system. In: Proceedings of the IEEE INFOCOM 2000. Annual joint conference of the IEEE computer and communications societies (Cat. No.00CH37064), vol 2, pp 775–784. https://doi.org/10.1109/infcom.2000.832252
16. Breiman L (2001) Random forests. Mach Learn 45:5–32. https://doi.org/10.1023/A:1010933404324
17. Hühn J, Hüllermeier E (2009) FURIA: an algorithm for unordered fuzzy rule induction. Data Min Knowl Discov 19: 293–319. https://doi.org/10.1007/s10618-009-0131-8

AMNA: Probe Agent Based Inter-Process Dependency Model for Wireless Sensor Network's Fault DiAgnosis

Amna Zafar, Ali Hammad Akbar, Bilal Wajid, Beenish Ayesha Akram and Talha Irfan

Abstract Process failures in communication protocol stack of wireless sensor networks (WSNs) are caused by underlying errors and faults. This work presents AMNA: probe Agent based inter-process dependency Model for wireless sensor Network's fault diAgnosis. The probe agent performs online process execution tracing to analyze the effect of errors, omissions and channel misbehavior on process execution at node level. An inter-process dependency model is proposed that is implemented through a decision tree for fault diagnosis. AMNA has been implemented and evaluated on Castalia simulator.

Keywords Probe · Process · Fault diagnosis · WSN

1 Introduction

Autonomous deployment of sensor nodes in unattended and hostile environments results in high frequency of faults in WSNs. A fault is an erroneous state of a hardware and/or software component that may cause operational failure [1]. Software layer of sensor nodes consists of processes executing the communication protocol stack. These correlated processes exchange information and thus, need to be monitored for potential failures. Therefore, to ensure reliable communication for data delivery, it is imperative to design efficient and effective fault detection and diagnosis techniques. Fault diagnosis schemes for WSN are generally investigated using these two methods (i) a known set of faults is tested for a given network under several parameters and

A. Zafar (✉) · A. H. Akbar · B. A. Akram
Department of Computer Science and Engineering, UET Lahore, Lahore, Pakistan
e-mail: amnazafar@uet.edu.pk

B. Wajid
Department of Electrical Engineering, UET KSK Campus, Kala Shah Kaku, Pakistan

T. Irfan
Department of Computer Science and Information Technology, Superior University Lahore, Lahore, Pakistan

A. Boyaci et al. (eds.), *International Telecommunications Conference*,
Lecture Notes in Electrical Engineering 504,
https://doi.org/10.1007/978-981-13-0408-8_11

channel behavior emulating the real world phenomena [2] (ii) gradual impact of the same parameters changes and channel behavior is analyzed to identify the likelihood of particular behavior as being anomalous, erroneous or faulty i.e., to perform fault classification [3].

This work presents AMNA: probe Agent based inter-process dependency Model for wireless sensor Network's fault diAgnosis that is based upon later of the two methods discussed. AMNA performs (i) probe based process execution tracing for fault detection and classification (ii) analysis of propagation effect of one process failure on stacked (vertical) processes execution to build an inter-process dependency model for fault diagnosis. Existing fault classification based diagnosis schemes do not incorporate these layered dependencies. The rest of the manuscript is organized as such. Section 2 gives a detailed survey of WSNs fault classification and diagnosis techniques. Section 3 describes AMNA system architecture in detail. In Sect. 4, experimental details and results obtained in Castalia simulator have been discussed. Section 5 discusses conclusion and future work.

2 Related Work

Existing fault diagnosis frameworks for WSNs artificially induce random faults to evaluate performance there off. Fault diagnosis is performed via periodic monitoring or reactive probing of network exceptions. Thereby, root cause analysis is conducted to identify failure source(s). The authors in [2] constructed a Belief Network based inference model for fault diagnosis at sink node. By observing network problematic symptoms, the inference model produces several candidate fault sets and their associated belief values. Incremental probing is performed to retrieve information about un-observed symptoms for final fault diagnosis. The proposed inference model was evaluated by random injection of faults in radio and sensing units.

Observation based method for fault diagnosis categorizes WSN anomalies based upon regularity and continuity into intermittent and transient faults. Intermittent faults occur sporadically and their effect is not generally obvious. Software malfunction is major source of these faults. Similarly, transient faults occur momentarily due to external factors such as environmental noise and interference. To diagnose intermittent hard and soft faults, the authors in [3] implemented a periodic network health sampling scheme through 'heartbeat' messages. The neighbors exchange these heartbeat messages and propagate local diagnostic views. Afterwards, the authors analyzed the effects of transient faults through simulations.

An on-site diagnostic system for 6LowPAN was proposed in [4]. It comprises of a traffic monitor, a frame decoder and a decision tree. The frame decoder excerpts and investigates network and Medium Access Control (MAC) layer headers to detect network failures including unresponsive node, network partition and intermittent disconnectivity. Subsequently, a decision tree is employed to perform root cause analysis. The diagnostic system lacks the ability to diagnose network partition due to problems along routing path. Additionally, the decision tree depends on examination of communication protocol control packets for correct diagnosis but protocol failures have been overlooked.

Researchers have investigated standard communication protocols mostly concentrating on their routing or MAC layers. In [5], the authors have utilized statistical model checking as a diagnostic technique to analyze performance of Ad hoc On-Demand Distance Vector routing protocol (AODV) [6] for wireless mesh networks. In another work [7], the chronological behavior of IEEE802.15.4 based wireless networks under communication errors caused by control field omissions was investigated. However, the study was confined to MAC layer only. Therefore, in all, there is need to further investigate inter-process dependency of routing, MAC and physical layer protocols [8]. To the best of authors knowledge, existing fault diagnosis schemes do not substantiate relationship between fault diagnosis and inter-process dependencies of functional peers. The research article [9] was our first attempt to propose a fault diagnosis system based upon intra-process communication and inter-process dependencies. The next section describes AMNA in detail.

3 Amna

For effective communication between sensor nodes, operating system implements protocol stack processes. AMNA identifies typical processes which run on any general network communication protocol stack. For practical considerations, AMNA is based upon processes on three layers i.e., network, MAC and radio layer. AODV routing [6] and IEEE 802.15.4 processes [10] have been selected. The execution flows of these processes serve as a basis for inter-process dependency model. AMNA performs probe based online tracing of process execution for failure detection and error classification. Fault diagnosis is performed using a decision tree based on inter-process dependency model as shown in Fig. 1.

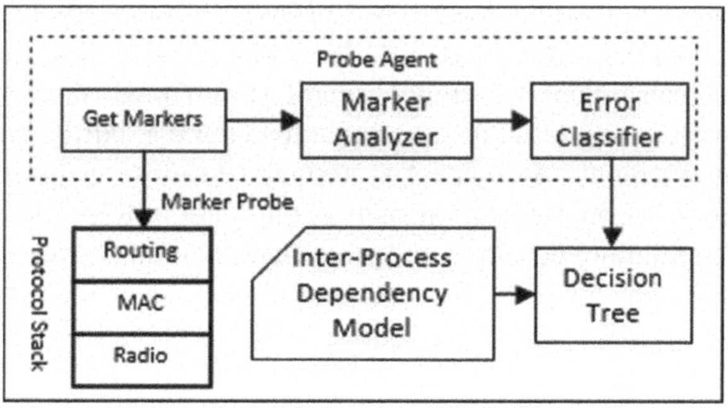

Fig. 1 AMNA system architecture: Probe agent collects markers through probe for analysis of process executing on protocol stack layers

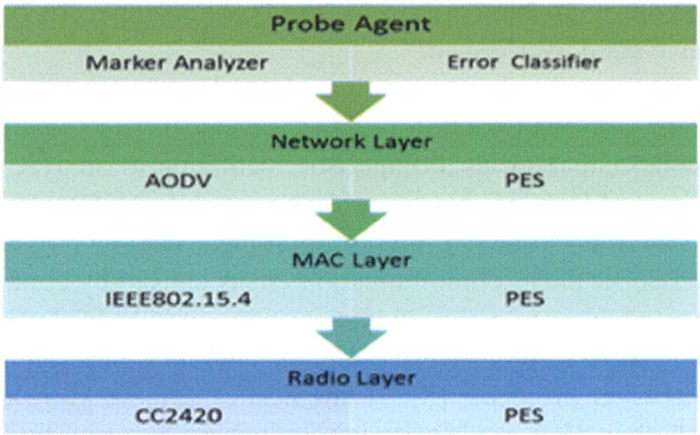

Fig. 2 Probe agent on each sensor node

3.1 *Probe Agent*

Probe Agent on each sensor node contains Marker Analyzer and Error Classifier modules, as shown in Fig. 2. The runtime execution state of each procedure within a process is stored as 'Procedure Execution Marker (PEM)'. Process execution order and PEM are stored in the form of 'Process Execution Stack' (PES). These error markers are generated on occurrence of runtime omissions through unified exception handling code implementation for multiple layers. Under normal circumstances, procedures run without any error and return normal marker. The probe agent periodically sends a marker probe that traverses routing, MAC and radio PES to collect respective PEMs. Marker Analyzer parses the received markers and generates 'Procedure Error Counter (PEC)' for each process.

Error Classifier module analyzes routing, MAC and radio PECs to detect process failures and categorize corresponding error levels against specific error threshold. The errors are categorized based upon frequency, regressive impact on process(es) and sensor node functional behavior. Error levels are classified as critical, warning and alert. Critical errors are major causes of process failures which interrupt communication and may lead to sensor node disconnectivity from rest of the network, for example synchronization failure. The errors which may have an ephemeral impact on intermittent occurrences might as well have a completely different impact such as process failure if these continue to occur e.g. buffer overflow. These may be labeled as warnings. Alert level is assigned to those errors that cause transient interruptions in communication such as radio not in RX mode for channel sensing process. After failure detection and error classification, the decision tree is employed for fault diagnosis that is based upon inter-process dependency model.

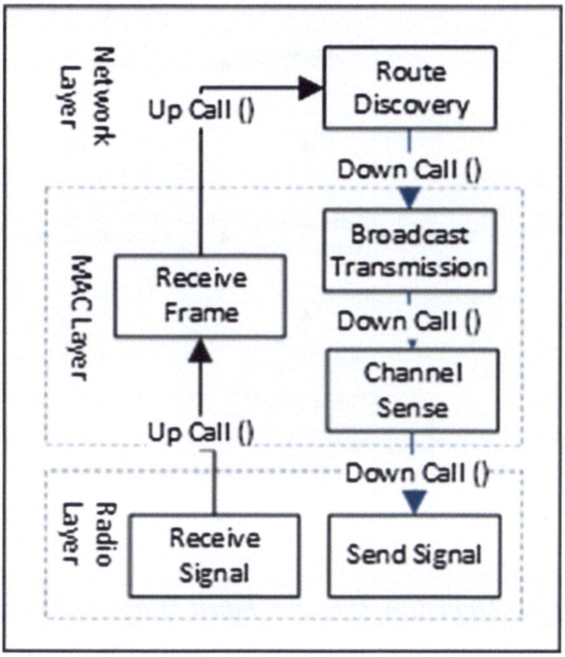

Fig. 3 Downcall/upcall sequence for route discovery process

3.2 *Inter-Process Dependency Model*

Process execution order is based upon inter-process communication in the form of up/down calls between stacked processes as shown in Fig. 3. The inter-process dependency is therefore extracted from these calls and is dynamically represented as a model. The components in the model are stacked processes on each node, as shown in Fig. 4. The directional links represent conditional dependency of these processes. Based upon downcall sequence, network layer route discovery process is dependent on MAC layer broadcast transmission process as shown in Fig. 3. Similarly, the broadcast transmission is further dependent on MAC association, channel sense and radio send signal processes. The association process relies on synchronization with coordinator node through beacons. To receive beacon packets, receive frame process is executed i.e., an upcall. Therefore, synchronization depends on the receive frame process. The dependent process may fail due to the failure of any of these processes. For a downcall, the failure effect is propagated upward the protocol stack. Consequently, route discovery process may fail.

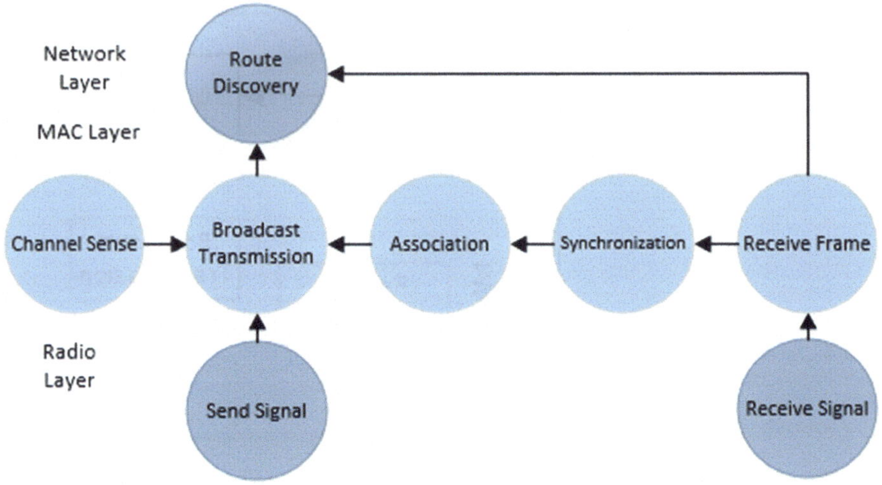

Fig. 4 Inter-process dependency model for node level stacked processes

3.3 Decision Tree for Fault Diagnosis

The inter-process dependency model forms a top down hierarchical structure that necessitate the usage of decision tree for root cause analysis. AMNA performs fault diagnosis using a decision tree that is based upon the inter-process dependency model as shown in Fig. 5. For each critical failure, decision tests are executed to detect potential root causes. Each decision test compares the downcall counter of the failed process with error counters of correlated stacked processes (critical, warnings and alerts). For example, in case of route discovery critical failure, the decision test compares the broadcast transmission, channel sense and send signal error counters with number of times route request (RREQ) packet was sent to MAC layer i.e., downcall counter. If these error counters are greater than the downcall counter for the route discovery process, the corresponding error sources are added to list of probable root causes. Subsequently, a failure report containing critical failures and primary root causes is generated.

4 Evaluation

AMNA has been implemented on Castalia-3.3 using OMNET++ver.4.6 under ubuntu 14.04 operating system. Performance has been evaluated through extensive simulations. The simulation setup consists of a sink node and a number of sensor nodes deployed uniformly. Sensor nodes periodically send data packets to the sink

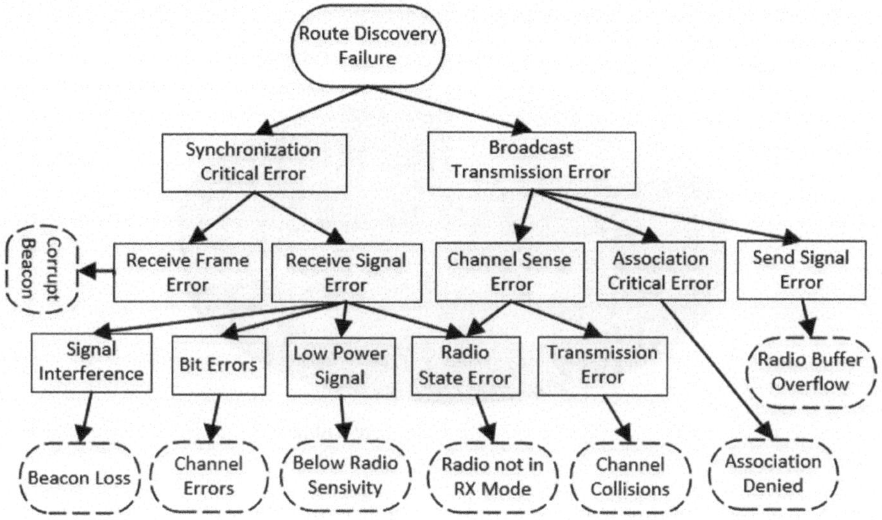

Fig. 5 Decision tree for fault diagnosis: dotted ovals represent warnings and alerts as possible route causes for process failure

Table 1 Simulation parameters

Number of nodes	10
Simulation time	10000 s
Deployment area	30 × 40 m
Traffic type	CBR
Probe period	300, 700 s
Radio TX power	−5 dbm
Interference model	Additive

through multi hop communication that triggers execution of routing processes. The simulation parameters are summarized in Table 1.

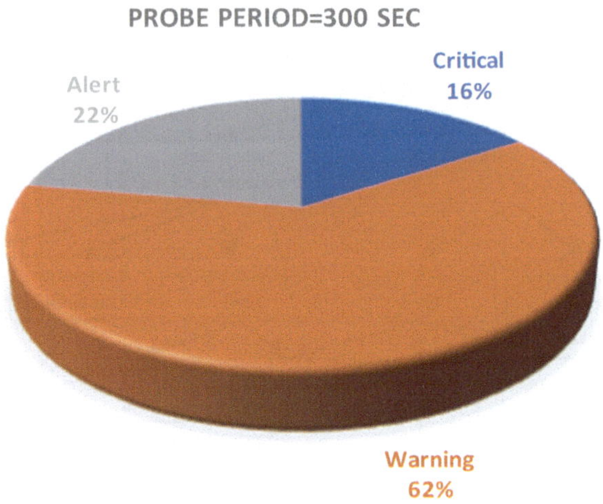

Fig. 6 Spatial aspects of error-warning-alert classification showing a higher proportion of warnings over shorter period

To evaluate performance of AMNA, probe period parameter has been varied. For each probe period, output of Error Classifier is analyzed. Error classification is an insight into long term (steady state) behavior of WSN under natural and, to an extent induced faults. The effect of probe period variation can be best explained by first understanding the relationship that exists between the anomalies declared, either as critical errors, warnings or alerts. The relationship between them is a function of the implementation of networking protocols and the exception handling code that reports errors. The spatio-temporal frequency of alerts and warnings with respect to the frequency of critical errors themselves shows the flexibility and adaptability of the code implemented through timers and counters. For a shorter probe period, a high number of warnings are reported network wide as compared to critical errors and alerts due to short term network stability as depicted in Fig. 6. However, for a longer probe period since the network restores to normalcy after erring within, warnings are lesser in number as compared to the previous case. Subsequently, the proportion of critical errors is reduced as shown in Fig. 7.

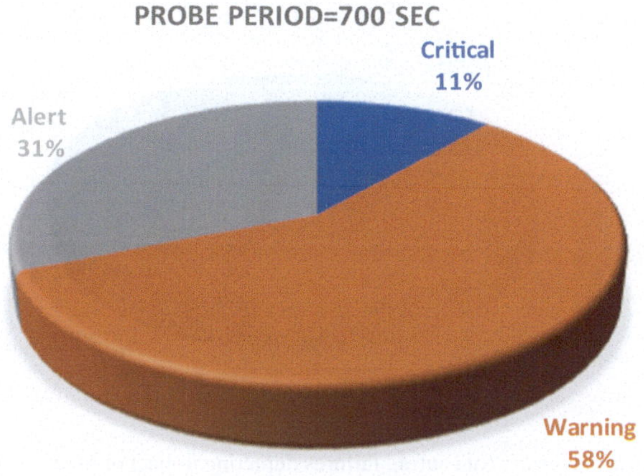

Fig. 7 Network wide error classification depicting a lesser proportion of critical errors over a longer period

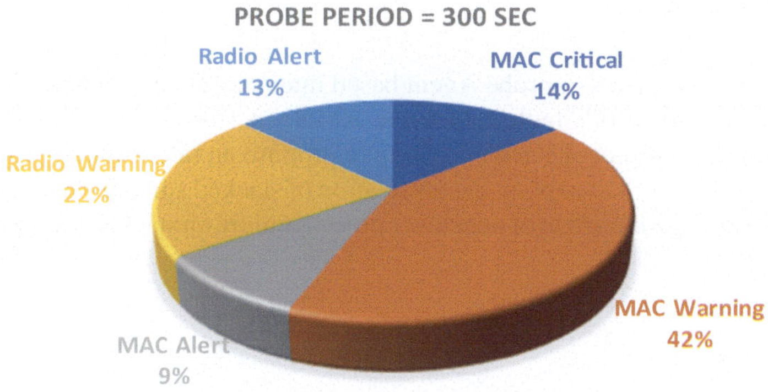

Fig. 8 Errors in stacked processes for routing failures, showing dominant impact of MAC warnings

To evaluate the inter-process dependency model based fault diagnosis, the impact of critical errors, warnings and alerts in stacked layer processes on routing process failure is analyzed. For a shorter probe period, the frequency of MAC layer critical errors is reduced due to the fact that network remains stable in a short-term interval. According to Fig. 8, decision tree infers a higher number of MAC and radio warnings as dominating root causes of the overall routing failures. It is because the effect of warnings on critical errors is cumulative. However, the impact of MAC layer critical errors on routing failures is profound for a longer probe period due to manifestation of warnings and alerts as critical errors over a longer interval as shown in Fig. 9. In all, alerts have lesser pronounced effect.

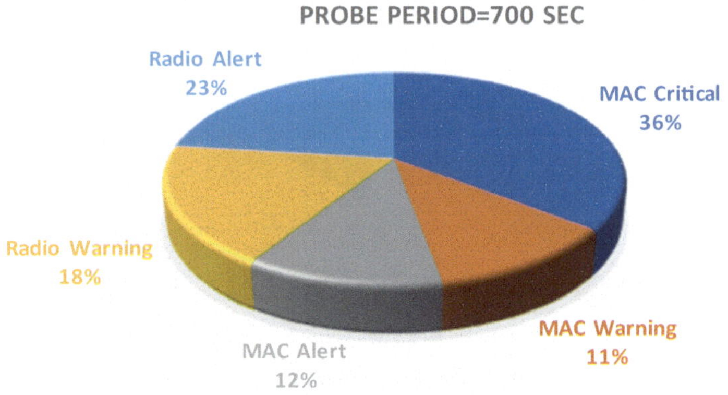

Fig. 9 Stacked processes errors for routing failures depicting impact of MAC critical errors caused by warnings on MAC and radio layer

5 Conclusion

This work presents AMNA: probe Agent based inter-process dependency Model for wireless sensor Network's fault diAgnosis. AMNA performs probe based analysis of communication protocol stack processes to propose an inter-process dependency model. In future, we shall explore the correlation of stacked and peer layer process to analyze the propagation effect of peer layer process failure on stacked layer processes.

References

1. Muhammed T, Shaikh RA (2017) An analysis of fault detection strategies in wireless sensor networks. J Netw Comput Appl (November 2016) 78:267–287
2. Gong W, Liu K, Liu Y, Member S (2015) Directional diagnosis for wireless sensor. Networks 26(5):1290–1300
3. Mahapatro A, Khilar PM (2014) Online fault diagnosis of wireless sensor networks. Open Comput Sci 4(1):30–44
4. Rodenas-Herraiz D, Fidler PRA, Feng T, Xu X, Nawaz S, Soga K (2017) A handheld diagnostic system for 6LoWPAN networks. In: Proceedings of the 2017 13th annual conference on wireless on-demand network systems services WONS 2017, pp 104–111
5. Peter H Quantitative analysis of AODV and its variants on dynamic topologies using statistical model checking. no. Mc
6. Perkins CE, Park M, Royer ME Ad-hoc on-demand distance vector routing
7. Souza JLR Improving NS-2 network simulator to evaluate IEEE 802. 15. 4 wireless networks under error conditions

8. Al-Anbagi I, Erol Kantarci M, Mouftah H (2014) A survey on cross-layer quality of service approaches in WSNs for delay and reliability aware applications. IEEE Commun Surv Tutor 18(1):1–1
9. Zafar A, Wajid B, Akram BA (2015) A hybrid fault diagnosis architecture for wireless sensor networks. In: International conference on open source systems and technologies, pp 7–15
10. Man LAN, Committee S, Computer I (2015) IEEE standard for low-rate wireless networks. In: IEEE Standard for Low-Rate Wireless Networks, vol 2015

Position at Response: A Novel Highly Accurate and Low Cost Laser Positioning System

Turhan S. Dervish, Caleb G. Teague, Fangyao Liu, Adam Coffey, Amit Kachroo, Sabit Ekin and Jerzy S. Krasinski

Abstract Global Positioning Systems (GPS) are widely used in many navigational applications such as vehicles, aircraft, and ships to find their speed and position. However, the main drawback of GPS is reliability, especially in an urban environment. Tall buildings reflect the signal and can make the receiver's location appear 10 m or more from its actual position. This paper details a low-cost system designed to overcome this accuracy problem. The proposed system, termed as Position at Response (PaR), utilizes a laser and a network of transponders to locate the vehicle position more accurately than a GPS. The system can work for an indoor application as well. It is easily adaptable to new routes, paths, or destinations. Our preliminary test results show an accuracy within three percent, relative to the distance of the transponders.

Keywords Laser navigation · Autonomous vehicle · Triangulation · Transponder localization · Positioning · Local positioning system

---

T. S. Dervish (✉) · C. G. Teague · F. Liu · A. Coffey · A. Kachroo · S. Ekin · J. S. Krasinski
School of Electrical and Computer Engineering, Oklahoma State University, Stillwater, OK, USA
e-mail: stuart.dervish@okstate.edu

C. G. Teague
e-mail: caleb.teague@okstate.edu

F. Liu
e-mail: fangyao.liu@okstate.edu

A. Coffey
e-mail: adam.coffey@okstate.edu

A. Kachroo
e-mail: amit.kachroo@okstate.edu

S. Ekin
e-mail: sabit.ekin@okstate.edu

J. S. Krasinski
e-mail: krasins@okstate.edu

A. Boyaci et al. (eds.), *International Telecommunications Conference*,
Lecture Notes in Electrical Engineering 504,
https://doi.org/10.1007/978-981-13-0408-8_12

1 Introduction

Today, Global Positioning Systems (GPS) are ubiquitous, being found on most cellular phones, equipped in many cars as a navigational guide, and in many other applications. The system uses a collection of satellites to trilaterate the position of a base station. Accuracy is guaranteed to be within 6 m 95% of the time [1]. In a study of over 10,000 smart phones in over 100 countries, GPS had an open sky mean average of 4.9 m [2]. GPS's inaccuracies only increase with obstructions like trees or buildings.

An autonomous vehicle (AV) needs to reliably determine its location with a better resolution than 6 m. There are two common methods for doing this, either triangulation, or trilateration, as illustrated in Fig. 1. Triangulation uses angles to determine distance, when two angles and the distance between them are known. Using the laws of sines and cosines, lines can be drawn from the base line at the angles known, and the intersection marks the location. Geometry then finds the distance. Trilateration measures distances not angles. The distance from three known points are used as the radius of three circles, and the circles intersect at the location.

There have been several studies based on the primary concept of trilateration and triangulation. Brooks et al. [3] surveys these navigation methods: Time of Flight, Time Difference of Arrival, and Angle of Arrival. Other methods such as Line Tracking [4] and RFID Mapping [5] are also defined and discussed below.

Time of Flight (ToF) When used with multiple base stations at known positions, ToF can be used to determine location. Having one base station can determine only distance. After a precise measurement of the time it takes to send a signal at a known velocity, the distance to the base stations are determined. Then trilateration is used to find position. One of the drawbacks to this system is the possibility of interference or multipath propagation of the signal [3].

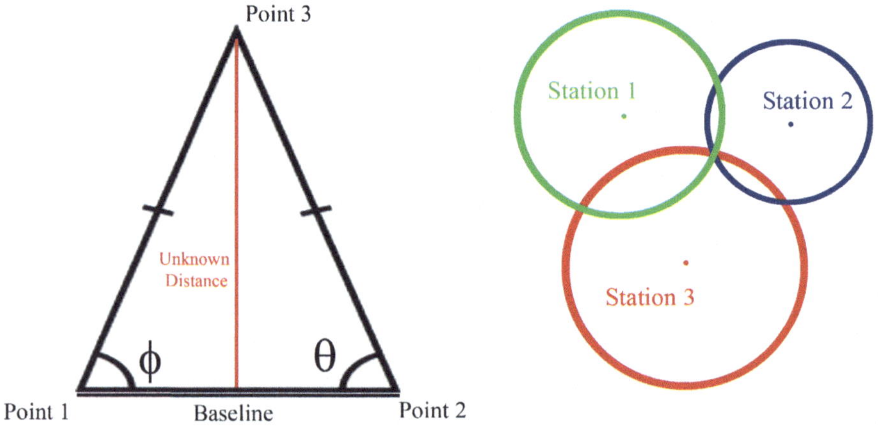

Fig. 1 Triangulation versus trilateration

Time Difference of Arrival (TDoA) This system uses multiple sources of regularly timed signals sent to one receiver. Since the signals locations are known, then by timing the arrival, it is possible to trilaterate position relative to the signal sources. GPS is a specific type of TDoA, using synchronized stationary orbit satellites. It can be completely localized as well, like the system written about by Cheng [6]. TDoA systems without a very accurate clock can have inaccurate distance and position calculations [3], and multipath propagation and atmospheric changes can also affect results.

Angle of Arrival (AoA) AoA finds position in one of two ways. In the first, the base station uses an array of antennas to find the angle of incoming transmissions. This information, used with ToF and trilateration, determines the position in polar coordinates. A second method uses two antenna stations to determine the angle of the incoming signal from each antenna, then uses triangulation to determine the location. One benefit of this second method lies in its angle based system, so no rigorous timing devices are needed. A second benefit is that more stations allow for reduced uncertainty through redundancy. However, the systems accuracy depends on the precision of the antenna's angular resolution [3].

Line Tracking (LT) One form of LT is achieved by burying a wire a few cm deep, and sending a signal through it. The path can be followed using the electromagnetic induction of the wire [4]. This system's main drawbacks are the disruption to existing infrastructure for installation and it's permanence. Optical line tracking is easier to implement, but requires a visible line along the intended path. However, a visual line can be obstructed (e.g. snow) or worn away over time. If the line is lost, the AV cannot move [5].

RFID Mapping (RM) This form of location is most commonly used in warehouses. RM embeds a grid of radio frequency identification (RFID) chips in the floor (or ceiling) [5, 7], to allow AVs to traverse to specific bins and collect materials, or have the bins move to desired locations [8]. This method while effective also requires drastic infrastructure disruptions to install or expand.

Position at Response (PaR) PaR is a new method for triangulation. Unlike the systems listed above, PaR is low cost and doesn't require any precision timing of signals (like ToF or TDoA), antenna beamforming (like AoA), or infrastructure changes (like LT and RM). Instead it uses an motor encoder to track the position of a laser it is rotating. As the laser sweeps across transponders with known locations, they return an identifying signal by transceiver. The position at the response is sent to a microcomputer to triangulate the location. Additionally, it is designed to be easy to setup, maintain, and determine or change existing routes.

The rest of the paper is organized as follows: In Sect. 2, the system overview and the triangulation method are detailed. In Sect. 3, testing and results are discussed, along with some problems of the PaR system, and how it compares to other methods. Section 4 discusses future work and conclusions.

2 System Description

This system was designed at the Oklahoma State University to provide accurate location data for an autonomous golf cart giving campus tours. However, the system will work any environment, indoors or out, with almost no infrastructure changes needed. Furthermore, it has user friendly software designed for easy adaptability in route changes; whether temporarily (i.e., for construction) or permanently. It was developed to be easily installed, maintained and adjusted by laypeople.

The system naturally divides into four parts: The laser base, the transponders, the location software, and the Graphical User Interface (GUI) as can be seen in Fig. 2.

1. **Laser Base**
   This central component contains a laser diode, an encoded motor, and processing hardware. A laser is decollimated into a vertical line and rotated about the z-axis. The motor's encoder data is sent to a Raspberry Pi. The base also has an Inertial Measurement Unit (IMU) to track magnetic north and the yaw (rotation about the z-axis), which it relays to the Pi. The laser base uses a transceiver to receive signals from the transponders.
2. **Transponders** As the rotating laser sweeps across photodiodes on each transponder, a small current is sent through a Trans-Impedance Amplifier (TIA). This triggers a PIC microcontroller to read the values from the user set ID for the transponder, and transmit this ID back to the laser base. The ID is user defined by rotary hexadecimal switches on the casing.
3. **PaR Location Software**
   The PaR system is elegant in its simplicity. The program's flowchart is detailed in Fig. 3. When a transponder broadcasts an ID signal, the PaR location software retrieves four elements of information: the angle of the laser from the encoder, the orientation from the IMU, the ID of the transponder, and the predetermined location of that transponder. That information is used to extrapolate a line. A second transponders angle, ID, and location, allows the software to calculate a second line. The lines intersect at the location of the laser base. Any additional transponders allow for error correction. It is intended to have at least three transponders in range at all times, but can work with two, as shown in Fig. 4.
   The software is fast and agile. Because trigonometry takes more clock cycles to process than linear algebra, a dictionary of the motor encoder's 3200 sine and cosine values was created. This way any value returned on the encoder can be referenced for a trigonometric value.
   This means in Fig. 4 that the positions of the transponders (c_1, d_1) and (c_2, d_2) and the angle of the transponders θ_1 and θ_2 are known values. By using linear algebra, finding the intersection of the lines and the position of the base station can be determined with minimal calculations.
   In Eqs. (1) and (2), a_1 and a_2 are values found in the dictionary, not solved:

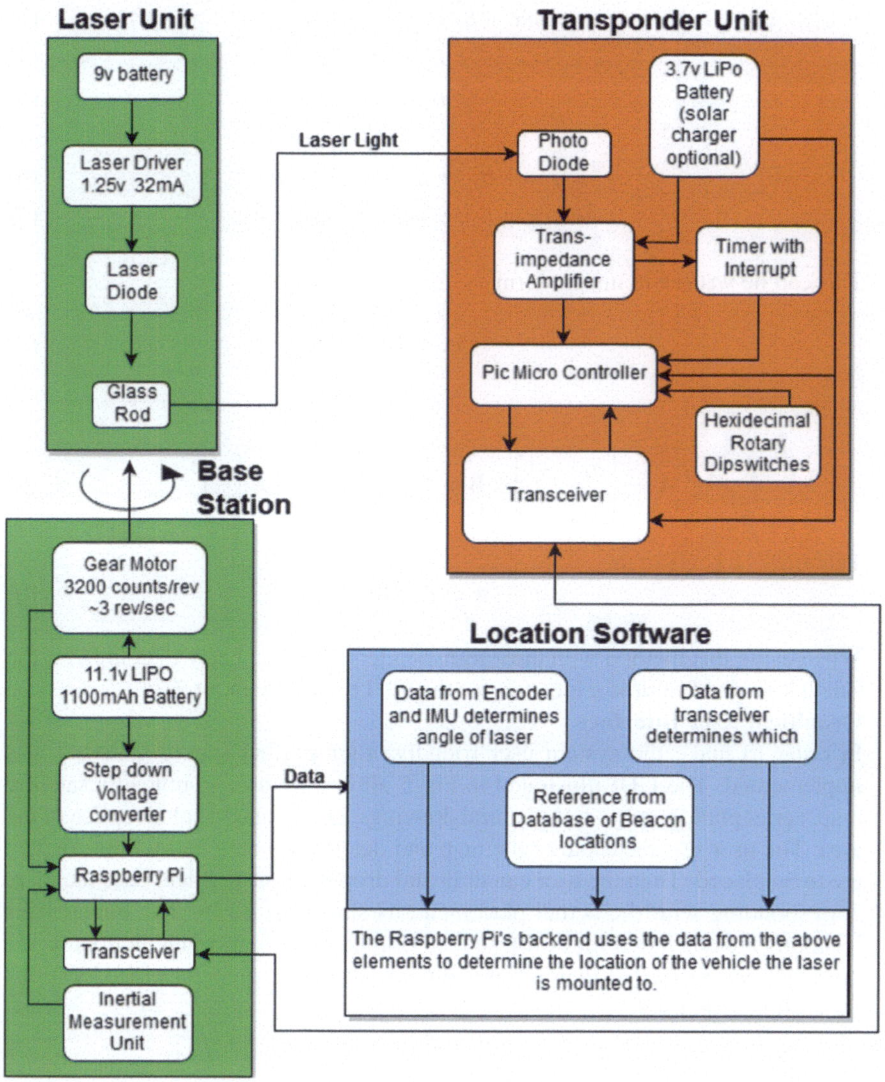

Fig. 2 System overview

$$a_1 = \tan(\theta_1) \quad (1)$$

$$a_2 = \tan(\theta_2) \quad (2)$$

Starting with the equation of a line, we can manipulate a system of equations:

$$y = mx + b, \Rightarrow b = y - mx \tag{3}$$

$$\therefore b_1 = d_1 - a_1 c_1 \tag{4}$$

$$\therefore b_2 = d_2 - a_2 c_2 \tag{5}$$

$$y = a_1 x + b_1 \Rightarrow a_1 x - y = -b_1 \tag{6}$$

$$y = a_2 x + b_2 \Rightarrow a_2 x - y = -b_2 \tag{7}$$

This can be written in matrix form as:

$$\mathbf{A}x = \mathbf{B} \tag{8}$$

where,

$$\mathbf{A} = \begin{bmatrix} a_1 & -1 \\ a_2 & -1 \end{bmatrix}, \ \mathbf{B} = \begin{bmatrix} -b_1 \\ -b_2 \end{bmatrix} \text{and} \quad x = \begin{bmatrix} x \\ y \end{bmatrix} \tag{9}$$

Therefore, x is given as:

$$x = \mathbf{A}^{-1}\mathbf{B} \tag{10}$$

When using this method with three transponders, the Raspberry Pi completes the calculations and returns a location for the GUI to plot in about 2 μs.

4. **Graphical User Interface**
In order to make the system user friendly, a simple but secure interface was implemented. The GUI illustrated in Fig. 5 allows the user to upload a satellite map, aerial photography, architectural drawings, or any other scaled replica of the area. The user sets the scale of the map and determines how many transponders are to be placed. Then the user can drag and drop the transponders onto the map, corresponding with the actual placement, as shown in Fig. 6. The transponder locations are converted into Cartesian coordinates and saved in a dictionary file on the Raspberry Pi. The Pi uses this to quickly access the predetermined locations for each beacon used.

3 Testing, Results, Problems, and Solutions

The positional information sent back during testing is accurate within approximately 3% (± 1.5%) relative to the distance to the transponders. Figure 7 shows the Cartesian coordinates found, while simultaneously Fig. 8 shows the angles being reported by encoder to the PaR software.

Occasionally, errors will jump to nearly 10%. These errors are sporadic, and the cause is unclear, but evidence suggests that the IMU is the cause, since they happen uniformly across all transponders. Since this was intended to be merely a proof of

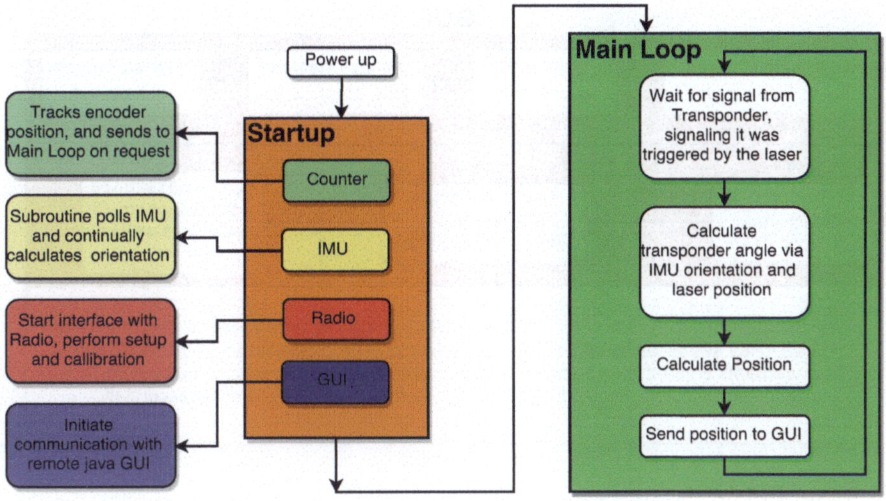

Fig. 3 The location system block diagram

Fig. 4 Triangulation using two transponders

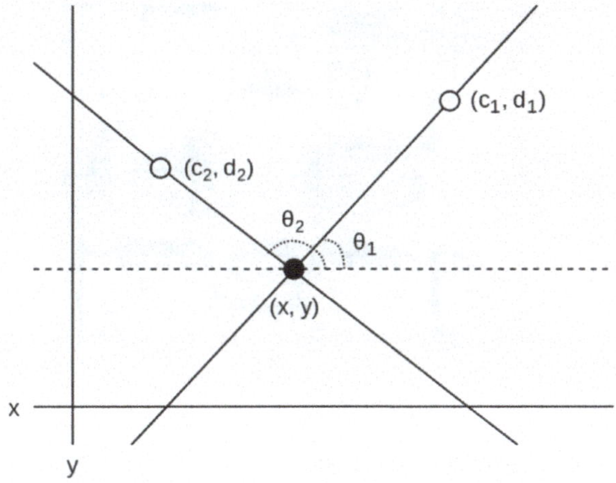

concept, the limitations of the system and variance shown in the resultant data were considered acceptable, but the addition of a more robust IMU and an absolute encoder would reduce the errors dramatically.

Non-line of sight (N-LoS), when something gets in between the transmitter and receiver, degrades the navigation performance. The reason the GPS loses resolution is the N-LoS with some of it's satellites. In addition, since the constellation of GPS sattelites is in constant change, the error varies constantly [9]. ToF and TDoA need a clear path, without it, multipath propagation adds time to the signal's flight, and causes error. If solutions to these problems are known, the expense to correct it is prohibitive.

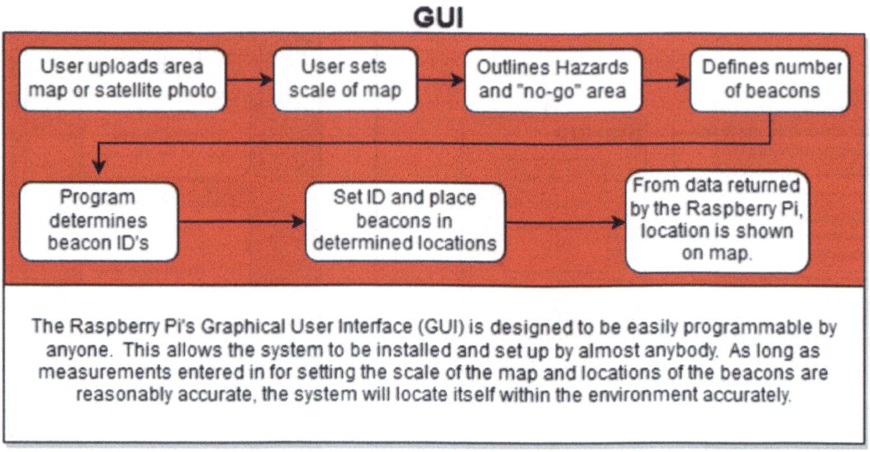

Fig. 5 GUI flowchart

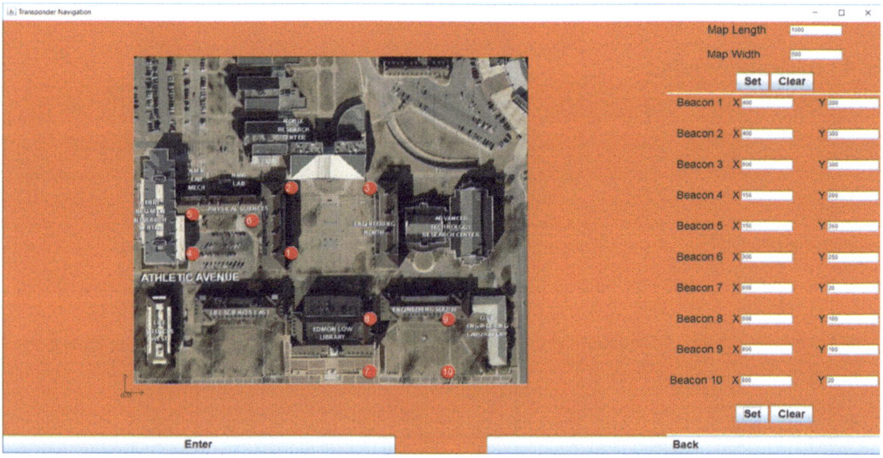

Fig. 6 A screenshot of the GUI

The same N-LoS issue affects the PaR system, but the cost of an additional transponder is currently under $15 (prototype cost is significantly higher than production cost will be), therefore putting more transponders in the system is an effective solution to the N-LoS issue. The system can have up to 65,536 unique transponders, which would blanket a very large area. Slight modifications allow over one million transponders, and exponentially greater coverage.

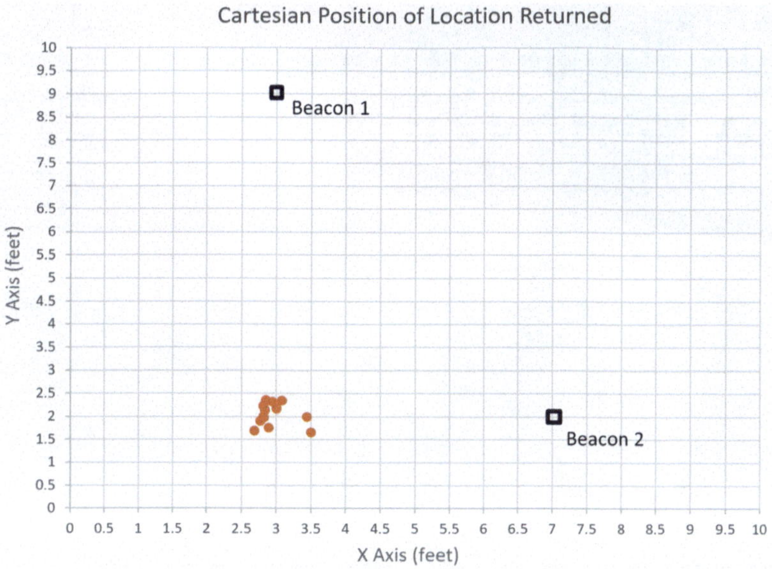

Fig. 7 A plot of the locations found over time

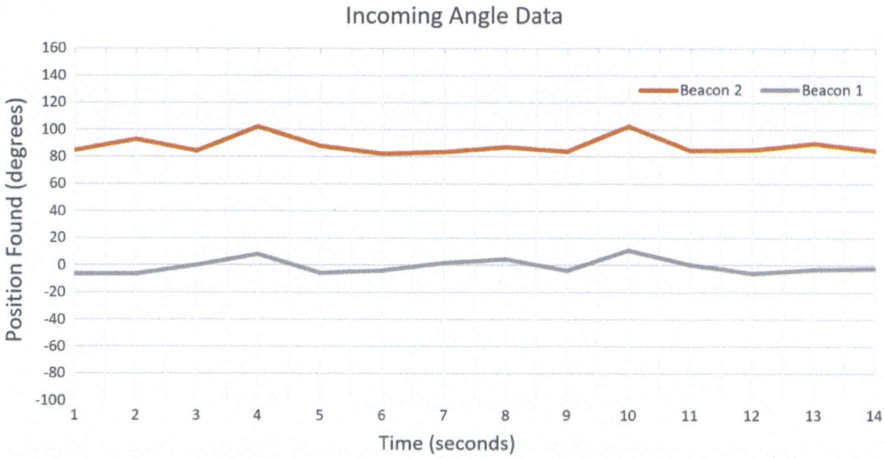

Fig. 8 The angles of beacons found over time

The physical system shown in Fig. 9 has a small footprint. The transponders being 90 × 70 × 40 mm and weigh less than 250g with the battery. The base is 200 × 75 × 120 mm and weighs less than 1kg. These small sizes make them unobtrusive, and adaptable to many sizes of AV. It connects via Wi-Fi to the computer running the GUI, making monitoring from a distance a feasible option (Table 1).

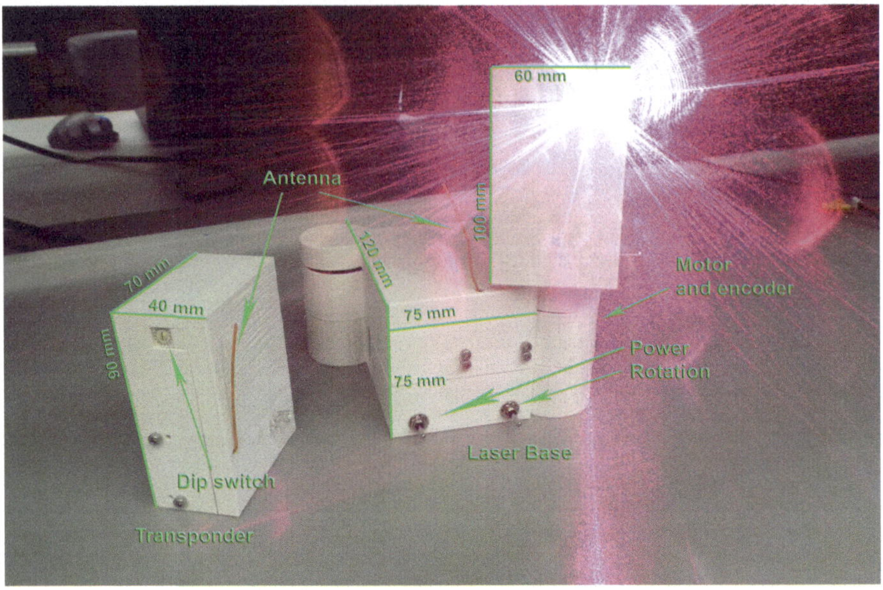

Fig. 9 Transponder and base station

Table 1 List of components[a]

Placement	Item	Cost
Base Station	Laser Diode	$47.40
	16 mm Glass Rod	$8.70
	9 DOF IMU	$15.95
	Raspberry Pi	$39.99
	Buck Converter	$6.99
	2000 mAh Battery	$12.99
	50:1 Gearmotor	$39.95
	Radio Transceiver	$9.95
Transponders	3 units	
	(x3) PIC microcontrollers	$5.40
	(x3) Op Amps	$8.64
	(x3) Radio Transceivers	$29.85
	(x3) Photodiodes	$3.95
	(x3) Rotary Dip Switch	$6.69
Total		$243.39

[a]Some items not listed (wire, resistors, capacitors, 3D printing filament, perf board, solder, etc.) would add a minimal cost, and were not accounted for in the budget

4 Conclusion and Future Work

The proof of concept detailed in this paper shows a new localized positioning technique, and a system implemented to run it. Preliminary testing shows it to be more accurate than GPS, especially indoors or outside in urban environments. Using a laser and a network of transponders, a new method of triangulation has been developed with an accuracy of approximately 3% relative to the distance to the transponders. It is designed to overcome the inaccuracy of GPS; but with a low cost of production and implementation, and can be used either indoors or outdoors. Furthermore, very little (if any) infrastructure change is needed to implement this system. The low cost and versatility of this system makes it a system worth fine tuning. We have plans to add more functionality to the software and upgrade the hardware in the next phase of our research.

References

1. US DoD (2001) Global positioning system standard positioning service performance standard. Assistant secretary of defense for command, control, communications, and intelligence
2. van Diggelen F, Enge P (2015) The worlds first gps mooc and worldwide laboratory using smartphones. In: Proceedings of the 28th international technical meeting of the satellite division of the institute of navigation (ION GNSS+ 2015), pp 361–369
3. Brooks TTJ, Bakker HHC, Mercer KA, Page WH (2005) A review of position tracking methods. In: 1st International conference on sensing technology, pp 54–59
4. Rosandich RG, Lindeke RR, Berg J (2007) Developing an automatic guided vehicle for small to medium sized enterprises
5. Burke EM, Ewing Jr DL (2014) Improving warehouse inventory management through rfid, barcoding and robotics technologies. Technical report, Naval Postgraduate School Monterey CA Graduate School of Business and Public Policy
6. Cheng X, Thaeler A, Xue G, Chen D (2004) Tps: a time-based positioning scheme for outdoor wireless sensor networks. In: INFOCOM 2004. Twenty-third AnnualJoint conference of the IEEE computer and communications societies, vol 4. IEEE, pp 2685–2696
7. Martinelli F (2016) Robot localization using the phase of passive uhf-rfid signals under uncertain tag coordinates. J Intell Robot Syst 82(3–4):577–593
8. Samad T, Annaswamy AM (2011) Success stories for control: Mobile-robot-enabled smart warehouses. The Impact of Control Technology, pp 1–2
9. Modsching M, Kramer R, ten Hagen K (2006) Field trial on GPS accuracy in a medium size city: the influence of built-up. In: 3rd workshop on positioning, navigation and communication, pp 209–218

Part IV
Measurement

Jamming Resilience of LTE Networks: A Measurement Study

Yağmur Coşkun, Mert Eygi, Gediz Sezgin and Güneş Karabulut Kurt

Abstract Focusing on LTE physical layer (PHY) and its channels, we investigate the resilience of an LTE network against jamming attacks. We conduct tests on an operational LTE network. Although LTE PHY jamming attacks have been studied, studies have not investigated the jamming attacks on a commercial LTE network. We examine the system performance of LTE PHY under two types of jamming attacks. Detailed measurement results are presented based on performance metrics of LTE networks. Finally, several counter measures against such jamming attacks are suggested.

1 Introduction

Mobile communication technologies are rapidly evolving and their standards are constantly being upgraded. The current target is to connect billions of devices with each other. As more devices become connected, the impact area of malicious intentions will increase. Searching possible attacks and their effects can help guard LTE networks and improve their resiliency against attacks.

There are many attack types in the literature. Among these, the denial of service (DoS) attacks are expected to be critical in mobile networks and they have to be considered. These attacks can be carried out at different LTE layers according to [1]. Attacks targeting core and backhaul networks would use vulnerabilities in authentication mechanisms to attack roaming interconnection or to steal International Mobile Subscriber Identity (IMSI). Similar to these attacks, signaling amplification attacks are also important for mobile networks, particularly in the signaling plane due to the scarcity of resources. Mobile networks suffer from limited radio resources to

Y. Coşkun (✉) · M. Eygi · G. Sezgin · G. K. Kurt
Wireless Communication Research Laboratory (WCRL), Istanbul Technical University, Istanbul, Turkey
e-mail: coskunyag@gmail.com

M. Eygi · G. Sezgin
TURKCELL İletişim Hizmetleri AŞ., Istanbul, Turkey

A. Boyaci et al. (eds.), *International Telecommunications Conference*, Lecture Notes in Electrical Engineering 504,
https://doi.org/10.1007/978-981-13-0408-8_13

provide service to every customer simultaneously. The scarcity of bandwidth requires advanced techniques to reuse idle resources in an efficient manner. If a large number of malicious user equipments (UE) guided by a botmaster which controls a botnet (a set of compromised devices) send to set up bearer connection, for example, the legitimate UEs would suffer depletion of resources. In addition to these, physical layer (PHY) attacks are also critical threats.

The common way to pollute the air interface is jamming attacks. For understanding the peril of the jamming attack issues experienced in LTE PHY, a comprehensive search has been conducted in this paper and possible different jamming attack cases are investigated by testing on a real LTE network. We present weaknesses and resiliency of the LTE PHY under several jamming attacks. Associated countermeasures are discussed.

In Sect. 2, we provide a summary on jamming attacks on LTE networks. In Sect. 3, a brief LTE air interface overview is given. Then, used jammer characteristics which was used during tests, testbed descriptions, performance metrics and test results are explained in detail in Sects. 3 and 4, respectively. Some mitigation methods are offered to mitigate jamming attacks in Sect. 5. We conclude the paper in Sect. 6.

2 Literature Overview

There exist multiple surveys focused on only PHY vulnerabilities and jamming attacks on LTE networks. As pointed out in [2], jamming attacks are the simplest but also effective attacks that LTE networks are exposed. Reference [3] is a tutorial paper addressing the PHY security concerns and resiliency of OFDM communications which is the de facto air-interface of most modern wireless broadband standards including 3GPP LTE. Reference [4] investigates the extent to which LTE is vulnerable to intentional jamming, by analyzing the components of the LTE downlink and uplink signals. The authors of [5] investigate the extent to which LTE is vulnerable to RF jamming, spoofing, and sniffing, and assess different PHY threats that could affect next generation critical communication networks. In [6] some potential ways of jamming attacks are discussed and an easy-to-implement method is analyzed in details. Authors of [7] propose an analysis on how an LTE based mobile broadband communication system would be able to survive and resist to intentional jammers as encountered in some military use cases. An investigation on the impact of jamming threats on the performance of LTE networks takes place in [8]. Reference [9] analyzes the vulnerability of LTE initial synchronization mechanism, and puts forward a LTE jamming method based on spoofing synchronization signals to verify this vulnerability. Finally, [10] focuses on investigating the effects of interference on the physical uplink control channel (PUCCH); the uplink control channel used in LTE.

Even though these studies have been conducted for demonstration of possible effects of a jamming attack, they were either simulation or a limited test system consisting of several equipment mimicking network components. In contrast to other researches, we had a cooperation with a mobile network operator giving us the opportunity to test on their commercial LTE network in a controlled environment. The reason why we are interested in jamming a commercial LTE network, is to observe attack results on a real-time network and also analyze the resilience of LTE networks. After understanding the impact of such attacks, effective countermeasures are discussed.

3 LTE PHY Overview and Description of Test Environment

LTE PHY is responsible of carrying both data and control info via radio interface between base station (eNB) and UE. A multicarrier approach is chosen for fourth generation in order to reach higher spectral efficiency, flexibility, to increase capacity and to enable efficient scheduling. Orthogonal frequency division multiple access (OFDMA) is employed for downlink. Single carrier frequency division multiple access (SC-FDMA) is preferred for uplink because of the fact that OFDMA produces a high peak-to-average power ratio (PAPR) and UEs are power limited [11–13]. Physical channels of LTE are responsible for carrying of user data and control messages and varies between uplink and downlink direction. Brief explanations of these channels are shown in Table 1.

3.1 Jamming PHY Channels and Its Implications

UE has to receive the master information block (MIB) message because MIB has significant information such as downlink bandwidth, system frame number and physical hybrid automatic repeat request (HARQ) indication channel (PHICH) configuration.

Table 1 LTE physical channels and their functions

Channel	Function	Direction
PUSCH	Carries uplink data, RRC signaling messages, uplink control information(optional)	UL
PUCCH	Carries uplink control information	UL
PRACH	Enables Random Access functions, has the only non-synchronized connection	UL
PDSCH	main data bearing channel includes SIB,paging, RRC connection messades	DL
PDCCH	Carries downlink control information (DCI) and scheduling information	DL
PBCH	Broadcast MIBs needed for accessing network	DL
PHICH	Carries HARQ ACK/NACK messages	DL
PCFICH	Carries control Format Indicator (CFI) and informs UE OFDM format	DL

Physical broadcast channel (PBCH) is transmitted at the center of the downlink (DL) frequency with 1.08 MHz bandwidth [5]. Thus, it is expected that jamming it by using a narrow-band continuous jammer operating in a specific frequency would be so vital for system performance. Physical downlink control channel (PDCCH) tells the UE reserved DL assignment and the uplink (UL) grant where the UE can receive or send their data, jamming this channel may cause that the UE cannot decide which data belongs to it and hence the UE cannot receive or send the data correctly.

Physical control format indicator channel (PCFICH) is a crucial channel for PDCCH because it determines the number of OFDM symbols belongs to PDCCH. Jamming PCFICH is not considered as an important threat since the channel is an extremely sparse channel whose data is carried by 4 resource element groups (REG) consisting of 4 resource elements and these 4 REGs are evenly distributed across the whole band regardless of the bandwidth. However PDCCH is not sparse as in PCFICH. Due to this property, jamming PDCCH would be easier than PCFICH, but it requires synchronization. Moreover, physical HARQ indicator channel (PHICH) provides reliable communication with the help of the HARQ mechanism which carries ACK-NACK messages for UL data transfers [12]. Therefore, jamming PHICH may cause an increase in the number of retransmissions and transmission failures in UL which degrades the UL data transfer performance.

Finally, UEs would not be able to receive or send data correctly if data channels (PDSCH or PUSCH) are jammed. However such an attack can be considered to be unlikely since it needs more complex jammers and system information. It is expected that UEs data throughput will be degraded and the degradation is directly proportional with jamming bandwidth. PUCCH that carries control information, scheduling requests and HARQ messages for DL data transmissions is placed at two edges of the UL bandwidth.

3.2 *Jammer Characteristics*

To investigate the jamming effect on the LTE PHY, we need to identify the jammer attacks first. According to [14], four different jammer attacks are put forwarded which are constant jammer (emits radio signal), deceptive jammer (injects regular packets to the channel), random jammer (a jammer which alternates between sleeping mode and jamming mode by sending radio signal) and lastly, reactive jammer. Reference [15] provides a different approach to the types of jamming attacks by postulating barrage jamming attack, pilot tone jamming attack, pilot tone nulling attack for OFDM systems which utilize pilot tone signals for channel estimation. A barrage jammer jams all subchannels of an OFDM system whereas pilot tone jamming and pilot tone nulling attacks are more complicated which are used to jam pilot tone signals or to avoid the reception of pilot tones, respectively. When an adversary targets a specific protocol, he or she could realize a very effective attack by jamming one of the subsystems that LTE has in forms of physical channels and signals. Reference [3] proposes that as long as those sub-channels including control channels, random access requests or HARQ, are vital for the smooth operation of the network system, with a high signal to jamming ratio, a denial of service (DoS) attack can be achieved. Reference [16] discusses the jamming type that targets specific channel instead of jamming the entire network bandwidth as smart jamming.

In our study, we consider two types of jammers having different characteristics used as the interference source during tests. They have been implemented with USRP N210 software defined radio nodes. First type is a conventional partial band jammer having an OFDM signal characteristics whose subcarriers are considered to be a Gaussian distributed signals with zero mean and parametric variance. A snapshot is shown in Fig. 1b. The second type of the jammer is referred to as the "chirp" jammer whose signal has an exclusive waveform having lower PAPR when compared to a Gaussian distributed signal. An example is shown in Fig. 1a. Reference [17] demonstrates that chirp jamming attacks may decrease the communication quality more severely than Gaussian noise based conventional jamming attacks. Considering limited output power of the antennas utilized for jammers, a chirp jammer having a better PAPR characteristics would be more effective since its impact on the system would be more severe than a conventional jammer.

3.3 *Testbed Descriptions*

Tests were carried out at the Ayazaga Campus of Istanbul Technical University as shown in Fig. 2. There were three active LTE base stations of the operator called AYAZA, ISPOR and INSAT during the test. The duplexing mode was frequency division duplex (FDD). LTE network specification of the operator is given in the Table 2. 1800 MHz LTE frequencies were jammed as an operational network test scenario. Although each base station has at least one 1800 MHz sector, the second sector of ISPOR was selected because of low number of instantaneous user relative to other sectors. The red circle shown in Fig. 2 indicates the location where tests were carried out. Table 3 also describes the test specifications.

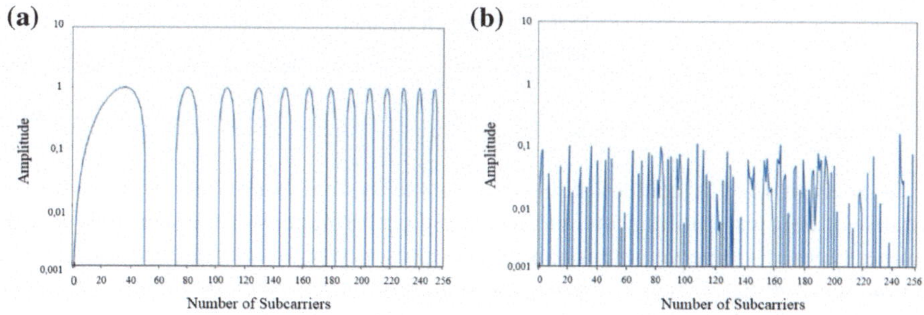

Fig. 1 **a** A realization of the chirp partial band jammer, **b** A realization of the conventional partial band jammer

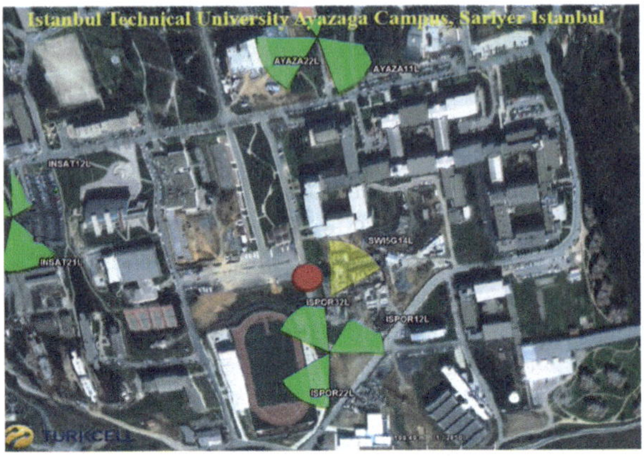

Fig. 2 A view of active LTE sectors of the operator and the testbed location (red circle) in the university campus

Table 2 LTE configuration.

Operating frequency	800 MHz	1800 MHz	2600 MHz
LTE bandwidth	10 MHz	20 MHz	20 MHz
eNode B antenna number	2	4	4
eNode B antenna gain	15.4	17.4	17.4
Duplexing mode	FDD	FDD	FDD
eNodeB output power	2×43 dBm	4×44.7 dBm	4×44.7 dBm
Reference signal gain	3 dB	0 dB	0 dB

Table 3 Test specifications

Center frequency of UL, f_{c1}	1755,1 MHz
Center frequency of DL, f_{c2}	1850,1 MHz
Jammer bandwidth	4 MHz
Jammer output power	27 dBm
Test terminal UE category	Category 6
Test terminal output power	23 dBm
Distance between UE and eNodeB	15 m
Distance between UE and Jammer	<1 m

As seen in Fig. 3, center frequencies of the system bandwidth were jammed because of hardware limitations. We compared performances of the aforementioned jammer types. To observe potential impact of presence of an interference source to a LTE network, 6 different test scenarios were logged as described in Table 4. DL1 and UL1 refer to the case when there is no jamming attack on DL or UL direction. The network's central 4 MHz bandwidth has been jammed with a conventional jammer in both direction during test cases which are labeled as DL2 and UL2. Finally, chirp

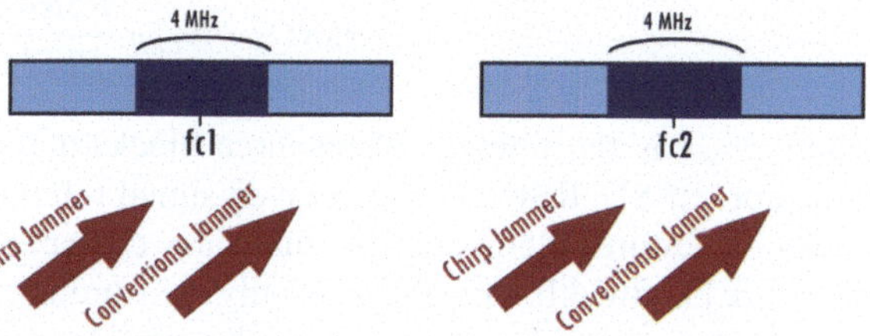

Fig. 3 Jammed frequencies of uplink and downlink bands with two types of jammers

Table 4 Test descriptions

Test Label	Jammed and measured frequency direction	Jammed frequency location	Jammer type
DL1	Downlink	None	None
DL2	Downlink	Center	Conventional
DL3	Downlink	Center	Chirp
UL1	Uplink	None	None
UL2	Uplink	Center	Conventional
UL3	Uplink	Center	Chirp

jammer is employed instead of conventional jammer for the test cases DL3 and UL3. The test labels also will refer to the figures in the following results and discussion section.

3.4 Performance Metrics

Several performance metrics are used to measure LTE system performance and logged by UEs and eNBs in order to monitor problems while providing several services to the users. To be able to interpret the test results, relevant metrics are briefly explained.

Most common notion used for wireless networks is *SINR*, signal to interference plus noise ratio, is a common performance metric used as an indicator for the channel quality and defined as:

$$SINR = \frac{S}{I + N}, \tag{1}$$

where S indicates the measured power of usable signals (which can be reference signals and PDSCH channels), I indicates interference power from other cells and N indicates the background noise. However, even though *SINR* is commonly used for expressing the signal quality for a wireless network, it is not defined in 3GPP specifications, thus do not reported by UEs. Instead, another performance metric, *RSRQ*, reference signal received quality is used which is defined as

$$RSRQ = N_{PRB} . \frac{RSRP}{RSSI}, \tag{2}$$

where N_{PRB} is the number of resource blocks across the received signal strength indicator (RSSI). RSRP is the received signal reference power which is the average power of resource elements (RE) that carry cell specific reference signals (RS) over the entire bandwidth, thus *RSRP* is only measured in the symbols carrying RS. *RSSI* is a parameter providing information about total received wide-band power (measure in all symbols) including all interference and thermal noise defined as

$$RSSI = S + I + N. \tag{3}$$

Those performance metrics depend on the RF conditions of the environment. It is obvious that any type of interference such as jammer sources will cause a deviation. For observation of the effects of jamming to the system, the variations in those performance metrics are compared and discussed.

4 Test Results and Discussions

The measurements have been conducted with the help of engineers from the network operator and the measurement reports containing performance metrics have been obtained from the serving cell's eNB and the used UE. Observing UE behavior and obtaining its statistics, LTE Category 6 TEMS terminal was used. The distance between terminal and jammer was under 30 cm. The measurements were taken for a duration of approximately 3 min. Accordingly, for convenience, average values of performance metrics are examined in this study for observation of impacts of presence of a jammer.

Figure 4a shows that the average values of RSRP, a performance metric used for expressing received signal's power level, were similar (nearly −58 dBm) for each test scenario since interference does not have a significant effect on total signal power transmitted by the eNB. In contrast to RSRP, SINR and RSRQ were effected by the jammer because these metrics are related to DL transmission. Figure 4b, c display measurement results. Firstly, we have to note that in any UL jamming scenario, RSRQ and SINR values are nearly the same as the DL1 scenario since the UE had already camped and serving on the eNodeB while UL frequencies were jammed. During the DL1 test, the network load was more than the load during all UL tests. Hence RSRQ values are different (−14 dB vs. −9 dB) as shown in Fig. 4b. From both figures, it is clear that chirp jammer has a more negative effect than the conventional jammer. For DL, jamming center frequencies with bandwidth more than 1 MHz causes more severe performance loss than UL center frequencies jamming since mostly DL control signals are located at the center of operating frequency. Since we have 4 MHz jammer bandwidth, in addition to PBCH channel, some part of PHICH channel, which is located to the near of center, were effected by the

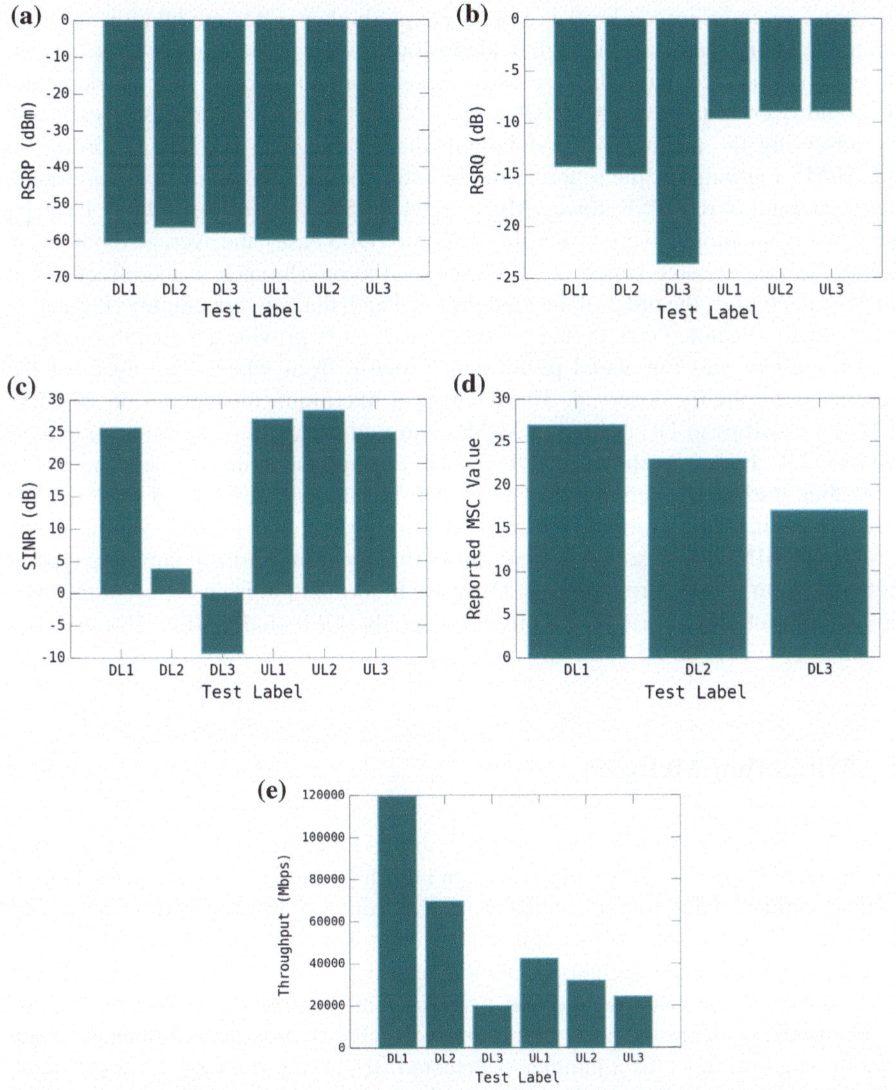

Fig. 4 **a** Average RSRP measurements, **b** Average RSRQ measurements. **c** Average SINR measurements, **d** Average modulation coding scheme number measurements, **e** Average throughput measurements

jammer. Consequently, during the DL3 test, the test terminal was disconnected from the jammed LTE network due to poor RF conditions (average RSRQ was 23.6 dB, average SINR was −9.26 dB).

Since LTE networks employ adaptive modulation coding schemes (MCS), the network switches the modulation and coding type as the signal quality varies. When the noise or interference level is low, it adapts higher order modulation whereas when the interference level is high it adapts lower order modulation schemes. Thus, it could be expected that if the resulting average modulation order graph is depicted, it should overlap the average SINR graph. When this phenomenon is investigated by observing the average of modulation schemes given in Fig 4d. According to the TEMS log outputs, the reported modulation coding scheme numbers are range between 0 and 28 (0:QPSK; lowest MCS, 28:64 or 256 QAM; highest MCS). It can be seen that in the most effective jamming scenario (DL3 case), the average modulation has the lowest average which is 17. Hence, as the interference or the effect of the jammer decreases, the order of the modulation which the network employs increases. It should also be noted that, higher modulation schemes provide a higher throughput.

Throughput was the actual performance metric from which the impact of the jammer is implicitly observed. The theoretical maximum throughput on 20 MHz BW is 150 Mbps in DL (with 2x2 MIMO and without carrier aggregation) and 50 Mbps in UL. Figure 4e shows the DL and UL throughput values on the same graph. Note that, during DL jamming test cases, DL throughput was logged from the UE whereas during UL jamming test cases, UL throughput was logged. Similar to the RSRQ and SINR figures, Fig. 4e also shows that the most harmful jamming type in both direction is the chirp jamming. Decrease in throughput is an expected result of any kind of interference in the spectrum especially if it includes UE's allocated data channels (PDSCH on DL, PUSCH on UL).

5 Mitigation Methods

It was shown that LTE networks are vulnerable to jammer attacks even though they are resource-constrained. An adversary can launch jamming attacks with the help of a smart jammer. Therefore, the network suffers from a significant performance loss, as shown during the test and may not be able to recover itself using current protocols. However our proposed mitigation methods can be used by the network to recover most of its performance loss and may even force an adversary to retract.

Primarily, with the help of regulations of eNB, its broadcast frequency could switch when presence of a jammer is detected. If jammer does not have the capability to adapt this change, this method could be enough to evade a possible attack. Also, when a jammer having capability of synchronization to the network is present, this method can be developed by changing the timing of eNB. Since more sophisticated attacks aiming to break transmission on PRACH and shared channels need to be synchronized to the system, they can be eliminated by breaking the synchronization of the jammer when the interference is detected or an attack starts. After detection, eNB can stop broadcasting any information for a length of time and starts transmitting again. Furthermore, data being transmitted on channels like UL control channel which can be easily located and jammed without synchronization or sniffing any information about system, may be transmitted alternatively on shared channels. Although this strategy can reduce the maximum uplink throuhput by sending control information on the shared data channels, which is spread over most of the available bandwidth, it would make the operation harder for adversaries. Thus, alternative transmission ways should be considered for unprotected, easy accessible and vital data of the network in order to keep network secure and supply uninterrupted communication for users.

6 Conclusions

This is the first paper which performed a jamming attack on a commercial LTE network with two different jammer types to the best of the authors' knowledge. According to the results, chirp partial part jammer is more effective than conventional partial band jammer. Additionally, DL center frequencies have control signals where as UL center frequencies have data signals. Hence, we get more severe result in DL than the UL. Jamming UL control channels would be more effective than the jamming of DL control channels since UE's power is limited with respect to the eNodeB. Due to hardware limitations the edge bands of UL, which consist of UL control channels, were not jammed at the same time. Thus we have considered the center 4 MHz of UL frequency. As a future work, more precise and effective jammers will be considered for protocol aware attacks on both UL and DL control channels and associated countermeasures will also be designed and tested.

References

1. Henrydoss J, Boult T (2014) Critical security review and study of DDoS attacks on LTE mobile network. In: Asia Pacific conference on wireless and mobile, pp 194–200
2. Talbot D (2012) One simple trick could disable a Citys 4G phone network. MIT Technol Rev
3. Shahriar C, La Pan M, Lichtman M, Clancy TC, McGwier R, Tandon R, Sodagari S, Reed JH (2015) PHY-layer resiliency in OFDM communications: a tutorial. IEEE Commun Surv Tutor 17(1):292–314
4. Lichtman M, Reed JH, Clancy TC, Norton M (2013) Vulnerability of LTE to hostile interference. In: Global conference on signal and information processing (GlobalSIP). IEEE, pp 285–288
5. Lichtman M, Jover RP, Labib M, Rao R, Marojevic V, Reed JH (2016) LTE/LTE-A jamming, spoofing, and sniffing: threat assessment and mitigation. Commun Mag 54(4):54–61
6. Krenz R, Brahma S (2015) Jamming LTE signals. In: International Black Sea conference on communications and networking (BlackSeaCom), pp 72–76
7. Philippe G, Montaigne F, Schiel J-C, Georgeaux E, Gruet C, Roy P-Y, Force P, Mege P (2013) LTE resistance to jamming capability: To which extend a standard LTE system is able to resist to intentional jammers. In: Military communications and information systems conference, pp 1–4
8. Bhattarai S, Wei S, Rook S, Yu W, Erbacher RF, Cam H (2015) On simulation studies of jamming threats against LTE networks. In: International conference on computing, networking and communications (ICNC), pp 99–103
9. Li X, Xie X, Zeng J, Wang Y (2015) Vulnerability analysis and verification for LTE initial synchronization mechanism. In: 36th Sarnoff symposium, pp 150–154

10. Lichtman M, Czauski T, Ha S, David P, Reed JH (2014) Detection and mitigation of uplink control channel jamming in LTE. In: Military communications conference (MILCOM), pp 1187–1194
11. 3GPP. LTE (2017) http://www.3gpp.org/technologies/keywords-acronyms/98-lte
12. 3GPP. LTE (2011) Evolved Universal Terrestrial Radio Access (E-UTRA); Physical channels and modulation. TS 36.211 version 10.0.0 Release 10, 2011
13. 3GPP. LTE (2009) Evolved Universal Terrestrial Radio Access (E-UTRA); Physical layer procedures. TS 36.213 version 8.8.0 Release 8, 2009
14. Xu W, Trappe W, Zhang Y, Wood T (2005) The feasibility of launching and detecting jamming attacks in wireless networks. In: Proceedings of the 6th ACM international symposium on Mobile ad hoc networking and computing, pp 46–57
15. Clancy TC (2011) Efficient OFDM denial: Pilot jamming and pilot nulling. In: International conference on communications (ICC). IEEE, pp 1–5
16. Aziz FM, Shamma JS, Stüber GL (2014) Resilience of LTE networks against smart jamming attacks. In: Global communications conference (GLOBECOM). IEEE, pp 734–739
17. Durmaz MA, Alakoca H, Kurt GK, Ayyıldız C (2017) Chirp subcarrier jamming attacks: an OFDM based smart jammer design. In: 16th Annual mediterranean Ad Hoc networking workshop (Med-Hoc-Net), pp 1–5

Impulsive Noise Modeling and Cancellation Strategies Over Power Line Channels

Waled Gheth, Khaled M. Rabie and Bamidele Adebisi

Abstract Impulsive noise is one of the main impairments over power line channels. In this respect, this paper presents the available impulsive noise models and then reviews and compare several existing mitigation techniques proposed in the literature. These methods include multi-carrier modulation, nonlinear preprocessors, multiple-input multiple-output, coding and iterative techniques. Interesting comparisons between these techniques and meaningful insights are presented and discussed.

Keywords Coding · Impulsive noise · Multi-carrier modulation · Power line communications

1 Introduction

The principle of power line communication (PLC) is based on the use of the existing power lines in buildings and utility grid to carry data and electricity signals simultaneously. This technology has been around for some decades, but it has only been used for limited applications such as, home automation, public lighting and narrow band tele-remote relay applications. Recently, PLC technology has attracted much attention as almost every single building in the world can be reached by the existing electrical wiring. PLC also plays an important role in home networking applications due to the spreading of Internet and modern communication technologies.

On the other hand, there are several challenges when using this medium for high-data rate transmissions. Firstly, the frequency-selective nature of the channel introduces inter-symbol interference (ISI) which can significantly degrade bit error rate (BER) performance [1, 2]. Secondly, there is the issue of electromagnetic compatibility (EMC) to other wireless systems in the surroundings. For this reason, there are regulatory limits for the maximum allowable power that a PLC modem can transmit [3] and this could result in very low signal-to-noise (SNR) values at the receiver.

W. Gheth (✉) · K. M. Rabie · B. Adebisi
School of Engineering, Manchester Metropolitan University, M1 5GD, Manchester, UK
e-mail: waled-ali.m.gheth@stu.mmu.ac.uk

A. Boyaci et al. (eds.), *International Telecommunications Conference*,
Lecture Notes in Electrical Engineering 504,
https://doi.org/10.1007/978-981-13-0408-8_14

Thirdly, noise over power lines is one of the most destructive parameters which can disturb communication signals over PLC channels [4]. According to [4], the PLC noise is categorized into five types namely, colored background noise, narrow-band noise, periodic impulsive asynchronous to the mains frequency, periodic impulsive synchronous to the mains frequency and asynchronous impulsive noise. The first three types vary slowly over time and remain stationary over periods of up to minutes and sometimes for hours; therefore, they are summarized as background noise. However, the last two types are classified as impulsive noise which has short duration, random occurrence and high power spectral density (PSD). Communication signals over power lines can be severely degraded by impulsive noise. In this respect, this paper is dedicated to review and compare the existing techniques used to model and mitigate the impulsive noise.

The rest of the paper is organized as follows. Section 2 describes the noise modeling methods over power line channels. The impulsive noise mitigation techniques are discussed in Sects. 3 and 4 conclude the paper.

2 Impulsive Noise Modeling

The statistical modeling of noise in power line channels has been an important interest to researchers recently. Plenty of research has been conducted to model impulsive noise [5, 6]. Below, we discuss the most common impulsive noise models used in analyzing and evaluating the performance of PLC systems [7].

2.1 *Two-Terms Gaussian Model*

For its simplicity, this model is widely used to characterize impulsive noise [7]; it is also known as two-component mixture-Gaussian model and is defined as

$$\eta(t) = \varepsilon(t) n_1(t) + (1 - \varepsilon(t)) n_2(t), \tag{1}$$

where $\varepsilon(t)$ is a Bernoulli random process, $n_1(t)$ and $n_2(t)$ are independent Gaussian processes with zero mean and σ_b^2 and $K\sigma_b^2$ variances, respectively. The power of $n_2(t)$ is K times greater than that of the background noise. The following formula describes the probability density function (PDF) of the total noise

$$p_\eta(v) = (1 - p) N\left(0, \sigma_b^2\right) + p\, N(0, K\sigma_b^2), \tag{2}$$

where $N(0, \sigma_b^2)$ denotes the Gaussian distribution and p is the probability occurrence of impulsive noise. The following formula is used to obtain the background noise variance

$$\sigma_b^2 = 2R_{IN} \int_{f_1}^{f_2} 10^{[R_{nb}(f)-30]/10} \mathrm{d}f \left[V^2\right], \tag{3}$$

where R_{IN} is the receiver input impedance and is usually equal to 50Ω, $f_1 = 1\text{MHz}$ and $f_2 = 100\,\text{MHz}$.

The following assumptions $\sigma_b^2 = 2.6456 \times 10^{-5}\left[V^2\right]$, $K = 100$ and $p = 0.01$ are usually used to simulate the two Gaussian model, see e.g., [8, 9].

2.2 *Middleton's Class-A Model*

Both the background and impulsive noise are incorporated in this model. It is worth mentioning that the communication channel that experiences class-A noise is also referred to as Additive White Class-A Noise channel (AWCN) [6]. The PDF of this noise model is defined as

$$p_\eta(v) = \sum_{k=0}^{\infty} \frac{e^{-A} A^k}{k!} \frac{e^{v^2/2\sigma_k^2}}{\sqrt{2\pi\sigma_k^2}} \tag{4}$$

where

$$\sigma_k^2 = (1 + \frac{1}{\Gamma})\frac{k}{A} + \frac{\Gamma}{1+\Gamma}\sigma_b^2, \tag{5}$$

Γ represents the background-to-impulsive noise ratio and A is called the impulsive index.

A small number of terms in this series can be used to give acceptable approximations to this distribution. It was found that the cumulative sum can be cut off at the third term and the results obtained were very identical to the one found in the simple two-terms Gaussian model. Also, note that Middleton class-A noise becomes very close to Gaussian noise as the impulsive noise statistical characteristic becomes continuous when A is large. On the contrary, highly structured impulsive noise is found when A values are small. Similarly, same comments can be made when Γ is changed while keeping A constant.

3 Impulsive Noise Cancellation Strategies

In this sections, several impulsive noise mitigation techniques over PLC channels are presented and compared.

3.1 Multi-carrier Modulation (MCM)

A lot of work has been carried out trying to reduce the effect of impulsive noise. For example, the effect of impulsive noise on orthogonal frequency division multiplexing (OFDM) systems in radio communications was first analyzed in [10] where a closed-form expression for the probability of error was derived for two different scenarios, namely, single-carrier (SC) and MCM. It was found that MCM outperforms SC systems, more specifically when the noise power is moderate and the probability occurrence of noise is not too high. This was justified as follows: the effect of impulsive noise gets spread over several data symbols in the case of MCM systems whereas in SC systems, the impulsive noise will affect only one symbol. Furthermore, the author added that this advantage can turn into a disadvantage if the energy of impulsive noise exceeds a certain level. However, this study did not consider the effect of the channel. In [11], the BER performance of OFDM systems and the effect of multi-path over PLC channels were investigated by means of computer simulations. In [12], the BER performance of such systems under the effects of impulsive noise and multi-path was analyzed theoretically and closed-form formulas were obtained. It was concluded that the adverse effect of multi-path over PLC channels can be more serious than that of impulsive noise.

3.2 Nonlinear Preprocessors

Another mitigation technique is proposed in [13] where a nonlinear device is placed at the front end of the receiver. This technique is simple and easy to be implemented; due to these two reasons, this method is widely used in practice [9, 14, 15]. There are three different types of nonlinear preprocessors namely: blanking, clipping and combined blanking and clipping (Hybrid) whose basic principles are discussed below. These are also referred to as time domain (TD) mitigation techniques.

- Blanking non-linearity

$$y_k = \begin{cases} r_k, & |r_k| \leq T_b \\ 0, & |r_k| > T_b \end{cases} \quad k = 0, 1, \ldots, N-1 \tag{6}$$

where T_b is the blanking threshold, y_k is the output of the nonlinear device and r_k is its input.

- Clipping non-linearity [16]

$$y_k = \begin{cases} r_k, & |r_k| \leq T_c \\ T_c e^{j\arg(rk)}, & |r_k| > T_c \end{cases} \quad k = 0, 1, \ldots, N-1 \tag{7}$$

where T_c is the clipping threshold.

- Hybrid (blanking/clipping) non-linearity [17]

$$y_k = \begin{cases} r_k, & |r_k| \le T_c \\ T_c e^{j\arg(rk)}, & T_c < |r_k| \le T_b \quad k = 0, 1, \ldots, N-1 \\ 0, & |r_k| > T_b \end{cases} \tag{8}$$

where $T_b \ge T_c$.

The assumption in this technique is that the received OFDM signal amplitude is conspicuously lower than the impulsive noise amplitudes. Therefore, any signals with amplitudes greater than a specified limit are assumed to be disturbed by impulsive noise [13]. However, using noise amplitude clipping or blanking techniques can enhance the communications performance. The received signal phase is not altered while the amplitude of the received signal is limited to a defined threshold by the non-linear preprocessor [18]. Any OFDM sample that has amplitude lower than threshold passes without modifying its amplitudes as it is considered a clear OFDM signal. On the other hand, signal samples with amplitude above the threshold are clipped to T_c in clipping technique or replaced with zero in blanking technique. Nevertheless, in the hybrid approach, signal samples that have amplitudes less than T_c can pass without clipping or nulling. As T_c is less than T_b, any sample with an amplitude above T_c and less than T_b is clipped to the clipping threshold while the OFDM sample whose amplitude is above both thresholds T_c and T_b is blanked. The concepts of the different TD techniques are illustrated in Fig. 1.

In [18], optimal T_c and T_b values were found and used to combat the impulsive noise over the narrow-band PLC systems. However, it was confirmed that the use of optimal threshold values for the hybrid scheme performs better than the use of either blanking or clipping separately. Specifically, a gain of 1.5 dB in the performance was achieved by combining clipping and blanking methods as was shown in [18]. Different impulsive noise scenarios were mitigated in [19], where the blanking and/or clipping techniques were used separately; improvement of 5 dB in the output SNR was achieved in the worst impulsive noise scenario when the hybrid method was used [19].

3.3 *Time and Frequency Domain (TFD) Techniques*

Unlike the aforementioned methods where the signal is processed before the OFDM demodulator, i.e., in the TD, the study in [20] proposed an algorithm that compensates for the impulsive noise and channel equalization after the OFDM demodulator, i.e., in the frequency domain (FD). This algorithm was investigated by means of simulation and a large performance improvement were achieved. Figure 2 shows where the TD and the FD mitigation takes place.

A very small improvement in the SNR was achieved when a combination of amplitude clipping and channel equalization TFD technique was used for impulsive

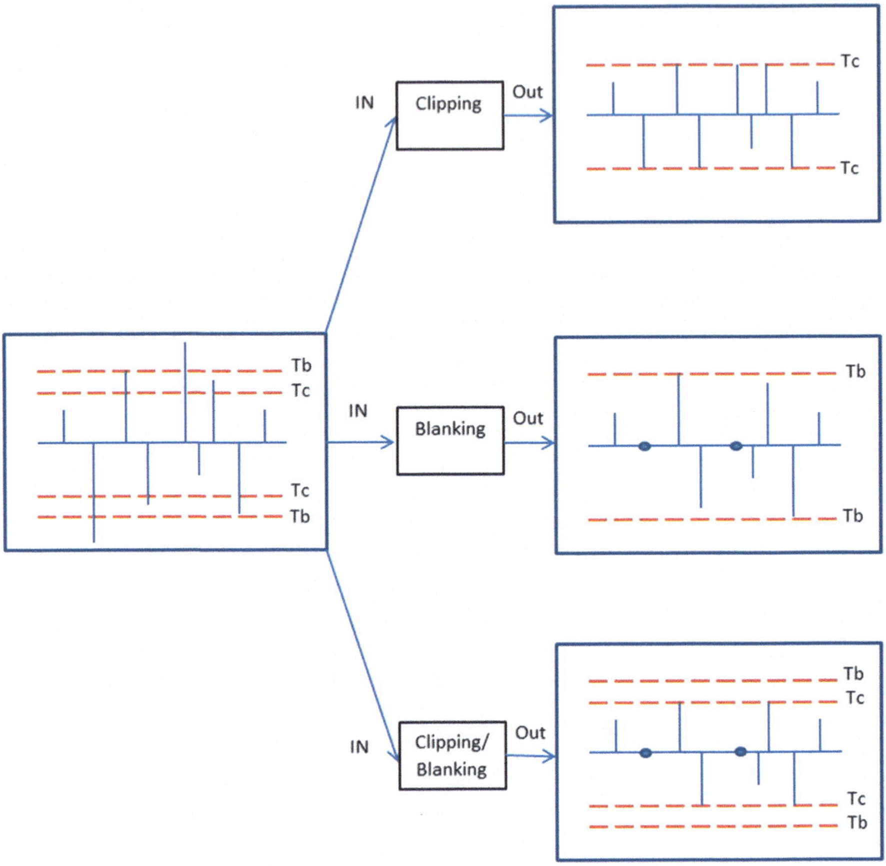

Fig. 1 TD impulsive noise mitigation techniques

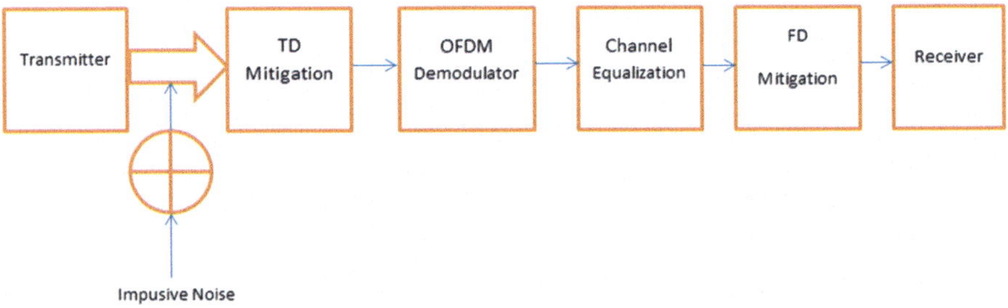

Fig. 2 TFD mitigation configuration

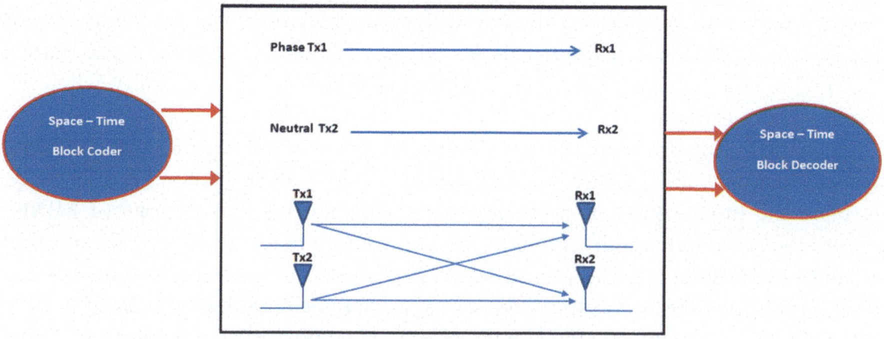

Fig. 3 MIMO technology in PLC and wireless channels

noise mitigation [21]. However, in this combined method, the amplitude clipping took place before the OFDM demodulator. The equalizer which was used as PLC channel compensator was located after the OFDM demodulator. The simulation results of this method showed that the system performance is enhanced by 0.7 dB as a maximum improvement in SNR [21]. Same impulsive noise mitigation technique was used in [19]. The OFDM signal is firstly passed through a clipping/blanking non-linearity TD processor then processed by the FD technique after it is passed through the demodulator and channel equalizer. A maximum of 5 dB SNR improvement was reached with this method. This study considered different impulsive noise scenarios, namely, heavily, medium and weakly [21].

3.4 Multiple Input Multiple Output (MIMO)

MIMO in PLC refers to the concept of using multiple cables between the transmitter and the receiver and can be used to reduce the impact of several PLC channel impairments. MIMO techniques are mainly divided into:

Spatial Diversity Fading can be combated by the use of diversity. It is called transmit diversity if it is at the transmitter side and receive diversity if it is exploited at the receiver side.

Spatial Multiplexing This method is not supposed to strengthen the transmission; the data rate is increased when spatial multiplexing is used. The data is divided into separate streams and different cables are used to transmit these streams. Figure 3 illustrates the main differences between MIMO in PLC and wireless channels.

The performance of differential phase shift keying (DPSK), in the presence Middleton class-A noise, with two different diversity techniques, namely, equal gain combining (EGC) and selection combining (SC) in Ricean fading channels is investigated in [22]. It was found that when the proposed system is combined with linear block codes, the system performance can be further improved. Whereas in [23], closed-form expressions for BER performance are derived for two diversity schemes, namely, maximum ratio combining (MRC) and post detection combining (PDC) over

Rayleigh fading channels using the two different impulsive noise models proposed in [24]. Unlike in Gaussian environments where MRC is the optimal, it was shown that PDC is more robust to impulsive noise than MRC scheme specially when the diversity order is high.

It is worth noting that there are some differences between the application of MIMO in power line channels and the wireless channel which can be seen in Fig. 3. Firstly, to obtain the input signal of every receiving antenna, the applications of MIMO in wireless communications combine the received signals from each transmitting antenna. On the other hand, as presented in [25], in order to simplify the transmission of the signals over the power lines, the wires ought to be completely isolated in PLC environments. Secondly, in PLC applications, the number of transmitting points must be equal to the number of receiving points [26]. In contrast, arbitrary number of antennas might be used in wireless applications. Thirdly, multiple transmitting and receiving antennas can be used in wireless communications whereas in PLC, a maximum number of three transmitting and receiving points are available. This means that for wireless communication with N_t transmitting antennas and N_r receiving antennas, a diversity order of $N_t N_r$ can be employed; while in PLC channels a maximum diversity order of three can be exploited.

Furthermore, many studies have investigated the use of space-time coding over PLC channels [27, 28]. In [27], the author examined the performance of space-time block coding (STBC) with block interleaving in PLC channel contaminated with Middleton class-A impulsive noise and was compared to that in wireless channels; it was found that STBC is as promising as it is in wireless channels. In [26], the BER performance and channel capacity of STBC2x2 system are evaluated in a typical PLC environment, i.e., frequency selective channel and in the presence of AWCN. It was found that a gain of about 16 dB at BER of 10^{-5} can be achieved in comparison with the conventional system. As this study assumed complete isolation between the wires, it was presented that there is no capacity improvement in the proposed system.

All the studies above considered SC systems. As was presented in the previous section, MCM is more suitable for PLC channels. Space-frequency block coded (SFBC) [29] OFDM over a 3-phase power line network was studied in [28]. It was demonstrated that the proposed system provides a significant performance improvement, a gain of about 10 dB at symbol error rate of 10^{-3}, compared to the conventional single-wire system.

STBC-OFDM over PLC channels was applied in [30] to improve the system reliability by applying symbol repetition technique. In [31], the performance of single-input single-output (SISO) and multiple-input single-output (MISO) techniques based on SC modulation was considered as well as the effect of feedback channels. It was demonstrated that MISO-SC systems with a feedback channel can outperform MISO-SC without a feedback channel and SISO-SC with and without feedback channels. In [32] MIMO-OFDM system was applied to PLC systems and the coupling effect between the conductors on the system capacity was evaluated experimentally.

3.5 Coding

Channel coding enables PLC nodes to detect and correct errors without the need for re-transmission, which may not be feasible in real-time based applications due to time delay restrictions. In general, convolutional and Reed-Solomon (RS) codes have been the main schemes adopted by narrow-band PLC standards. However, researchers have also evaluated the performance of other coding techniques such as the generalized array codes, Luby transform codes, bit-interleaved coded modulation with iterative decoding, row and column array codes. These codes have shown more robustness against power line channel impairments in comparison to the convolutional codes. Channel coding systems generally add redundancy bits to the original information to detect and correct errors. The performance of coded OFDM systems over power lines was first investigated in [33] by means of simulation. In this study, the authors examined two different channel coding methods namely, convolutional coding and RS coding. The author of [34] suggested adding bit-interleaving to the convolutionally coded OFDM and the BER performance was examined under three different impulsive noise scenarios based on the measurements of [4]. The simulation results showed that the combination of interleaving and convolutional coding can significantly minimize the effect of impulsive noise and a gain of up to 15 dB can be achieved at BER of 10^{-3}.

Turbo decoding for turbo codes in impulsive noise environments is discussed in [35] whereas the performance of double binary turbo coded OFDM system over PLC channels is analyzed in [36] where it was concluded that as the number of decoding process iterations increases, the BER performance is significantly improved. The study in [37] proposed using RS codes combined with M-ary modulation and showed that this can combat burst errors caused by impulsive noise even without applying interleaving. Quasicyclic low-density parity check (QC-LDPC) codes are implemented in PLC systems in the presence of Middleton class-A noise in [38].

It is true that the more robust the coding scheme employed is, the more link reliability is achieved; however, this will be attained at the expense of more computational complexity at the PLC modem resulting in more expensive chip-set requirements.

Iterative impulsive noise reduction Several iterative algorithms have been proposed in the literature to mitigate impulsive noise [39]. For instance, Fig. 4 shows the iterative algorithm used in [39] which is based on the findings of [10] that MCM can combat the impulsive noise. In this algorithm, the output of the OFDM demodulator is re-modulated by the OFDM modulator and $\bar{b}_k$ is given. After that $\bar{b}_k$ is subtracted from the received signal r_k and the result defines the estimated noise samples $\bar{n}_k$ which have large amplitudes. The estimated noise samples are subtracted from r_k. This process is repeated n times depending on the number of iterations performed. Finally, after n iterations, the resulting signal is passed through the OFDM demodulator. It was shown that when the number of iterations is 10, this algorithm provides a gain of about 7 dB at BER of 10^{-3} compared to the conventional OFDM system, i.e., when the number of iteration = 0.

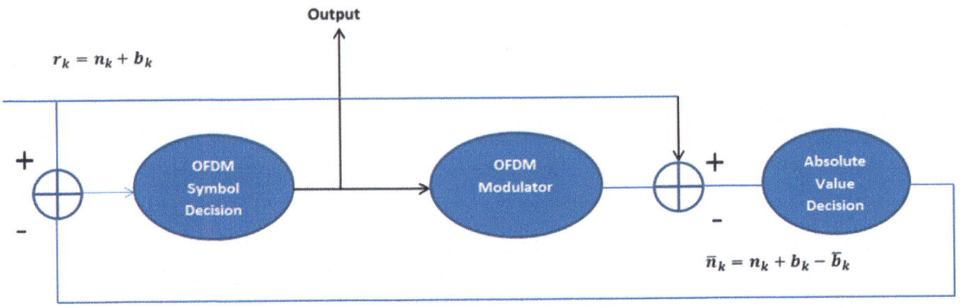

Fig. 4 Iterative OFDM impulsive noise cancellation

Table 1 Comparison between the various impulsive noise mitigation techniques

	Complexity	BER performance	Data rate	Transmission power
MCM	Medium	Good	Medium	Medium
Nonlinear preprocessors	Low	Good	Medium	Medium
MIMO	Medium	Good	Medium	Medium
Coding	High	Very good	Low	Medium
Iterative	High	Good	Medium	Medium

Based on the above review of impulsive noise mitigation techniques, we can demonstrate a comparison between these techniques in terms of complexity, performance and data rate as in Table 1.

4 Conclusions

In this paper, different impulsive noise modeling and mitigation techniques have been presented. The performance of the conventional communication schemes designed for Gaussian noise usually degrades drastically in non-Gaussian environments such as the PLC one. In general, MCM systems have better immunity to impulsive noise than SC systems. When the noise is very impulsive, nonlinear preprocessors become a very effective way of reducing the noise pulses. Furthermore, PLC systems can be made more resistant to such channel impairments by applying channel coding schemes. Increasing the transmit power to reduce the effect of such channels could solve the impulsive noise problem; however, this would create another problem, i.e. electromagnetic radiations. Therefore, there is a trade-off to which careful attention must be paid.

Acknowledgements This research has been carried out within the "CityVerve: IoTs and Smart Cities Demonstrator" project funded by Innovate UK (102561).

References

1. Anastasiadou D, Antonakopoulos T (2005) Multipath characterization of indoor power-line networks. IEEE Trans Power Del 20:90–99. https://doi.org/10.1109/tpwrd.2004.832373
2. Rozman M, Ikpehai A, Adebisi B, Rabie KM (2016) Channel characterisation of cooperative relaying power line communication systems. In: Proceedings of the international symposium on communication systems, networks and digital signal processing (CSNDSP), pp. 1–5. https://doi.org/10.1109/csndsp.2016.7574000
3. Adebisi B, Honary B (2006) Comparisons of indoor PLC emissions measurement results and regulation standards. In: Proceedings of the international symposium on power line communications (ISPLC), pp. 319–324, https://doi.org/10.1109/isplc.2006.247482
4. Zimmermann M, Dostert K (2002) Analysis and modeling of impulsive noise in broad-band powerline communications. IEEE Trans Electromagn Comput 44:249–258. https://doi.org/10.1109/15.990732
5. Liu D, Flint E, Gaucher B, Kwark Y (1999) Wide band AC power line characterization. IEEE Trans Consum Electron 45:1087–1097. https://doi.org/10.1109/30.809186
6. Middleton D (1977) Statistical-physical models of electromagnetic interference. IEEE Trans Electromagn Compat EMC-19:106–127. https://doi.org/10.1109/temc.1977.303527
7. Miyamoto S, Katayama M, Morinaga N (1995) Performance analysis of qam systems under class-a impulsive noise environment. IEEE Trans Electromagn Compat 37:260–267. https://doi.org/10.1109/15.385891
8. Rabie KM, Adebisi B (2017) Enhanced amplify-and-forward relaying in non-gaussian PLC networks. IEEE Access 5:4087–4094. https://doi.org/10.1109/ACCESS.2017.2680599
9. Adebisi B, Rabie KM, Ikpehai A, Soltanpur C, Wells A (2014) Vector OFDM transmission over non-Gaussian power line communication channels. IEEE Syst J 99:1–9. https://doi.org/10.1109/jsyst.2017.2669086
10. Ghosh M (1996) Analysis of the effect of impulse noise on multicarrier and single carrier QAM systems. IEEE Trans Commun 44:145–147. https://doi.org/10.1109/26.486604
11. Kuri K, Hase Y, Ohmori S, Takahashi F, Kohno R (2003) Powerline channel coding and modulation considering frequency domain error characteristics. In: Proceedings of the international symposium on power line communications (ISPLC), pp. 221–225
12. Ma YH, So PL, Gunawan E (2005) Performance analysis of OFDM systems for broadband power line communications under impulsive noise and multipath effects. In: IEEE Trans Power Del 20:674–682. https://doi.org/10.1109/tpwrd.2005.844320
13. Liu S, Yang F, Ding W, Song J, Han Z (2016) Impulsive noise cancellation for MIMO-OFDM PLC systems: a structured compressed sensing perspective. In: IEEE global communications conference (GLOBECOM), pp 1–6. https://doi.org/10.1109/GLOCOM.2016.7841891
14. Anoh K, Adebisi B, Rabie KM, Hammoudeh M, Gacanin H (2017) On companding and optimization of OFDM signals for mitigating impulsive noise in power-line communication systems. IEEE Access 99:1–1. https://doi.org/10.1109/access.2017.2747629
15. Ikpehai A, Adebisi B, Rabie KM, Fernando M, Wells A (2017) Energy-efficient vector OFDM PLC systems with dynamic peak-based threshold estimation. IEEE Access 5:10723–10733. https://doi.org/10.1109/access.2017.2709254
16. Suraweera HA, Chai C, Shentu J, Armstrong J (2003) Analysis of impulse noise mitigation techniques for digital television systems. In: Proceedings of the 8th International OFDM workshop, pp 172–176
17. Zhidkov SV (2008) Analysis and comparison of several simple impulsive noise mitigation schemes for OFDM receivers. IEEE Trans Commun 56:5–9. https://doi.org/10.1109/tcomm.2008.050391
18. Korki M, Hosseinzadeh N, Vu HL, Moazzeni T, Foh CH (2011) Impulsive noise reduction of a narrowband power line communication using optimal nonlinearity technique. In: Australasian telecommunications networks and applications conference (ATNAC), pp 1–4

19. Al-Mawali K, Sadik AZ, Hussain ZM (2008) Joint time-domain/frequency-domain impulsive noise reduction in OFDM-based power line communications. In: Australasian telecommunications on networks applications conference, pp 138–142
20. Zhidkov SV (2003) Impulsive noise suppression in ofdm-based communication systems. IEEE Trans Consum Electron 49:944–948. https://doi.org/10.1109/tce.2003.1261179
21. Kim Y-H, Kim K-H, Oh H-M, Kim K-H, Kim S-C (2008) Mitigation of effect of impulsive noise for OFDM systems over power line channels. In: Proceedings of the IEEE International symposium on power line communications (ISPLC), pp 386–390
22. Weng JF, Leung SH (1934) On the performance of dpsk in rician fading channels with class a noise. IEEE Trans Veh Technol 49:1934–1949. https://doi.org/10.1109/25.892596
23. Tepedelenlioglu C, Gao P (2004) Performance of diversity reception over fading channels with impulsive noise. In: IEEE International conference on acoustics, speech, and signal processing, 2004. Proceedings. (ICASSP '04), vol 4, pp 389–392
24. Delaney PA (1995) Signal detection in multivariate class-a interference. IEEE Trans Commun 43:365–373
25. Giovaneli CL, Yazdani J, Farrell P, Honary B (2002) Application of space-time diversity/coding for power line channels. In: Proceedings of international symposium on power-line communications and its applications 2002 (ISPLC 2002), pp 101–105
26. Papaioannou A, Papadopoulos G, Pavlidou F (2005) Performance of space-time block coding in powerline and satellite communications. IEEE J Commun Inf Syst 20(3):174–181. https://doi.org/10.14209/jcis.2005.24
27. Papaioannou GDPA, Pavlidou FN (2004) Performance of space-time block coding over the power line channel in comparison with the wireless channel. In: Proceedings of the international symposium on power line communications (ISPLC), pp 362–366
28. Giovaneli CL, Farrell P, Honary B (2005) Space-frequency coded ofdm system for multi-wire power communications. In: Proceedings of the international symposium on power line communications (ISPLC), pp 50–55
29. Lee K, Williams D (2000) A space-frequency transmitter diversity technique for ofdm systems. In: IEEE Global telecommunications conference 2000, vol 3, pp 1473–1477. https://doi.org/10.1109/glocom.2000.891885
30. Quan Z, Ribeiro M (2011) A low cost stbc-ofdm system with improved reliability for power line communications. In: Proceedings of the IEEE international symposium on power line communications (ISPLC), pp 261–266. https://doi.org/10.1109/isplc.2011.5764404
31. de Campos F, Machado R, Ribeiro MV, de Campos M (2009) MISO single-carrier system with feedback channel information for narrowband PLC applications. In: Proceedings of the IEEE international symposium on power line communications (ISPLC), pp 301–306
32. Hao L, Guo J (2007) A MIMO-OFDM scheme over coupled multi-conductor power-line communication channel. In: Proceedings of the ieee international symposium on power line communications (ISPLC), pp. 198–203
33. Babic M, Bausch J, Kistner T, Dostert K (2006) Perfomance analysis of coded OFDM systems at statistically representative PLC channels. In: Proceedings of the IEEE International symposium on power line communications (ISPLC), pp 104–109. https://doi.org/10.1109/isplc.2006.247445
34. Al-Mawali KS, Hussain ZM (2009) Performance of bit-interleaved coded OFDM in power line communications with impulsive noise, pp 49–53
35. Umehara D, Yamaguchi H, Morihiro Y (2004) Turbo decoding in impulsive noise environment, vol 1, pp 194–198. https://doi.org/10.1109/glocom.2004.1377938
36. Kim EC, Il SS, Heo J, Kim JY (2010) Performance of double binary turbo coding for high speed PLC systems. IEEE Trans Consumer Electron 56:1211 –1217. https://doi.org/10.1109/tce.2010.5606248
37. Chuah TC (2009) On reed solomon coding for data communications over power-line channels. IEEE Trans Power Del 24:614 –620. https://doi.org/10.1109/tpwrd.2008.917667

38. Andreadou N (2010) Mitigation of impulsive noise effect on the PLC channel with QC-LDPC codes as the outer coding scheme. IEEE Trans Power Del 25:1440–1449
39. Matsuo H, Umehara D, Kawai M, Morihiro Y (2002) An iterative detection for OFDM over impulsive noise channel. In: Proceedings of the International Symposium on Power Line Communications (ISPLC)

An IEEE 802.11x Implementation for V2X Communications Towards IoT and Big Data

Dilara Albayrak Serin, Ali Boyaci, Alper Özpınar and Serhan Yarkan

Abstract Green transportation is an integral part of green environment concept. Next generation transportation systems are desired to achieve high performances with reduced fuel consumption and carbon emission. In this regard, vehicle maintenance status along with some other critical diagnostic data should steadily be checked and tracked. It is known that vehicles that are poorly maintained or not maintained in a timely manner lead to both emissions exceeding the standards and low performance. Vehicle telematics along with some other conveniences such as infotainment systems, location–based services and applications are expected to improve safety, availability, and reliability of next generation transportation systems. From this perspective, intelligent transport systems (ITS) seems to be a promising solution candidate which encompasses all of the aforementioned topics as well as vehicle-to-vehicle (V2V), vehicle-to-infrastructure (V2I), and vehicle-to-cloud (V2C) opportunities. Therefore, in this study, a conceptual model that links vehicle telematics to the cloud along with V2V communications facility is proposed and a prototype based on IEEE 802.11x protocol suite is implemented. Mobile data collection and measurements are obtained. Results are presented along with relevant discussions as well as the end point storage and usage of the data are introduced. In the proposed model, the mobile data are transferred to cloud computing platform to create the big data for

D. A. Serin
Department of Electrical–Electronics Engineering, Istanbul Commerce University, Istanbul, Turkey
e-mail: dilaraalbayrak@ticaret.edu.tr

A. Boyaci
Department of Computer Engineering, Istanbul Commerce University, Istanbul, Turkey
e-mail: aboyaci@ticaret.edu.tr

A. Özpınar
Department of Mechatronics Engineering, Istanbul Commerce University, Istanbul, Turkey
e-mail: aozpinar@ticaret.edu.tr

S. Yarkan (✉)
Center for Applied Research on Informatics Technologies (CARIT), Istanbul Commerce University, Istanbul, Turkey
e-mail: syarkan@ticaret.edu.tr

A. Boyaci et al. (eds.), *International Telecommunications Conference*,
Lecture Notes in Electrical Engineering 504,
https://doi.org/10.1007/978-981-13-0408-8_15

further research opportunities for car manufacturers, policy makers, and researchers with the concern of ethics and security issues.

Keywords Communications network · Green communications · Network security · Vehicle-to-vehicle communications

1 Introduction

Global warming along with climate change is considered by many to be one of the most serious concerns for the upcoming decade. Especially human activities such as carbon dioxide emissions due to burning fossil fuels or to deforestation trap the escaping heat leading to the phenomenon called greenhouse effect. Because transportation is currently heavily relies on fossil fuel, drastic changes are required both to reduce the pollutant emissions and to improve the energy efficiency while achieving high performances. In this regard, intelligent transport systems (ITS) seems to be a promising solution to city of tomorrow concept with efficient vehicle and motor designs and efficient use of renewable energies.

Green environments are not only supported by ITS but also by next generation wireless network (NGWN). An immediate consequence of these concepts are Internet of things (IoT). Furthermore, it is believed in the near future that wireless revolution will lead to smart cities, environments, and infrastructures which are driven by cyber–physical systems. Such an envision points out the following key points. It is clear that wireless communications allow one to stay online almost anywhere–anytime implying ubiquitous communications. Considering the fact that number of wireless nodes will increase more in the near future, amount of data of numerous types to be generated, stored, and processed will increase dramatically as well. Such an anticipation entails big data. Of course, coping with such a massive information flow of various types could only be possible with non–traditional approaches including parallel processing, extensive data storage, and virtualization. Therefore, cloud computing should be included into the picture as well.

In light of the discussions above, one could conclude that ITS should be fortified with NGWN which is supported by cloud computing in order to reach the point where green environments are envisioned to be. There are various ITS implementations in the literature including communications to some extent. For instance, an intelligent speed adaptation mechanism is proposed and vehicle data are transferred to an online platform via Global Packet Radio System (GPRS) in [1]. Similarly, a sensitivity analysis is carried out in [2] and identified that gear choice and the distance from the traffic light at which vehicles are informed via a road network improve the performance. In [3], average traffic fuel economy, carbon dioxide–monoxide and several other emissions are estimated using a computer-vision-based methodology. In [4] writers present an ITS mechanism to detect and manage traffic congestion to reduce average trip time, CO2 emissions and fuel consumption. On the other hand in [5], a dynamic roadway network database is proposed for eco–routing navigation system based on energy/emissions factors available. There are several other studies and implementations in the literature which strive to merge several aspects of ITS and NGWN. Interested readers could refer to [6] for a very detailed and a more elaborative list of studies in this field of research with a comprehensive comparative discussion.

Although there are numerous studies and implementations available in the literature, it is observed that an ITS equipped with NGWN is still not complete without big data and cloud computing. Therefore, in this study, a conceptual model that links vehicle telematics to the cloud along with vehicle-to-vehicle (V2V) communications facility is proposed and a prototype based on IEEE 802.11x protocol suite is implemented. Mobile data collection and measurements are obtained. Results are presented along with relevant discussions as well as the end point storage and usage of the data are introduced. In the proposed model, the mobile data are transferred to cloud computing platform to create the big data for further research opportunities for car manufacturers, policy makers, and researchers with the concern of ethics and security issues.

The organization of the rest of the paper is as follows. Section 2 provides the details of the proposed conceptual model. Section 3 discusses cloud computing and big data including several concerns related to security and privacy issues. Section 5 briefly provides the implementation, validation, and test procedures carried out. Finally, Sect. 6 concludes the paper by outlining the future works.

2 Conceptual System Model

The proposed conceptual system model consists of the following modules: vehicle computer interface (VCI), IEEE 802.11x suite, an RF frontend. The modules except for the RF frontend are considered to have cross-layer capabilities with a robust security support on both PHY and upper layers. In the proposed system model the key module is the IEEE 802.11x since it paves the way of vehicle data to the cloud. It is important to keep in mind that cross-layer functionality could both be employed on IEEE 802.11x suite and VCI; therefore, security support could be established both on VCI and IEEE 802.11x suite. From this perspective, the proposed system model provides PHY-layer security on VCI, whereas MAC (and possibly upper) layer security on IEEE 802.11x suite.

Main task of VCI is to establish a fast, reliable, and secure short range wireless communications between the vehicle and the IEEE 802.11x suite. Considering the fact that majority of the on-board diagnostics data are fed to VCI without a substantial security support, VCI remedy these shortcomings. Furthermore, VCI provides a vast variety of both wireless and wired communications interfaces to IEEE 802.11x suite. As will be discussed subsequently, IEEE 802.11x suite is used as a communications hub for the vehicle; therefore, a versatile VCI is required.

IEEE 802.11x suite is the key component of the proposed system model. It enables vehicle data to reach the cloud via VCI; establishes communications between

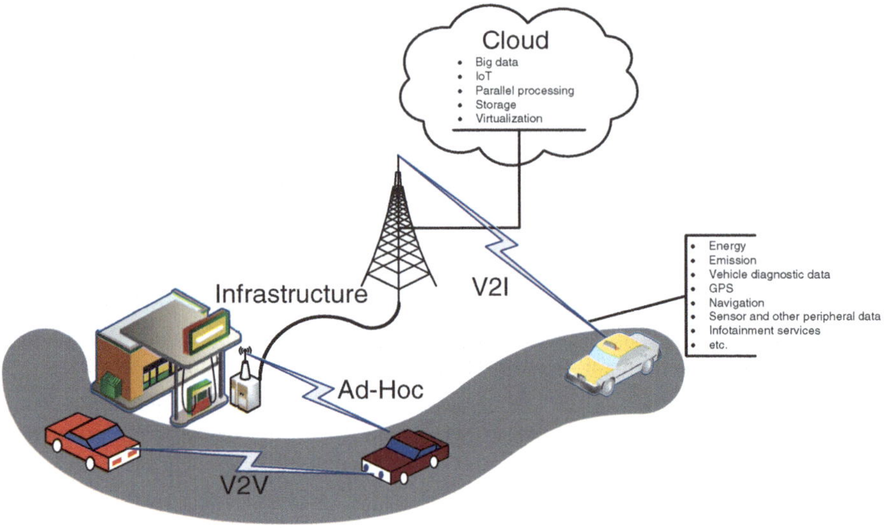

Fig. 1 The proposed conceptual model which extends ITS with NGWN and cloud computing

various peripherals and extensions; and more importantly allows the vehicle to enter a vehicle-to-vehicle (V2V) network and maintain it. Furthermore, IEEE 802.11x suite could be modified and tailored to specific requirements, applications, and/or services.

In contrast to VCI and IEEE 802.11x suite, RF frontend is the least adaptable and/or flexible component among these three. Its only task is to provide interface between the physical propagation medium and the digital domain. Of course, it is expected RF frontend to support multi-band operations with high performance with a possible MIMO support. An overview of the proposed conceptual system model is given in Fig. 1.

3 Cloud Computing and Big Data

Cloud computing is the latest step of the computer evolution includes massive flexibility, scalability, and complex computing. The common idea behind the cloud computing is delivering all suitable types of information technologies via Internet as a service–based business model namely X as a Service (XaaS) or Everything as a Service (EaaS). The X or E can be replaced by well–known services like software (SaaS), infrastructure (IaaS), and platform (PaaS) as well as numerous unique or extended possibilities such as network (NaaS), communication (CaaS), and education (EaaS). The core technology behind the computational power of cloud computing is based on parallel processing, extensive storage and virtualization.

There is a tremendous increase in the volume and the detail of data captured by different sources, organizations, such as the rise of social media, devices forming the IoT, and multimedia which has already produced an overwhelming flow of data in either structured or unstructured format [7]. The amount of data collected, generated, and captured require rapid processing. Moreover, the data is of mostly unstructured form or generated by different sources. Furthermore, they are massive as compared to the traditional database applications and structures such as ERP, MRP, CRM or sales automation system. Being massive and generated in bulks cause classical relational database systems or old fashioned hierarchical database systems to be unstable. This new form of data has a new name "Big Data". Big data receive a significant attention from many communities including especially academicians, engineers, finance and business companies, government institutions, and healthcare industry. Usage of big data for vehicle industry can be very advantageous as well. As stated in [8] novel-deep learning through big data provides a traffic flow predicton based on deep architecture models with big traffic data.

The main characteristics of big data are studied formerly in the literature in various ways. A technology research company, Gartner, starts with the definitions in terms of volume, velocity, and variety in 2001 but not directly calling it big data [9]. Meanwhile other characteristics such as veracity, variability, value, and complexity are added by different contributors [10].

Those characteristics of big data transform and adapt the big data storage and analysis processes to cloud computing with key technologies including NoSQL database systems, map reduce algorithms, and Hadoop. Big data provides users with the ability to use commodity computing to process distributed queries across multiple datasets and return resultant sets in a timely manner [11].

Cloud computing provides the underlying engine and service through the use of open source software for reliable, scalable, distributed computing framework namely Apache Hadoop. Hadoop is ready to distribute and operate on the cloud platform by using Hadoop Distributed File System (HDFS) storage. HDFS provides a class of distributed data processing platforms for large data sources from the cloud. It forms the big data in a distributed fault–tolerant database and processes it through a programming model for large datasets with a parallel, distributed algorithm on a cluster.

4 Analysis of Vehicle Big Data, Its Characteristics, and Related Issues

Total number of cars and vehicles has already reached and passed the 1 billion barrier in the previous years. The idea of exploiting vehicles as mobile sensors for data generators with different aspects is not new. It already started with acquisition of road traffic data, which is a crucial and necessary activity for any traffic management information system. Furthermore, it paves the way of creating the big data for different purposes such as planning, tracking, modeling [12]. The term floating car data (FCD) refers to the data being collected (continuously) by a fleet of vehicles. Nowadays it can be considered as a huge distributed network of sensors providing massive amounts of data without the problem of energy management for the sensors.

Sensors and VCI data include engine status, fuel consumption, geolocation, shifts and gears, service and maintenance data, motor heat, tire pressure, and exhaust greenhouse gas emissions such as carbon dioxide (CO_2), methane (CH_4), nitrous oxide (N_2O) emitted during combustion of fossil fuels. Reference [13] states that collection of floating car data is essential to improve traffic flows and for using efficient road capacity.

4.1 Analysis of Big Vehicle Data

There are several ways of analyzing the big vehicle data and each analysis can be used for different purposes. The following three categories of analysis are more prominent in the literature: descriptive data analysis, predictive data analysis, and prescriptive data analysis. Descriptive analytics use historical vehicle data to identify and set the patterns, trends, statistics and create decision making reports for the future projections based on the behaviors in the past. Predictive vehicle data analytics attempt to predict the future by using the past and contemporary data in a dynamical manner. Prescriptive data analysis of vehicle data helps decision makers in forming their decisions by determining actions and assessing their impact with respect to policies, legal infrastructures, social and political aspects as well as business models, requirements, and constraints for companies.

4.2 Main Characteristics of Vehicle Big Data (Multi-V Model)

Volume Volume refers to the amount of all types of data generated by different sources and continue sto expand. The benefit of gathering large amounts of data includes the creation of hidden information and patterns through data analysis.

Variety Variety refers to the different types of data collected via different vehicle models. Vehicle type and fuel type are two example dimensions of variety.

Velocity Velocity refers to the speed of data transfer. The contents of data constantly change because of the absorption of complementary data collections, introduction of previously archived data or legacy collections, and streamed data flowing from multiple sources.

Value Value is the most important aspect of big data. It refers to the process of discovering huge hidden values from large datasets with various types and rapid generation.

Veracity Veracity represents the quality of the data obtained from sensors.

Complexity Since the big data is obtained from different manufacturers, sources, and nodes; management and normalization become a critical tasks and require multiple sub–tasks including processing and evaluating the data, and making inferences. This intriguing structure of the management and normalization implies the complexity of the big data.

4.3 Data Management, Privacy, and Security Issues

There are several major concerns about the cloud computing: security, privacy [14, 15]. In [16] it is explained that there should be a trusted cloud identity ecosystem to promote user confidence and provide assurances. Moreover, there is a legal aspect which deals with who would be the owner of the big data in case it is hosted and processed in hardware which is provided as a part of service [17]. Considering the immense competition between the vehicle manufacturers, one could easily infer that the data collected from the vehicles are vital since they include several customer experience related information, service and maintenance data, real fuel consumption, and economy data. Similar concerns have already been addressed in the literature especially for gathering the medical records [18] and storing them [19]. As a result, studies take place in order to improve the big data architecture so that security level is increased as in the Hadoop architecture [15, 20]. In Sect. 6, a hybrid cloud structure will be discussed as well.

5 Implementation, Validation, and Test of the Prototype

5.1 Implementation

In implementation, x86–based Alix 3D2 development board is used with a mini–PCI IEEE 802.11a/b/g/n wireless network interface. OpenWrt is installed on the development board as the operating system. OpenWrt later on is customized according to the V2V and vehicle-to-cloud (V2C) needs. BATMAN protocol is employed to establish the mesh network layout between vehicles and infrastructure. As extensions, both Global Positioning System (GPS) and Third Generation Long Term Evolution (3GLTE) peripherals are included into the main development board. Vehicle diagnostic data are obtained via a CAN–bus module included into the overall architecture. Using Python scripting language, server and client architectures are formed. Data obtained from peripherals such as CAN–bus and GPS are read, processed, buffered, and sent periodically to the cloud via 3GLTE.

Note that data exchange between the vehicles and the infrastructure is not limited solely to 3GLTE. Especially local stations such as gas stations and parking zones are planned to contain mesh network infrastructure to provide a connection between the vehicle and the cloud. This way, 3GLTE becomes an option rather than a mandatory design requirement for the proposed architecture.

5.2 PHY–Layer Validation and Verification

Validity of the implementation is checked first by setting two nodes where one of them being in transmit mode sending a short text message at 5.9 GHz and the other being in receive mode. It is known that generic IEEE802.11 protocol suite employs a synchronization mechanism before the data transmission starts. Synchronization is established in two steps. First coarse synchronization takes place. Next, fine carrier frequency offset (CFO) estimation and adaptive channel equalization are invoked. Therefore, it is appropriate to check whether synchronization data could be captured to validate and verify the functionality of the implementation.

Data capture is carried out by Anritsu MS2830A vector signal analyzer (VSA) at 5.9 GHz. The parameter settings of the VSA are adjusted in such a way that a complete 20 MHz span is captured at Nyquist rate. The VSA is programmed to provide baseband in-phase/quadrature I/Q samples. I/Q samples are then transferred to a computer on which MATLAB analysis scripts are run. Several statistics along with plots are obtained on MATLAB. Based on the setup explained above, synchronization data are captured and plotted, which is seen in Fig. 2.

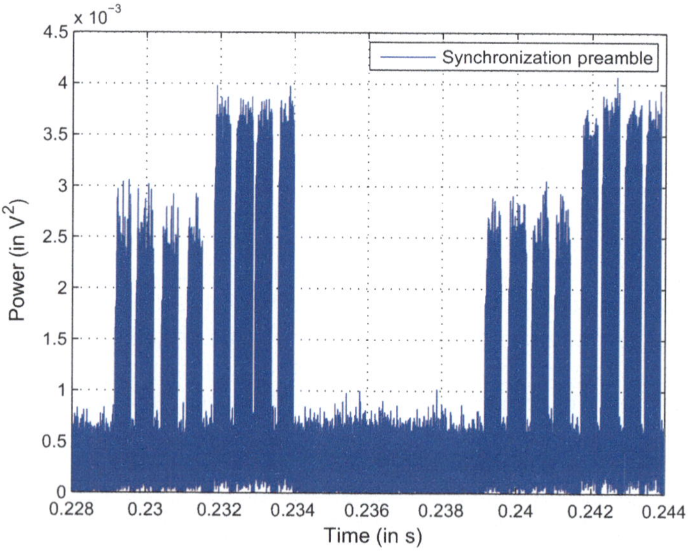

Fig. 2 Synchronization data captured at 5.9 GHz between the transmit and receive nodes prior to the message signaling

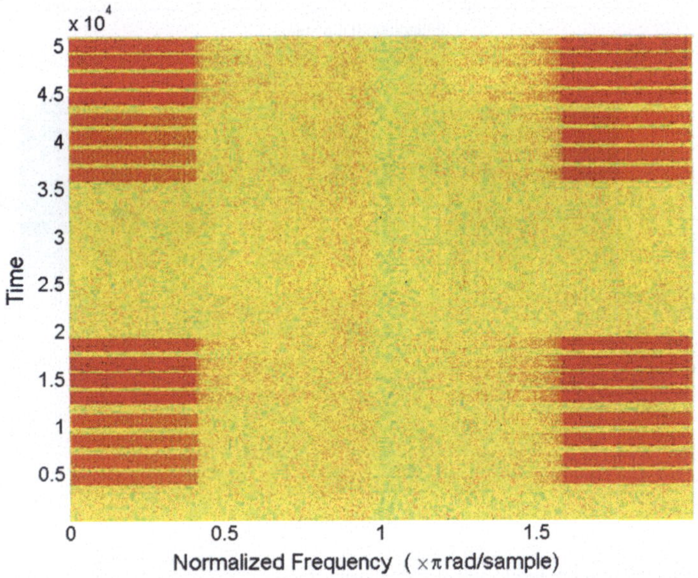

Fig. 3 Spectrogram of synchronization preamble. Recall that folding frequency implies the DC component, which is left vacant in IEEE802.11 protocol suite

Beside time–domain behavior, frequency–domain behavior of the physical layer (PHY) data is critical as well. Generic IEEE802.11x protocol suite sends synchronization preamble with a relatively narrower bandwidth as compared to the bandwidth reserved for the data. Because IEEE802.11 protocol suite is based on orthogonal frequency division multiplexing (OFDM) technology, subcarrier formation and allocation are two crucial tasks. Hence, it is appropriate to glance at the snapshot of the synchronization data, which is given in Fig. 3. Note that in Fig. 3 all of the subcarriers are distributed around the folding frequency, which points out the direct current (DC) component. In Fig. 4, subcarrier distribution can closely be investigated. As can be inferred from Fig. 4, subcarriers manifest themselves as uniformly distributed spikes all across the baseband. It is important to bear in mind that Fourier transform comes at the expense of some sort of averaging over frequency–domain; therefore, spikes come about in a uniformly distributed fashion at baseband. Subcarrier organization and deployment might differ for different bursts. However, it is outside the scope of this study.

5.3 *Test and Measurement*

In order to test the prototype and collect some data, a test drive is performed in Istanbul, Asia Minor side in real traffic with a C–segment medium size car. The

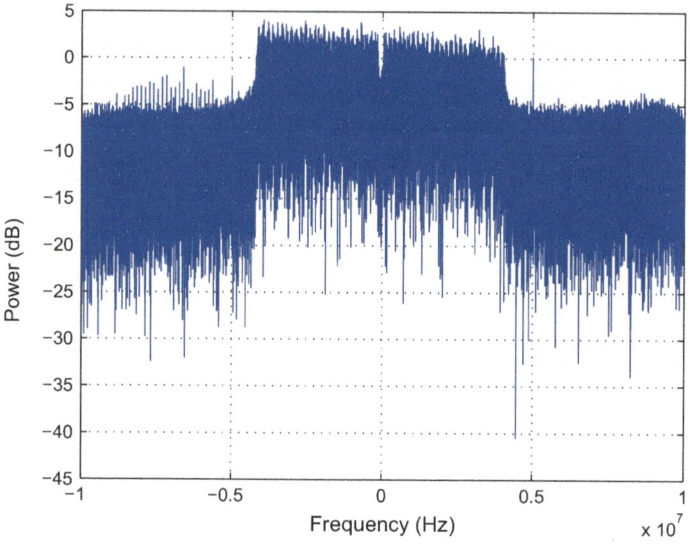

Fig. 4 Synchronization preamble subcarrier distribution

prototype and all its peripherals including a GlobalSat BU-353-S4 universal serial bus (USB) GPS along with 3GLTE extension are set in the vehicle up. It is worth mentioning at this point that the prototype is programmed to acquire GPS coordinates at a rate of 1sa/s. In addition, the prototype queries the car area network (CAN)–bus in a continuous manner so that several types of vehicle data including instantaneous gas consumption are recorded and processed. A continuous one hour test drive data are formed. Data collected are processed at a central location. Results are marked on Google maps and collected data are tracked on 2D plane.

6 Conclusion and Future Work

ITS seems to provide a promising solution towards green environments with optimized designs and systems, energy efficiency, and reduced carbon footprints. It is clear that NGWN is an integral part of ITS since explosive growth of the use of wireless communications along with emerging new technologies, standards, and services are anticipated to continue further. Merging ITS with NGWN immediately points out cyber–physical systems, IoT, and big data. Considering the opportunities provided by cloud computing such as parallel processing, high–volume storage, and virtualization, it is evident that ITS will gain newer dimensions and horizons.

First and foremost, extension of ITS with V2V, vehicle-to-infrastructure (V2I), and NGWN is expected to give rise to a very different, heterogeneous, and extremely dynamic network topology including both wired, wireless, long–, and short–range communication modes simultaneously. Furthermore, the data characteristics will have several intriguing overall network traffic behavior when examined with the ad–hoc structure of V2V and V2I. Note that the models which have been used in cellular networks up until now might not be sufficient to deal with the diversity and complexity of this behavior.

Mobile data which are transferred to cloud computing platform create the big data. It is expected that big data along with the hybrid cloud give rise to further research opportunities for all car manufacturers, policy makers, regulatory bodies, and researchers with the concern of ethics and security issues.

References

1. Servin O, Boriboonsomsin K, Barth M (2006) An energy and emissions impact evaluation of intelligent speed adaptation. In: IEEE intelligent transportation systems conference. Toronto, Ontario, Canada, pp 1257–1262
2. Tielert T, Killat M, Hartenstein H, Luz R, Hausberger S, Benz T (2010) The impact of traffic-light-to-vehicle communication on fuel consumption and emissions. In: Internet of things (IOT), Tokyo, Japan, pp 1–8
3. Morris BT, Tran C, Scora G, Trivedi MM, Barth MJ (2012) Real-time video-based traffic measurement and visualization system for energy/emissions. IEEE Trans Intell Transp Syst 13(4):1667–1678
4. Brennand CARL, da Cunha FD, Maia G, Cerqueira E, Loureiro AAF, Villas LA (2016) Fox: a traffic management system of computer-based vehicles fog. In: 2016 IEEE Symposium on Computers and Communication (ISCC), pp 982–987
5. Boriboonsomsin K, Barth MJ, Zhu W, Vu A (2012) Eco-routing navigation system based on multisource historical and real-time traffic information. IEEE Trans Intell Transp Syst 13(4):1694–1704
6. d'Orey PM, Ferreira M (2014) Its for sustainable mobility: a survey on applications and impact assessment tools. IEEE Trans Intell Transp Syst 15(2):477–493
7. Hashem IAT, Yaqoob I, Anuar NB, Mokhtar S, Gani A, Ullah Khan S (2015) The rise of big data on cloud computing: review and open research issues. Inf Syst 47:98–115
8. Lv Y, Duan Y, Kang W, Li Z, Wang FY (2015) Traffic flow prediction with big data: a deep learning approach. IEEE Trans Intell Transp Syst 16(2):865–873
9. Laney D (2001) 3d data management: controlling data volume, velocity, and variety. Application delivery strategies
10. Assunção MD, Calheiros RN, Bianchi S, Netto MA, Buyya R (2014) Big data computing and clouds: trends and future directions. J Parallel Distrib Comput
11. Andreolini M, Colajanni M, Pietri M, Tosi S (2014) Adaptive, scalable and reliable monitoring of big data on clouds. J Parallel Distrib Comput
12. Messelodi S, Modena CM, Zanin M, De Natale FG, Granelli F, Betterle E, Guarise A (2009) Intelligent extended floating car data collection. Expert Syst Appl 36(3):4213–4227
13. Tiedong W, Jingjing H (2014) Applying floating car data in traffic monitoring. In: 2014 IEEE international conference on control science and systems engineering, pp 96–99
14. Wei L, Zhu H, Cao Z, Dong X, Jia W, Chen Y, Vasilakos AV (2014) Security and privacy for storage and computation in cloud computing. Inf Sci 258:371–386
15. Kshetri N (2013) Privacy and security issues in cloud computing: the role of institutions and institutional evolution. Telecommun Policy 37(4–5):372–386
16. Adjei JK (2014) Explaining the role of trust in cloud service acquisition. In: 2nd IEEE international conference on mobile cloud computing, services, and engineering, pp 283–288

17. Kemp R (2014) Legal aspects of managing big data. Comput Law Secur Rev 30(5):482–491
18. Lin W, Dou W, Zhou Z, Liu C (2014) A cloud-based framework for home-diagnosis service over big medical data. J Syst Softw
19. Yang J-J, Li J-Q, Niu Y (2015) A hybrid solution for privacy preserving medical data sharing in the cloud environment. Future Gener Comput Syst 43–44:74–86
20. Zhao J, Wang L, Tao J, Chen J, Sun W, Ranjan R, Kolodziej J, Streit A, Georgakopoulos D (2014) A security framework in g-hadoop for big data computing across distributed cloud data centres. J Comput Syst Sci 80(5):994–1007

EMC Measurements in Indoor Power Line Communication Environments

Waled Gheth, Matjaz Rozman, Khaled M. Rabie and Bamidele Adebisi

Abstract The transmit power of communication signals over power lines is limited due to restrictions imposed by regulatory bodies and standards. In this paper, the electromagnetic emissions from two different power line communication (PLC) modems are measured in different indoor environments in the frequency range between 1–30 MHz. The obtained measurements are compared with the current available limits and regulations. Also, a comparison between the emission measurements of the two modems is made. Results have shown that, in some scenarios, the measured emissions do not comply with some of the available standards. Based on this, recommendations are made towards the end of the paper.

Keywords Electromagnetic comparability (EMC) · Measurements · Power line communication (PLC) · Regulations and standards

1 Introduction

Power lines have always been a means to distribute electrical power. However, with the rising dependence on communications, power line networks were seen as a possible medium for delivering data. This technology is commonly known as power line communication (PLC). PLCs exploit the existing power cables in buildings or utility grid to simultaneously carry communication signals and deliver electricity. PLCs can play an important role in home networking applications due to the spreading of internet and modern communication technologies. However, the power line network was initially designed to transmit alternating current (AC) to consumers at frequencies 50 and 60 Hz; hence, several challenges are faced when using this medium for high-data rate transmissions [1–3]. Among these is the electromagnetic interference (EMI) issue due to the fact that at high frequencies power lines tend to act as radiating antennas [4]. There are two types of the EMI, namely, inductive and

W. Gheth (✉) · M. Rozman · K. M. Rabie · B. Adebisi
School of Engineering, Manchester Metropolitan University, Manchester M1 5GD, UK
e-mail: waled-ali.m.gheth@stu.mmu.ac.uk

A. Boyaci et al. (eds.), *International Telecommunications Conference*,
Lecture Notes in Electrical Engineering 504,
https://doi.org/10.1007/978-981-13-0408-8_16

In [4], the measurements were taken within the frequency range from 1 to 30 MHz in an indoor power line environment. The radiated emission measurements were carried out at a distance of 3m away from the power line using an active loop antenna and the results were then compared to the Federal Communications Commission (FCC) standard. The differential and common modes of the conducted emission readings were taken using a current probe connected to a spectrum analyzer. The busy state of the common mode was compared to the EN55022 (main port and common mode). However, the comparison between the test outcomes and the standards showed that the measured results do not totally comply with the specified limits.

The possibility of extending the frequency band limits above 30 MHz considering the EMC standards was investigated in [10]. A bi-conical antenna was used to measure the electric field at a distance of 3m away from a 15m power line installed in an open area site. Different signals in the frequency bands 2–30 MHz and 30–100 MHz were injected to the PLC network and EMC measurements were conducted. Results showed that the power spectral density (PSD) should be at a certain level in the frequency range 30–100 MHz in order to meet the EMC limits [10].

3 Overview of EMC Standards

EMC certificate is a topic that has been and will always be discussed in PLC systems. A different set of EMC standards exist for PLC technologies according to their operating frequency bands. Technologies that operate at frequencies 500 kHz and below are known as narrow band PLC, whereas those technologies using frequencies above 1.6 MHz are referred to as broadband PLC (BB-PLC) systems [11]. However, frequency bands from 1.6 to 30 MHZ are recommended by European Telecommunications Standards Institute (ETSI) for PLC systems [4]. This framework concentrates mostly on the BB-PLC technology and its relevant standards.

The issue of the EMC standards is still discussed by several standardization bodies. The International Special Committee on Radio Interference (CISPR) is an organization that plays a fundamental role in the context of EMC. It is a non-governmental organization and was established in 1934 to set standards for controlling EMI. The CISPR is a part of the International Electro-technical Commission (IEC). However, the standards that have been set by CISPR should be amended by local authorities in order to have legal effects [11].

Furthermore, European Committee for Electro-technical Standardization (CENELEC) is the organization that is responsible for European standardization in this area. This organization started from CISPR standards and produced the European Norms (EN), which is called harmonized standards. These standards are then amended by local committees in EU and also have been adopted by the US standardization body, i.e., FCC. FCC Part 15 and CISPR 22, which were published in 1979 and 1985 respectively, are the two EMC standards that have an impact on PLC. The former is valid for commercial products marketed in the US and the latter is for outside the US. Nevertheless, CENELEC adopted the specific standards CISPR 22 with the norm EN55022 for the EU [11].

4 Methods of Measurement and EMC Limits

The conduced and radiated emissions are two classes of the EMC. The interference that is conducted through the power line is called the conducted emissions whereas those caused by electric and magnetic fields produced by appliances and that propagate outside the power conductors are referred to as the radiated emissions. According to EN55022, the radiated emissions are only measured above 30 MHz, while the conducted emissions ought to be measured between the frequencies 150 KHz and 30 MHz. Recently, measurements below 150 KHz have been specified by the norm EN50065. However, FCC in the US includes that radiated emissions should be measured below and above 30 MHz [11].

Each of the standardization bodies has its own limits and measurement setups. For instance, FCC specifies radiated and conducted limits in μV/m and μV, respectively, and the measurement is made at distance of 30m using quasi-peaks level. NB30 standard uses a peak detector at distance of 3m and the radiated limits are specified in dBμV/m. Modem operating mode during the measurement is another crucial factor. While EN55022 specifies a normal operating mode for measurement setup, both FCC and NB30 specify maximum emissions mode for measurement setup.

The conducted emission is distinguished among the terminal port and the telecommunication port in the EN5022 standard, where the former indicates the power supply and the later refers to the data transmission port. However, two modes are defined and considered in the conducted emission, namely, differential mode and common mode. Differential mode is related to the flow of the generated current from the phase conductor (P) to the neutral conductor (N) whereas the common mode is associated with the current flow through the phase conductor (P) and neutral conductor (N) synchronously then returns through the Protective Earth (PE) conductor [11]. In Fig. 1 we present these modes as used in our experiments (more details will be provided later).

It can also be observed that in EN5022, devices are split into two classes: class A and class B. The devices that are used in industrial or commercial environments are specified as class A while the ones used in residential environments are called class B. However, EN5022 specifies class A limits with less strength than that of class B. For example, the class B quasi peaks (QP) main port limit for the frequency range 5–30 MHz is 60dBμV, while the same condition limit in class A is 73dBμV. For better understanding and clarity, we present in Table 1 (shown at the top of the next page) some differences in the limits of class A and class B devices [11].

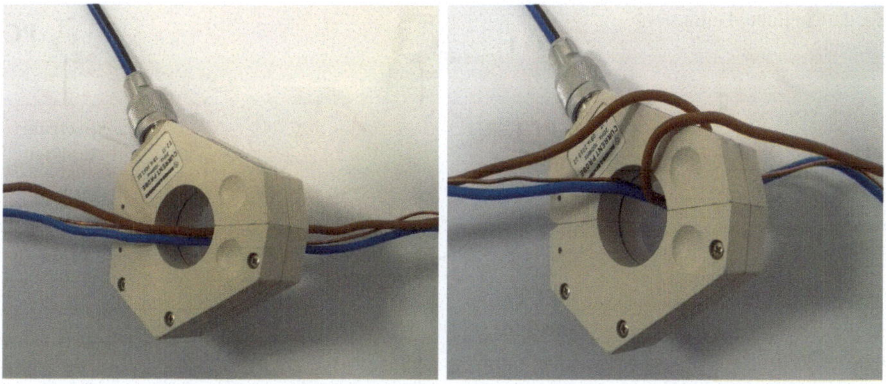

(a) Common mode measurement (b) Differential mode measurement

Fig. 1 The real setups of the common and differential modes used in our measurements

Table 1 EN55022 conducted emission limits for class A and class B devices.

	Class A				Class B			
	Main port limits		Telecom limits		Main port limits		Telecom limits	
Frequencies (MHz)	QP (dBμV)	AV (dBμV)	QP (dBμV)	AV (dBμV)	QP (dBμV)	AV (dBμV)	QP (dBμV)	AV (dBμV)
0.15–0.5	79	60	97–87	84–74	66–56	56–46	84–74	74–64
0.5–5	73	66	87	74	56	46	74	64
5–30					60	50		

5 EMC Analysis

In this section, we present the setup and procedures of our experiments. In order to establish a PLC network, two computers were connected to two modems which are plugged into two sockets attached to the power line. Three measurement scenarios are involved, namely,

1. PLC modems are powered off to measure the noise.
2. PLC modems are on with no communication between the computers (i.e., Quiescent state),
3. PLC modems are switched on and data is active on the network (i.e., Busy state).

5.1 Conducted Emission Measurements

Common mode and differential mode measurements were taken using a current probe which is connected to the spectrum analyzer. Firstly, the common mode is measured

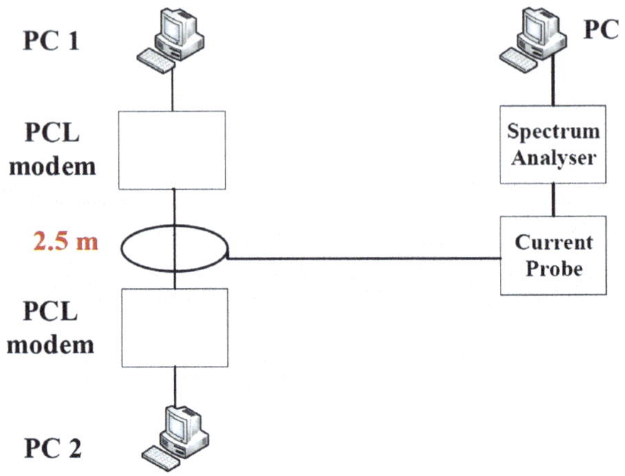

Fig. 2 Conducted emission test setup

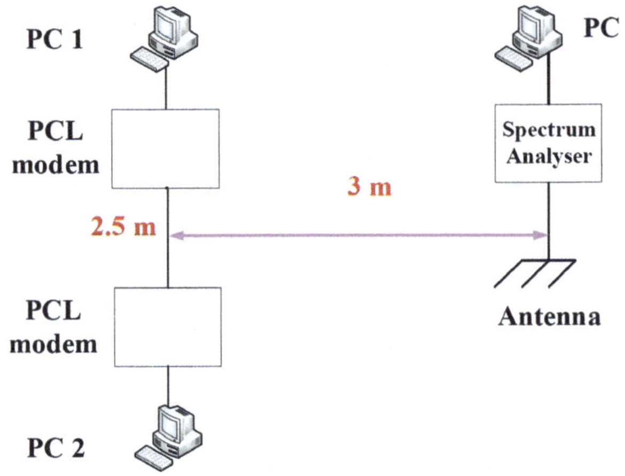

Fig. 3 Radiated emission test setup

in both states quiescent and busy (see Fig. 1a) and, similarly, the differential mode is measured in quiescent and busy states (see Fig. 1b). A simplified diagram of the conducted emission test setup is illustrated in Fig. 2.

Radiated Emission Measurements Radiated measurements are carried out in this paper by a magnetic tracking antenna at a distance of 3 m from a 2.5 m power cable which connects the two computers as represented in Fig. 3. The data is collected and recorded on a PC which is already connected to the spectrum analyzer that is connected to the antenna.

6 Results and Discussions

In this section we present and discuss the conducted and radiated emission results.

6.1 Conducted Emissions

To begin with, Fig. 4 presents the common and differential mode results over the frequency band 1–30 MHz for a TP-LINK modem. We can clearly see that the differential mode current is the dominant signal as it carries the desired date along the power cable. Although, the common mode current should not exist, we still can see its presence in Fig. 4.

Figure 5 illustrates the emission for busy state of the common and differential modes for both modems: TP-LINK and Homeplug Turbo. From Fig. 5a and b it is evident that in some cases, the readings of the TP-LINK are higher than the Homeplug Turbo measurements and vice versa. The conducted emission measurements in this figure are compared to the EN55022 main port limits and, as shown, there are three different values of the EN55022 limits depending on the operating frequency. However, the EN55022 limits are 32dBμA for frequency range from 0.15 to 0.5 MHz, 22dBμA for frequencies between 0.5 and 5 MHz and for the frequencies from 5 to 30 MHz, it is 26dBμA. The common mode measurements indicate that the tested PLC modems do not exceed the EN55022 limits throughout the frequency spectrum

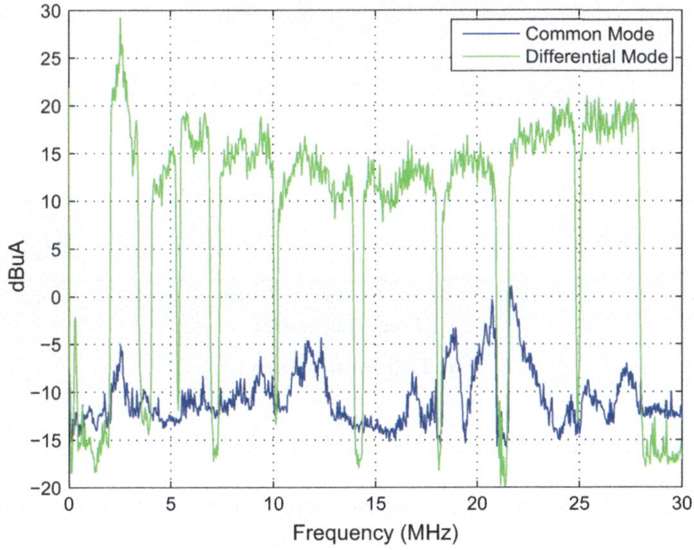

Fig. 4 Conducted current measurements for both the differential and common modes

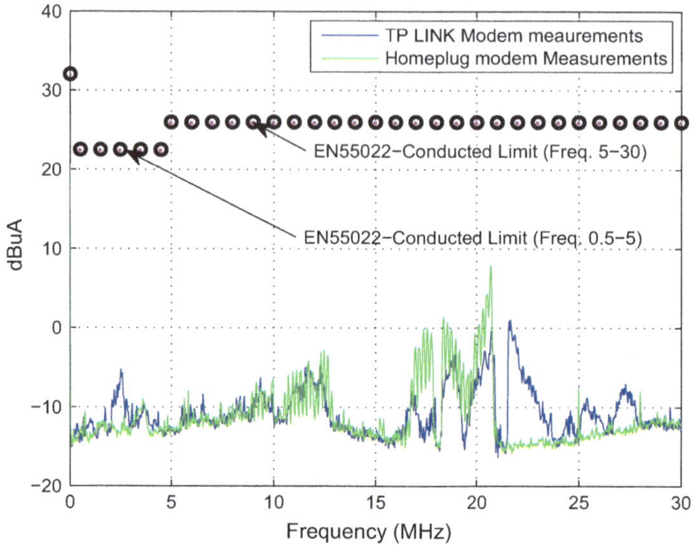

(a) Common mode

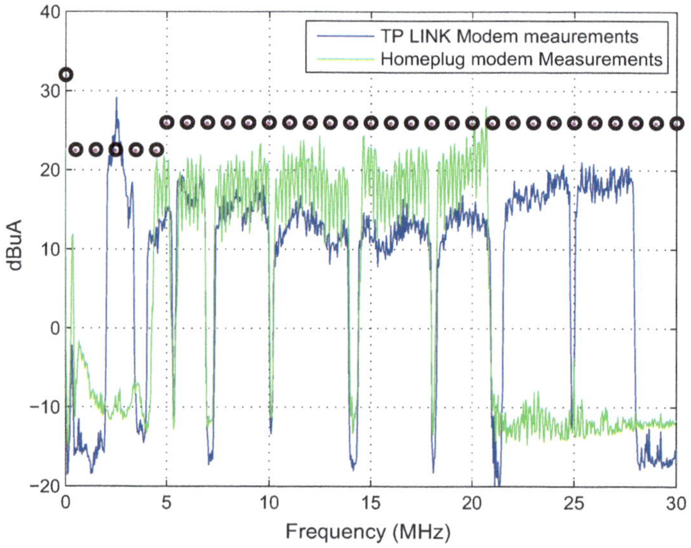

(b) Differential mode

Fig. 5 Conducted emissions during busy state of the common and differential modes for TP-LINK and Homeplug turbo modems compared to the EN55022 limits

under consideration. On the other hand, the differential mode measurements show that the TP-LINK exceeds the EN55022 limits between 2.5 and 3 MHz. Furthermore, the EN55022 limits are exceeded by the differential mode readings of the Homeplug Turbo modem at frequency 22 MHz.

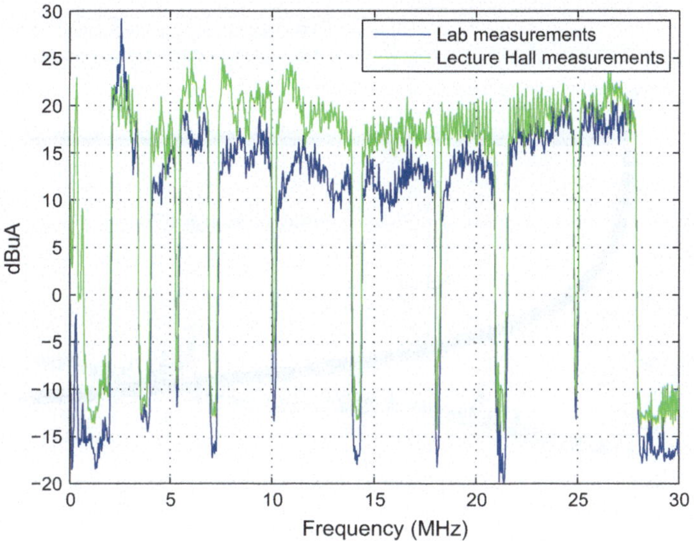

(a) Differential mode

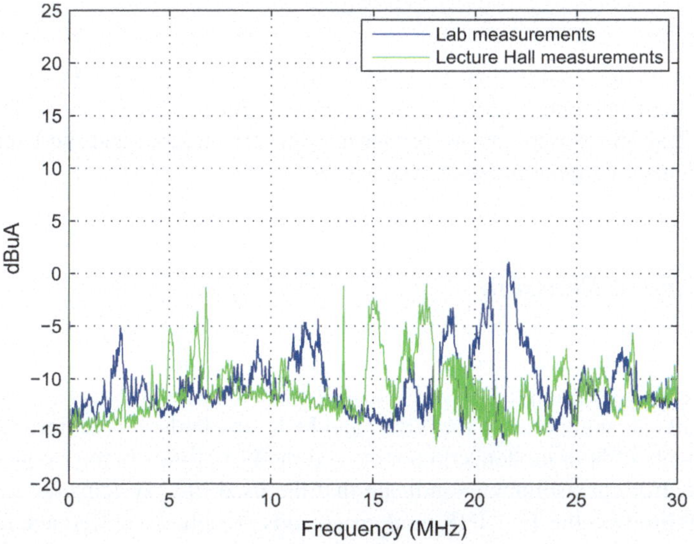

(b) Common mode

Fig. 6 Conducted emission measurements for the TP-LINK modem in two different indoor environments

Moreover, the conducted emissions were measured in two different environments: a medium-sized Lab and a large lecture hall. In this respect, differential and common modes are compared in Fig. 6. From Fig. 6a, it is clear that, in most cases, the Lecture

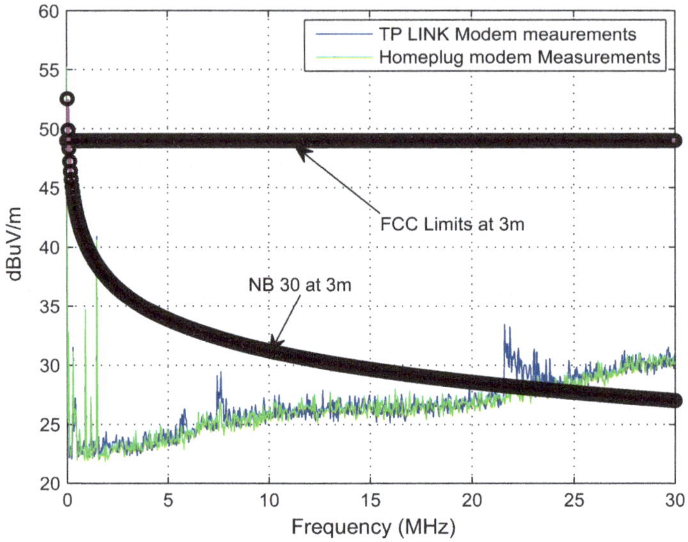

Fig. 7 Radiated emission measurements and their comparison to the standards

Hall measurements are dominant in the differential mode. Figure 6b shows the highest difference between both readings in common mode measurements is about 12dBμ*A* where the Lab measurements are higher at some frequencies and the Lecture Hall measurements are higher at other frequencies.

6.2 Radiated Emission

Figure 7 depicts the radiated emission for both modems under consideration. Clearly, the peak of the field strength is about 45dBμV/m which occurs at very low frequency. These results are compared to the NB30 and FCC standards. It is noticeable that the measurements of both modems do not exceed the FCC limits in the frequency band 0–30 MHz. Both measurements indicate that the used PLC systems do not comply with NB30 limits as the TP-LINK modem exceeds these limits at frequencies around 2 MHz and above 23 MHz. Moreover, all results taken at almost 22 MHz and above exceed the NB30 limits for Homeplug Turbo modem. It should be pointed out that these measurements are compared to the standards without considering the antenna calibration factor which should be added to the measurements before comparing them to the standards. This factor is called antenna factor; it is treated as a loss in dB. Also, a cable losses ought to be considered and added to the readings.

The final set of results is shown in Fig. 8. This figure shows a comparison between the radiated emissions of TP-LINK modem in two different environments, i.e., lab and lecture hall. It is clear that the lecture hall measurements are higher than the

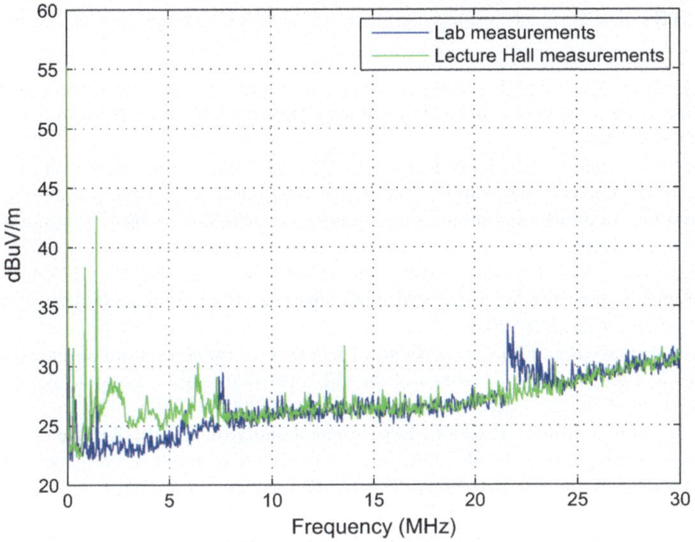

Fig. 8 Radiated emissions in two different indoor environments

lab readings for the frequency range 0–7 MHz whereas the emissions in the lab are dominant in the frequency range 22–24 MHz.

7 Conclusion

From an overall perspective to the measurements taken in this frame work, it was confirmed that the tested PLC modems totally comply with FCC limits for the radiated emissions and with the EN55022 in most cases for the conducted one. In fact, there was at least a minimum of 10dB margin between our radiated measurements and the FCC regulation standards; this margin reaches as high as 35dB at some frequencies. On the other hand, the comparison between measured radiated emissions and the standards showed that the former exceeds one of the proposed regulation limits. In some measured radiated emissions, the field strength curve exceeds the NB30 limits. Comparisons of both conducted and radiated emissions in different environments were presented. It was clear that emissions were affected by the place and surrounding areas where the measurements took place.

Acknowledgements This research has been carried out within the "CityVerve: IoTs and Smart Cities Demonstrator" project funded by Innovate UK (102561).

References

1. Meng H, Guan YL, Chen S (2005) Modeling and analysis of noise effects on broadband power-line communications. IEEE Trans Power Deliv 20:630–637. https://doi.org/10.1109/tpwrd.2005.844349
2. Rozman M, Ikpehai A, Adebisi B, Rabie KM (2016) Channel characterisation of cooperative relaying power line communication systems. In: 10th International symposium on communication systems, networks and digital signal processing (CSNDSP), pp 1–5. https://doi.org/10.1109/CSNDSP.2016.7574000
3. Adebisi B, Rabie KM, Ikpehai A, Soltanpur C, Wells A (2017) Vector OFDM transmission over non-gaussian power line communication channels. IEEE Syst J PP(99):1–9. https://doi.org/10.1109/jsyst.2017.2669086
4. Adebisi B, Honary B (2006) Comparisons of indoor PLC emissions measurement results and regulation standards. In: Proceedings of the IEEE international symposium power line communication, pp 319–324. https://doi.org/10.1109/isplc.2006.247482
5. Adebisi B, Stott J, Honary B (2006) Experimental study of the interference caused by PLC transmission on HF bands. In: IET international conference on ionospheric radio systems and techniques (IRST), pp 326–330. https://doi.org/10.1049/cp:20060295
6. Widmer H (2000) On the global emc aspect of broadband power line communications using the ldquo; hf rdquo; frequency band. In: 2000 International Zurich seminar on broadband communications. accessing, transmission, networking. proceedings (Cat. No.00TH8475), pp 179–184. https://doi.org/10.1109/IZSBC.2000.829249
7. Middleton D (1977) Statistical-physical models of electromagnetic interference. IEEE Trans. Electromagn. Compat. EMC–19:106–127. https://doi.org/10.1109/TEMC.1977.303527
8. Hansen D (2002) Review of emc aspects in recent european PLC development. IEEE Power Eng Soc Summer Meet 3:1200–1216. https://doi.org/10.1109/PESS.2002.1043488
9. Grassi F, Pignari SA, Wolf J (2011) Channel characterization and EMC assessment of a PLC system for spacecraft dc differential power buses. IEEE Trans Electromagn Compat 53:664–675. https://doi.org/10.1109/TEMC.2011.2125967
10. Tlich M, Razafferson R, Avril G, Zeddam A (2008) Outline about the EMC properties and throughputs of the PLC systems up to 100 mhz. IEEE Int Symp PLC Tech Appl, pp 259–262. https://doi.org/10.1109/ISPLC.2008.4510435
11. Girotto M, Tonello AM (2017) EMC regulations and spectral constraints for multicarrier modulation in PLC. IEEE Access 5:4954–4966. https://doi.org/10.1109/ACCESS.2017.2676352

Part V
Security

Security Problems and Attacks on Smart Cars

Ş. Okul, Muhammed Ali Aydin and Fatih Keleş

Abstract This study provides information about smart cars and some of the technologies and researches on smart cars. It then describes the modeling of security attacks in smart cars in terms of aggressive profile, attackable objects, attack requirements and security requirements. It also discusses the attack pattern and risk analysis related to vehicle speed ups and the stealing of personal information.

Keywords Smart car · Security attacks · Security attacks on smart cars · Attacker tree model

1 Introduction

Smart Cars integrates Internet Of Things (IOT) components to bring value-added services to drivers and travelers. These components communicate with each other and with the outside of the vehicle. Over the last few years, there have been many publications about the attacks on automotive systems. Some of these have been shown to be cheaper and easier to show, especially as they are a teenager who opens up and launches a car that remotely connects by using simple equipment, especially under the eyes of the media, especially at $ 1,500 a few dollars, turns out [1].

In addition to prestigious damage, the safety cost of cyberspace is becoming a problem for automobile manufacturers. Security vulnerabilities have been found in recent years and an increasing number of recalls have been made [2]:

Ş. Okul
TUBITAK BILGEM, Kocaeli, Turkey
e-mail: sukru.okul@tubitak.gov.tr

M. A. Aydin (✉) · F. Keleş
Computer Engineering Department, Istanbul University, Istanbul, Turkey
e-mail: aydinali@istanbul.edu.tr

F. Keleş
e-mail: fkeles@istanbul.edu.tr

A. Boyaci et al. (eds.), *International Telecommunications Conference*,
Lecture Notes in Electrical Engineering 504,
https://doi.org/10.1007/978-981-13-0408-8_17

- Charlie Miller and Chris Valasek were great prospects for remote attacks, taking over a Jeepin control and driving 1.4 million cars out of the way [3];
- Security researchers moved away from cutting off the BMW Connected Drive and managed to unlock vehicles from cutting Miller/Valaisek (2.2 million cars had more industrial use than expected) [4];
- More recently, more vehicles (including most Volkswagen cars since 1995) have been shown vulnerable to an attack against a remote keyless entry, thus increasing the size of the affected fleet once more. This last number showed that the number of potentially affected cars increased rapidly with a vehicle size of 100 million vehicles [5].

2 Technologies and Researches

As technology continues to progress, the tools will be smart and have more connections. According to ABI Research, less than 40 percent of old vehicles around world are anticipate to be linked by 2017 [6].

According to a study published by the Kelley Blue Book, 62% of US consumers thought that connected cars would be attacked, while 42% of respondents said they wanted cars to be more connected. This number has increased to 60% over the course of the millennium. Finally, more than a third of all participants said that a car did not have the desired technology and would then buy a different car [7].

Some automakers add the items of Apple's operating system to automakers cars, allowing their drivers to integrate an Apple appliance into the emulated system. Tools that use the Drive and the Android operating system can access and use traffic, maps, web services, etc.; They are all integrated with all the car's devices for full. At Google's goal, the car is becoming a "connected Android appliance" [8].

Moreover, the automobile infotainment system claims to be developed faster than it can analyze the effects of independent researchers [9].

3 Models of Security Attacks in Smart Cars

3.1 Offensive Profile

Different groups of attackers are attracted to attack vehicles. These groups range from the owner of the vehicle to a specialized hacker with advanced tools. Each of these groups typically has their own motivations:

- Falsification: An attacker may wish to misrepresent actual vehicle information, such as a tachograph exchange, to sell the vehicle by reading the tachograph or miscalculation.

- Illegal profit: An attacker can profit by selling the ability to steal or attack an entity to a different entity. Some attacks can be directed by a commercial opponent of target vehicle sellers to sabotage their products and gain share in the market.
- Insane fun and vandalism: Revenge and vandalism can motivate some attacks, such as an employee casualty attempt to punish an old company by deriving from the company the cars are sold [10].
- Research and test purposes: Attack and penetration testing can be done by security experts or test teams. The assailants, in this case, have benign motivations. Before they are used by third parties, they are trying to discover security flaws within the different components of the vehicle systems.
- Accidental: In some cases, there may be an attack without any intention. Such an attack can be carried out while upgrading an existing system or while reading unwanted malicious data, such as in the case of GPS, climate control and fault in the front console radio system in Toyota Lexus vehicles [11].
- Overlap: Sometimes there are multiple causes behind a single attack.

However, motivation alone is not enough. An attacker needs sufficient technical skills and different sets of hardware to reach its goals. The diversity of skills, abilities, technical equipment and financial resources can be used as indicators to separate attackers into different groups [12]:

1. Unscrupulous attackers (script kiddie): This group includes attackers with limited financial resources and insignificant knowledge of vehicle architecture. Such attackers lack the ability to use complex tools. Regular thieves are owners of those who want to install or change a component in their vehicle, and good offender of this group of offenders who give highway signals to gain credibility in their community.
2. Hacker: This group includes highly skilled specialists with enough equipment and equipment to perform the attack. Using the experience of this group of members, they can profit like land hat hackers. Mechanics and security researchers are involved in this group.
3. Organization: There are several members of the above group that these organizations work with. Generally, gigantic financial support enables them to take advanced tools and attract experts. Security research groups can be an example of this class.

3.2 Attackable Objects

Attackers can focus on different parts of the vehicle components:

- Data: Attackers can target data stored on some ECUs; This may be cryptographic private keys, digital certificates or special vehicle and driver activities (e.g. vehicle location, navigation destination, etc.). Or, the transmitted wired/wireless data may threaten the vehicle. This data includes the following:

- Inter-vehicle data exchange between different components and sensors with one component. Fraud between the transmitted data, the system on the vehicle, and the pressure sensors on the tires is an example of the safety margin of such data [13].
- Data transferred between the vehicle and the outside world; V2 V communication data, V2I communication data, etc.

- In-car hardware: In general, attacking the hardware infrastructure (in other words, ECUs, sensors and Car-top Units) requires direct access to the target devices. Attacking an in-vehicle navy can occur by replacing a device with a malicious device or by installing new hardware that is performed in error. Sometimes the attacked hardware may not be part of the vehicle. Your drive may be a 3rd party device, such as a mobile phone, that is plugged in. [14] An attacker may aim to reduce the performance of the vehicle's component or even deliberately produce misleading results (e.g. Volkswagen's Emission Scandal [15]).
- Media infrastructure: Some attacks may target the environment of the vehicle. A typical example of such an attack is a change in electronic road signs such as "Zombies Ahead", where an attacker finds how to change the text on electronic road signs to warn of a Zombies attack. Even such a ridiculous attack can create public safety problems for drivers on the road [16].
- Software and framework: The enormous amount of integrated software on each vehicle and the different levels of security control between different vendors make them more vulnerable to attack. The frame that controls the ECU can be the target of various attacks; Some attackers may interfere with this aspect of the ECU to achieve superior performance [17]. Malicious updates to the inner parts of an applet or framework can open doors that can be exploited by an attacker.

3.3 Attack Requirements

To be able to attack, it must be a smart car, direct, remote or mixed access to be attacked:

- Direct Access: Some attacks are based on direct access from the target vehicle. A vehicle can be accessed directly when parked. Subsequently, attackers may have the option of linking a GPS device to track the vehicle later, or determining the destination of the vehicle [18]. In some cases, taking a car to the service station to check it may be a way for direct access by attackers. In such cases, an attacker has full access to the vehicle. On-board Diagnostic port (OBD-II) is a physical interface already used for many attacks [19].
- Remote access: Other attacks do not require direct access to the target. Attackers can target from a distance. Such attacks exploit the integrated wireless capabilities of modern automobiles. These include Bluetooth, a cellular connection, wireless tire pressure monitoring, etc.. Entertainment system is another point that is inter-

rupted at a distance. For example, playing a song linked with Malware could give malicious messages to the Control Area Network [19].

- Mixed access: Direct access to the car, remote intrusion logging. Indeed, some attackers can install some devices into the vehicle (for example, USB, malicious DVDs, malicious components connected via OBD-II port, etc.), even though some attackers can access the car quickly. Then, these interference devices can be used to target the vehicle remotely. Attackers can use others to wear these tools, such as a surrogate who parks the victim's car, or a mechanic at a service station [19].

4 Attack Tree Modelling and Calculating Risk

In this section, there is a model of attack and risk analysis related to increasing vehicle speed and stealing personal information.

4.1 Attack Tree Model

Attacker modeling was done after dangers and safety vulnerabilities were expressed and separated by categories. It was used with scenario of the attack used in increase of the vehicle speed studies for modeling [19–22]. Increased vehicle speed includes threats arising from vulnerabilities, software updates, use and use of the Bluetooth Control Area Network. Three attack scenarios consist of OR nodes for increased vehicle speed.

Figure 1 shows an attack on the speed of the vehicle using a security update to update the firmware. When using the update of firmware, the attack first adds code to the Engine Control Unit, and then adds malicious software to send the car's acceleration message to the firmware. Thus, if updates the OBD (in-vehicle diagnostic) terminal and the firmware on a wireless or wired connection by the driver, the Engine Control Unit generates an error. Thus increases of the speed of the car. The OBD controls the electrical operating the vehicle status. Although initially used to protect electronic components such as motors, the way of reporting various vehicle conditions through the driver plays the role of computer user interface [23].

Figure 2 shows the increase in vehicle speed using Bluetooth security challenge. To use Bluetooth, you first need to have an infotainment system privilege. This can be done by connecting to Bluetooth via a Bluetooth communication interface or by means of a mobile appliance through which a software can be downloaded. Bluetooth is scanned to obtain the Bluetooth communication privilege and the PIN number is acquired with brute force attack [22].

Figure 3 shows the attack rate at a vehicle speed using the Control Area Network security policy. When a attacker uses a Control Area Network, the attacker analyzes the Control Area Network message first, thus generates an automobile self-diagnosis software that can send a malicious Control Area Network message using the analyzed

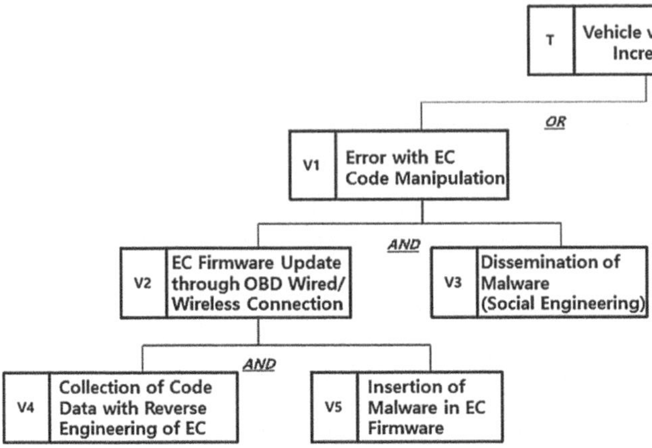

Fig. 1 Firmware updates threaten to increase vehicle speed using a security obligation [14]

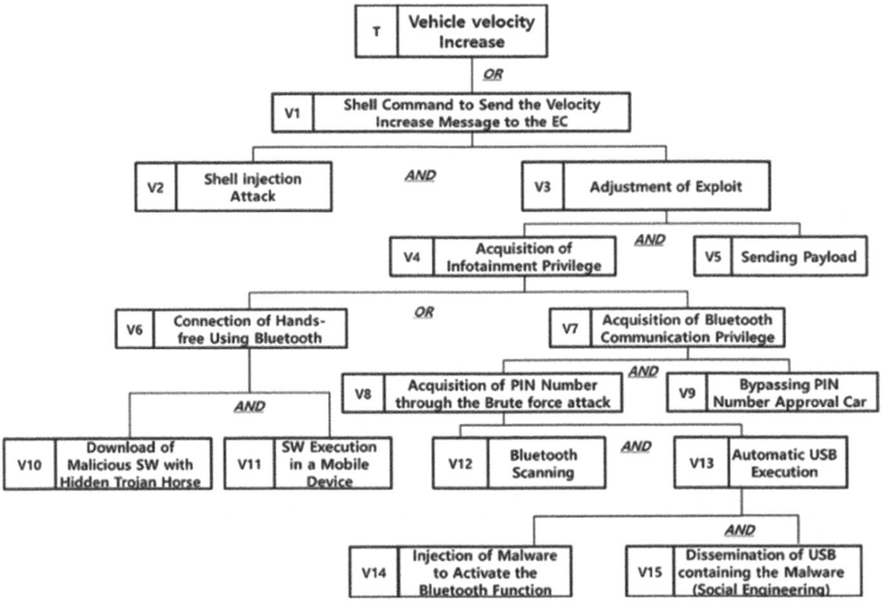

Fig. 2 Vehicle speed increase using Bluetooth weakness [23]

message. When a driver downloads a malicious software file and then hitches to a mobile appliance over a wired or wireless network, the attacker uses a message for increasing car speed through a pre-built car hacking program [23].

The modeling of the particular knowledge leak threat was formed with attack scenarios from previous works. The infotainment system uses for this scenario (like USB, Bluetooth). The infotainment system can be hitched to various external net-

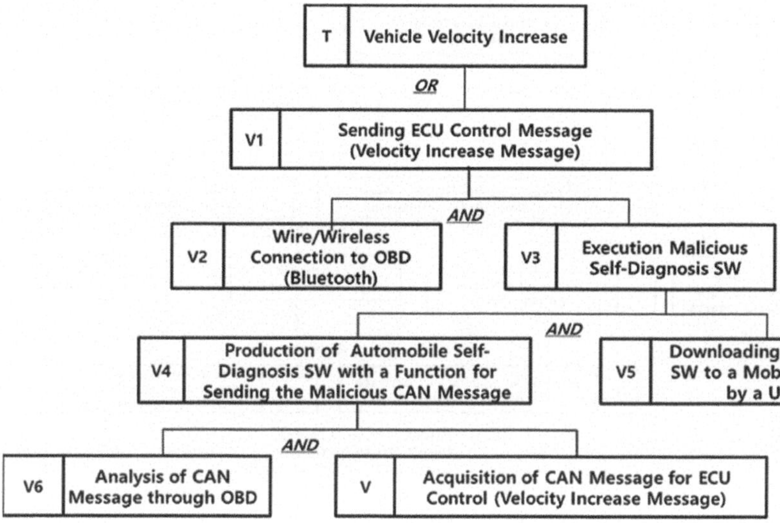

Fig. 3 Vehicle speed increase using the security implication of the CAN network [23]

works so that particular knowledge such as location data and special call records can be leaked into the system for it. Figure 4 shows the final of the modeling of the attack tree of the particular knowledge leak danger and shows that the attacks are possible with double weakness. For this reason Bluetooth is used for a Trojan horse and loads the mobile application onto an application or network that can be downloaded and sent to the driver. After driver connect via Bluetooth, the attacker uses the software to communicate with the car's infotainment system and access particular knowledge such as location information, vehicle identification and destination. When there is a Bluetooth weakness with automatic USB playback, a malicious software designed to activate the Bluetooth function can be added to an MP3, and USB with MP3 can cause the information of a driver to be stolen using a social engineering method. The drive activates USB to listen to MP3 s that enable Bluetooth information without the knowledge of the driver. The attacker controls the Bluetooth on the car with Bluetooth scan and receives the brute force attack and the PIN number. This PIN number can be used to obtain information about the driver.

4.2 *Risk Level Calculation*

R is defined for Risk Level. R is found by assessing the significance of assets and the dangers and weaknesses identified by the analysis procedure. That led to a R of increase of car speed, a major threat to intelligent automobiles. The risk levels calculated in this way are presented in Table 1.

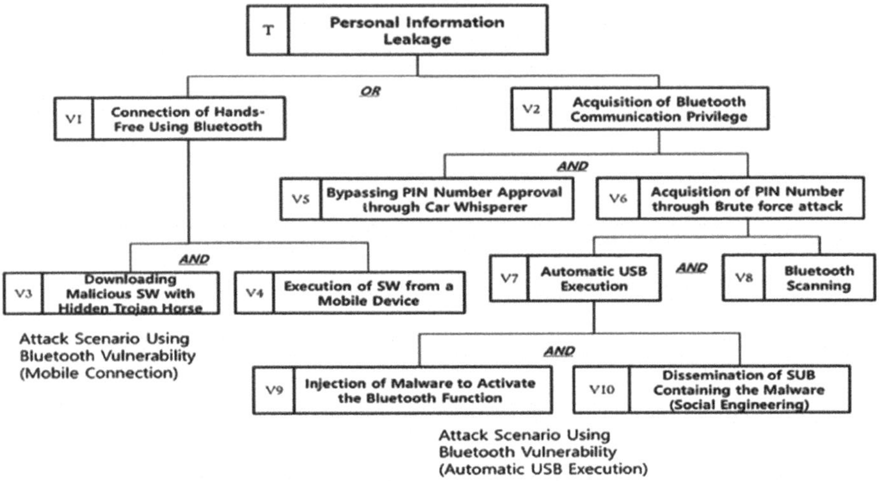

Fig. 4 Particular knowledge infiltration using Bluetooth open [23]

Table 1 For intelligent vehicle security risk [20]

Type			Value				
Asset (A)	Threat (T)	Vulnerability (V)	A	T	V	R	Rule
Powertrain	Increase of car velocity	Update for Fimware	S = 3	0.94	0.75	2.11	H
			P = 1				
			O = 2				
		Bluetooth	S = 3	0.91	0.515	1.44	M
			P = 1				
			O = 0				
		CAN	S = 3	0.93	0.687	1.94	M
			P = 1				
			O = 0				
Infotainment	Personal infor-mation leakage	Bluetooth (mobile connection)	S = 1	0.83	1	2.49	H
			P = 3				
			O = 1				
		Bluetooth (USB execution)	S = 1	86	0.687	1.77	M
			P = 3				
			O = 1				

The rating of security risk of a intelligent vehicle is presented in the framework of assessment of security risk as shown in Table 1. R was scored between 0.0 and 3.0. In the range of $1.0 < R \leq 2.0$, this means the medium risk level M values and if $0.0 < R \leq 1.0$, the low risk is categorized as L evaluation. Finally, if $2.0 < R \leq 3.0$, high risk H was considered. Estimates of assets, threats and weak points are multiplied, and then the increase in vehicle speed is rated as threat, and particular knowledge is graded on an H, L or M basis as the last calculation of the threat of leaking. It reflects the significance of being. In one case, the level of risk calculated to give the most high score. This reason for only reflecting the most high score is that the security dangers in the case of an automobile are handled from various angles; Security measures should focus on efficiency, with some of the highest score taking into account the much significant of each asset when calculating of security are established. For instance, the safety (S) direction received the most high score in the case of a power transmission system; this indicates that focusing on security should be on security (S) instead of privacy (P) or availability (O) [20].

The transponder system is related to the presence of a motor vehicle acceleration is due to the presence of three security vulnerabilities. In evaluating the importance of the entity, security (S) earned three points, privacy (P) one points and operational (O) two points. To assess the entity, only the highest rated security (S) is reflected. This means that safety measures must focus on safety, as it is an important vehicle presence that greatly influences the safety of the vehicle, rather than the power transmission organ's privacy [20].

In the infotainment system, the existence of particular knowledge leakage threat is due to two security reasons. In evaluating the importance of the entity, security (S) earned one point, privacy (P) three points and operational (O) one points. To assess an entity, only the highest score receiving privacy (P) is reflected; this means that the information system requires that the security measures for the IT containing the particular knowledge of the chauffeur focus on the confidentiality aspect [20].

5 Result

By using the information in this study, modeling of security attacks that take place in intelligent vehicles, modeling of two examples of attack types and modeling of attack trees and risk analysis have been expressed. It is aimed to perform related studies by modeling in other types of attack in future works that can be written by moving from here, or by adding new parameters to the attacking species in the attack types exemplified here.

Acknowledgements This work is also a part of the M.Sc. thesis titled Security Attacks Analysis For Smart Cars at Istanbul University, Institute of Physical Sciences.

References

1. http://www.forbes.com/sites/leoking/2015/02/23/14-year-old-hacks-connected-cars-with-pocket-money/. Accessed 19 July 2017
2. Anthony Foxx, Secretary, U S Department of Transportation and Mary Barra, the chairwomen and CEO of General Motors Company, stress the importance of these issues in a keynote talk at the Billington Cyber summit 2016
3. http://www.wired.com/2015/07/hackers-remotely-kill-jeep-highway/, 19.07.2017
4. http://www.heise.de/ct/artikel/Beemer-Open-Thyself-Security-vulnerabilities-in-BMW-s-ConnectedDrive-2540957.html. Accessed 19 July 2017
5. http://arstechnica.com/cars/2016/08/hackers-use-arduino-to-unlock-100-million-volkswagens/. Accessed 19 July 2017
6. Woodyard C, O'Donnell J (2013) Your car may be invading your privacy. USA Today. www.usatoday.com. Accessed 25 March 2013
7. http://www.businessinsider.com/smart-car-hacking-major-problem-for-iot-internet-of-things-2016-3. Accessed 19 July 2017
8. FAQs, Open Automotive Alliance. www.openautoalliance.net. Accessed 19 July 2017
9. Hands-Free Infotainment Isn't The Solution To Distracted Driving, Researchers Warn Webcast News Room (Feb 5, 2014) www.webcastnewsroom.com
10. Poulsen K (2010) Hacker disables more than 100 cars remotely. https://www.wired.com/2010/03/hacker-bricks-cars/
11. Bogage J (2016) Scary glitch affects luxury cars. https://www.bostonglobe.com/lifestyle/2016/06/09/scary-glitch-affects-luxury-cars/kj4wg2lhphlJDC3gATGuPM/story.html
12. Camek AG, Buckl C, Knoll A (2013) Future cars: Necessity for an adaptive and distributed multiple independent levels of security architecture. In: Proceedings of the 2Nd ACM international conference on high confidence networked systems, ser. HiCoNS '13
13. Rouf I, Miller R, Mustafa H, Taylor T, Oh S, Xu W, Gruteser M, Trappe W, Seskar I (2010) Security and privacy vulnerabilities of in-car wireless networks: A tire pressure monitoring system case study. In Proceedings of the 19th USENIX conference on security, ser. USENIX Security'10. USENIX Association, Berkeley, CA, USA
14. Izosimov V, Asvestopoulos A, Blomkvist O, Torngren M (2016) Security-aware development of cyber-physical systems illustrated with automotive case study. In: 2016 Design, automation & Test in Europe conference & exhibition, DATE 2016, Dresden, Germany
15. Guilbert G, Jack E, Karl R, Deerek W (2016) Explaining volkswagens emissions scandal. http://www.nytimes.com/interactive/2015/business/international/vw-diesel-emissions-scandal-explained.html
16. Olofsson J (2014) zombies ahead!a study of how hacked digital road signs destabilize the physical space of roadways. Vis Commun 13(1):75–93
17. Wasicek A, Andre W (2015) Recognizing manipulated electronic control units. In: SAE 2015 World congress & exhibition, April 2015
18. Verdult R, Garcia FD, Ege B (2015) Dismantling megamos crypto: Wirelessly lockpicking a vehicle immobilizer. In: Supplement to the 22nd USENIX security symposium (USENIX Security 13), 2015
19. Koscher K, Czeskis A, Roesner F, Patel S, Kohno T (2010) Experimental security analysis of a modern automobile. 2010 IEEE Symposium on Security and Privacy; 16–19 May. IEEE, Oakland, pp 447–462
20. EVITA (2009) Security Requirements for Automotive on-board Networks based on Dark-side Scenarios EVITA Deliverable D2. 3. EVIPA Project

21. Cho AR, Cho HJ, Son YD, Lee DH (2012) A message authentication and key distribution mechanism secure against CAN bus attack. J Korea Inst Info Sec Cryptol 22(5):1057–1068 (in Korean)
22. Miller C, Valasek C (2013) Adventures in automotive networks and control units.. http://illmatics.com/car_hacking.pdf. Accessed 13 Dec 2016
23. Kong H, Hong M, Kim T (2017) Security risk assessment framework for smart car using the attack tree analysis

A Survey on White Box Cryptography Model for Mobile Payment Systems

Öznur Şengel, Muhammed Ali Aydin and Ahmet Sertbaş

Abstract The technology is showing rapid development and these developments are changing our lives, our habits, and our needs. As electronic devices, which are indispensable for our daily lives, continue to be intelligent, we are able to do our every operation through these devices. Mobile payment technologies and services are one of the innovations. Consumers all over the world and in our country have started to use their mobile devices as a means of payment as well as communication services. With rapidly developing technology, one of the most important needs of many systems such as electronic, mobile and bank is to move and store the data safely. In addition to data security in electronic transactions, the speed of the system operations is becoming very important. Developing a mobile payment system whether by installing an application or using existing hardware, the most important issue in both cases is the creation of a reliable system based on the protection of the current situation of the consumer and the confidentiality of their information.

Keywords Mobile payment · Cryptography · Cryptography model · White box cryptography · Data security

1 Introduction

New applications and payment channels are emerging with the demands and needs of the payment systems of the Y and Z generations adopting communication with the Internet, E-mail and smartphones. Y and Z generations, which make up the

Ö. Şengel (✉)
Computer Engineering Department, İstanbul Kültür University, İstanbul, Turkey
e-mail: o.sengel@iku.edu.tr

M. A. Aydin · A. Sertbaş
Computer Engineering Department, İstanbul University, İstanbul, Turkey
e-mail: aydinali@istanbul.edu.tr

A. Sertbaş
e-mail: a.sertbas@istanbul.edu.tr

A. Boyaci et al. (eds.), *International Telecommunications Conference*, Lecture Notes in Electrical Engineering 504,
https://doi.org/10.1007/978-981-13-0408-8_18

vast majority of the population of Turkey, want to make purchases faster and easier than any other device, anytime, anywhere. In systems that are developed with these demands, we can now make payments through mobile phones, watches, glasses, body chips, skin that we have not separated from us at any moment. Today, in many countries, operations are carried out with naked nipples placed under the skin, especially opening the entrance doors. With the name of "Circet" we can realize all the operations that we made via our Android-based smartphone through our wristband, which is reflected in our wristband. We can make payment through application integrated with intelligent eyeglasses is named "intelligent eyeglasses payment" developed by BKMLab. All developments indicate that mobile payment applications will increase even more.

According to "Bankalararası Kart Merkezi (BKM)" as shown in Fig. 1, the number of card payments made via internet in Turkey increases rapidly every year and approximately 365 million transactions have been realized in 2016. According to the Card Monitor 2014 report, which is published by BKM, 49% of consumers will be preferred mobile phone, 44% of consumers will be preferred using credit card, 42% of consumers will be preferred using internet, 28% of consumers will be preferred using cash and 9% of consumers will be preferred [1]. The research shows that we make all our payment transactions, including small payments, by credit card. Developing technology and demands of consumers shows that even a cardholder will have a date to make a payment in the near future.

The widespread use of mobile payment systems has brought about the question of safely storing data in these applications. In particular, the use of payment systems in common Wi-Fi areas, the use of anonymous codes, the use of simple passwords and verification methods allow data to be captured by attackers. When payment systems

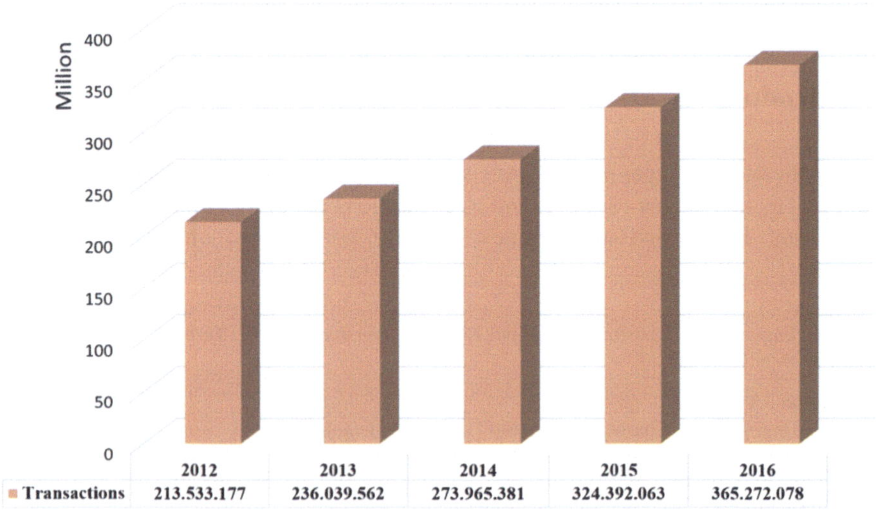

Fig. 1 Order by letter/phone and card payment transactions made on the internet [2]

conductive. A number of standards and limits for the two types have been set by different standardization bodies, for more details the reader may refer to [4, 5].

Power lines radiate EMI to the surrounding as communication signals travel along the wires and electronic equipment malfunction might occur [6]. It is known that the non-shielded electrical cables act as an effective transmission antenna at higher frequencies (>1MHz spectrum) utilized in PLC networks [7]. However, the interference caused by such signals is low in power, yet local consumers still have their own reason to complain. This local interference can be criticized by the amateur radio, who live or work near PLC systems [7]. They believe that sending data via power lines will interfere with the sensitive wireless systems being used.

As signals travel over power lines, they cause unacceptable risks in many areas of technology due to their radiation emissions. This interference should be controlled and kept at acceptable levels. Many studies on the topic of electromagnetic compatibility (EMC) have been reported in the literature, see e.g., [4, 6, 8]. In simple terms, EMC describes the capability of system devices to function in the surrounding electromagnetic environment with no error. Ensuring the reliability and safety of systems when they are used in electromagnetic environments is the main aim of EMC regulations. Practically, the electromagnetic disturbances caused by every single appliance on the network should be limited, also an appropriate level of immunity for each device to these disturbances must be considered [9].

Motivated by the above, in this paper, we will review and compare the common EMC regulations and standards. In addition, some measurements will be conducted in different indoor environments to measure electromagnetic emissions from two different PLC modems, unlike [4, 5] which considered only one environment and one modem. Experimental measurements are compared with the limits and regulations. Results show that the PLC modems under study mostly comply with the EMC standards but in some scenarios the measured radiated emissions were found to exceed the limits at some frequencies.

The rest of the paper is organized as follows. Section 2 presents and reviews the previous work done on this topic. Section 3 compares and discusses the common EMC standards. In Sects. 4 and 5, we discuss the measurement methods and analysis of the approaches adopted in our work. Results are presented in Sect. 6 and, finally, conclusions are drawn in Sect. 7.

2 Previous Work

A number of radiated and conducted emission measurements were carried out in previous years to find out whether the PLC links comply with EMC standards. Different frequency ranges and environments were investigated. Some of the outcomes of these studies were promising and showed that PLC systems comply with the standards while others showed that there is non compliance existed [4]. Other studies investigated the possibility of extending the frequency range of PLC systems such as [10].

are being developed different authentication schemes of security systems are used to keep all information of the user confidential. Just as in any system, authentication is performed in order to keep information securely in mobile payment systems.

Authentication is strengthened based on the importance of the information that needs to be kept confidential. The confidentiality of information is a secret knowledge that attackers should not acquire while the intention is not hidden for the person to whom it is communicated. The data that is important in the systems being built is not entirely confidential. The most important thing is to send important data to target without unaltered and untouched by attackers. In every system we use, we need authentication to ensure that information is kept secure. We need to authenticate while dealing with our debit cards, registering any web site, logging into our e-mail addresses as well as payment systems.

Different authentication methods are used, such as knowledge based, belonging based, and biometric information based. Knowledge-based authentication requires a password to access the system. It will prevent the information of the person not to share the password that he or she specifies with anyone, not to forget, not to be easily solvable. If the person specifies password that not to forget, not to be easily solvable, not to share with anyone, it will prevent the information of the person. Belonging based authentication uses its own identification card to access the system. The use of the card for entry into the system threatens the confidentiality of the information in cases such as being stolen, forgotten or lost. Biometric-based authentication is used by the person himself to login to the system.

The paper is organized as follows. Section 2 deals with how the mobile payment systems work and important points of mobile payment applications. Section 3 gives information about what is cryptography and types of cryptography. Section 4 describes white box cryptography and recent studies.

2 Mobile Payment Systems

Mobile payment is a service that allows you to make payments easily and quickly without the need for credit card information or cash over the application you use on our mobile phone. Any service or product that we purchase with our mobile phone is paid by the GSM bill or from the balance that is defined on our phone line. In this case, we can define the mobile payment as a payment system, in which all payment data and transaction are transmitted by the approved acknowledgment receipt accepted by the mobile device.

Mobile payment systems, which are increasing all over the world, are confronted in our country as systems used by domestic and foreign companies as Online Wallets, Mobile Wallets, and Short Messaging Service (SMS) Based Payment Systems [3]. Payment tools, which are usually renewed by banks, are now beginning to lead innovations offered by mobile operators in the case of mobile payment methods.

There are two methods for mobile payments: Proximity Payments and Mobile Remote Pay. If the mobile device has the required features, it is possible to make

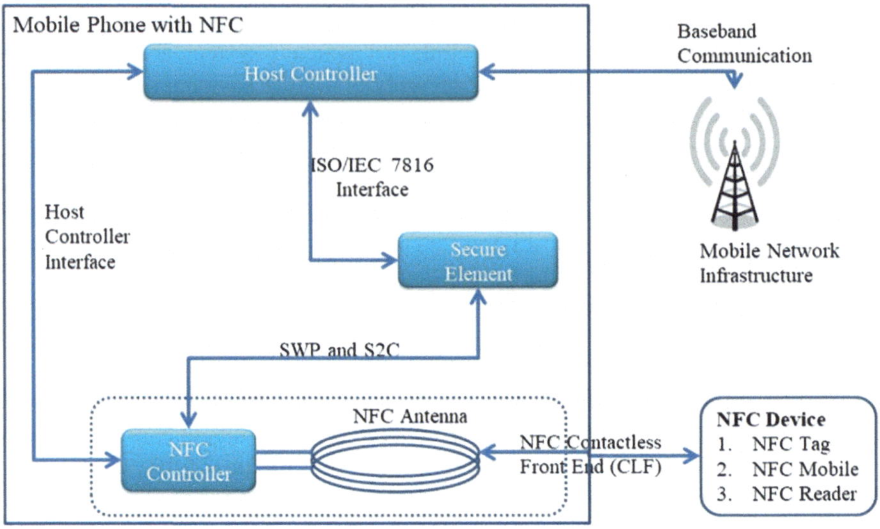

Fig. 2 Mobile phone with Near Field Communication (NFC)

payments in both types. For example, a payment system using Mobile Remote Pay while using the text message service on the mobile phone, another system application may request to be installed on the device. It is not necessary to use a secure element in the Mobile Remote Payment model. Because the system is configured according to the method used by the authentication payment service provider, the consumer (the paying party) authenticates directly from the payment server or uses the security features found on the SIM cards. There must be a Secure Element as well as a Near Field Communication Controller and interfaces that guarantee the secure operation of the application on Proximity Payments (see Fig. 2). Proximity Payments is software that fulfills the function of a payment card that can directly access Near Field Communication (NFC) and communicate with the transaction point.

Near Field Communication (NFC) Technology is a wireless technology that has been in the ISO RFID standard since 2003 and can be used for low-power data exchange, providing reliable access to short distance electronic devices and making reliable contactless transactions. When the customer wants to pay the payment through the mobile point of sale, he transmits the encrypted data by moving the mobile phone with the necessary hardware to the POS (Point of Sale) device. Payment is made after verifying the encrypted data transmitted.

RFID (Radio Frequency Identification) technology is a technology that uses radio waves to make mobile POS payments such as Near Field Communication Technology much longer and less secure than NFC technology in which RFID tags are read and transferred. Near Field Communication Technology is safer than RFID technology, but the security of communication between the buyer and the seller around the mobile payment center, such as secretly listening and watching the network to gain unfair earnings, gains importance.

Important points to consider in mobile payment services and systems are as follows.

- Protecting the confidentiality of data was given directly from the consumer or through the device for payment processing.
- Taking payment security precaution.
- Provide enough space on the handset to store data securely.
- Designing a secure interface that allows personal and payment agent information to be edited on the mobile device.
- Clearly reporting the cost of the system (the application to be used).
- Taking precautions for payment transactions that children can make.
- Determining the scope of automatic payment instructions within the applications used.
- Creation of an interface through which the user can specify payment options.

In mobile applications, different security parameters are used for login and payment transactions. Mobile banking systems where payment transactions are heavily used have begun to use various interfaces to increase security. New authentications systems have begun to be introduced to the mobile banking systems that are being used in our country. Many banks in the mobile systems using iris readers without requiring any password entry for customer's payment operations.

There are various authentications that are used in mobile applications such as One-time password [4–6], Voice-Based Authorization [7], Image-Based Authentication [8–12], Biometric Authentication [13–17].

3 Cryptography and Cryptography Types

The encryption process is the whole process of preventing modification of the data during transmission. If we look at traditional cryptographic logic, it consists of two parts, encryption and decryption. The computer A wants to send the data to the computer B over a secure channel that an attacker can listen the network. Data with a secret key, which are known from computers A and B, are encrypted by computer A with an encryption algorithm. Encrypted data is sent to computer B over a secure channel. The computer B obtains plain text from the cipher text by decrypting it in the decryption algorithm with the secret key. Encryption systems are divided into public key (asymmetric) and private key cryptography (symmetric) according to the key type used.

Two separate keys are used in public key cryptography (see Fig. 3): public key for encryption and private key for decryption. Everyone can know public key while the secret key can only be known in the person who decrypts it. Public key cryptography algorithms are used for digital signatures are used in the fields of authentication, information integrity and to securely identify and exchange the key to be used by

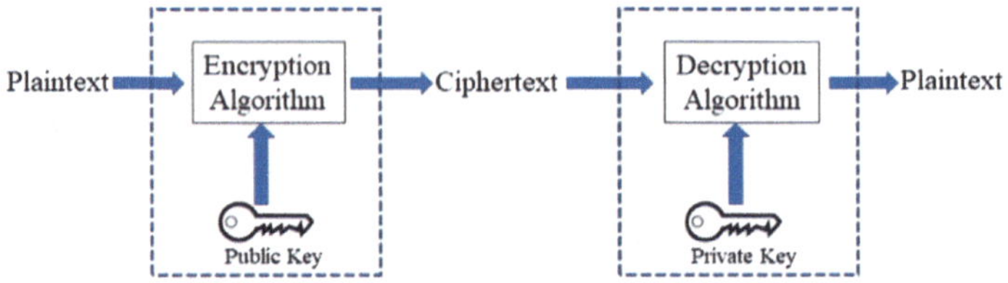

Fig. 3 Public key cryptography (asymmetric) system

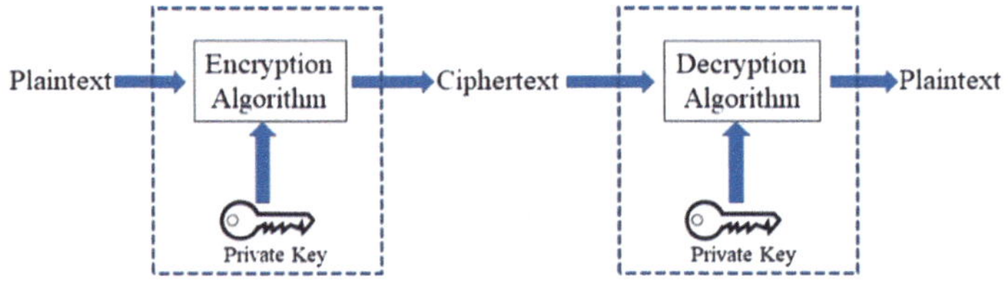

Fig. 4 Private key cryptography (symmetric) system

the two parties. Deffie-Helman (DH), Rivest-Shamir-Adleman (RSA), ElGamal and Paillier are well-known public key cryptography algorithms.

Private key cryptography (see Fig. 4) uses a single key for encryption and decryption. The sender also sends the agreed key along with the cipher text, so the key must be communicated securely to the destination. It is more secure than public key cryptography as long as the key is securely communicated to the other side. Caesar, Vigenere, Data Encryption Standard (DES), Triple DES (3DES), RC5, Blowfish, IDEA, SAFER, Advanced Encryption Standard (AES) are well-known private key cryptography algorithms.

In order to secure a system, the key must be stored securely. Attackers usually are intended to find the keys. The aim of the attackers is to change the application after finding the key with information from different cryptology models. There are three cryptography models; Black Box Model, Gray Box Model and White Box Model.

The Black Box Cryptography Model is the traditional encryption model that the attackers do not have physical access to any information in the system such as key or internal structure. This model just allows input and output information to obtain. As shown in Fig. 5a, the black box model provides access to only the plain text (input of the system) and the cipher text (output of the system), but does not have access to the encryption algorithm and other operable codes used by the system.

In the Gray Box Cryptography Model, a person who attacks the system has partial physical access through side channel information, as shown in Fig. 5b. It is possible to attack the system thanks to the information obtained from the physical application of the encryption system. The attacker knows side channel information such as power consumption, timing, fault analysis and uses this information to attack the system.

In the White Box Cryptography Model, the attackers have full control over the structure running on the system. The attackers can monitor executable dynamic codes

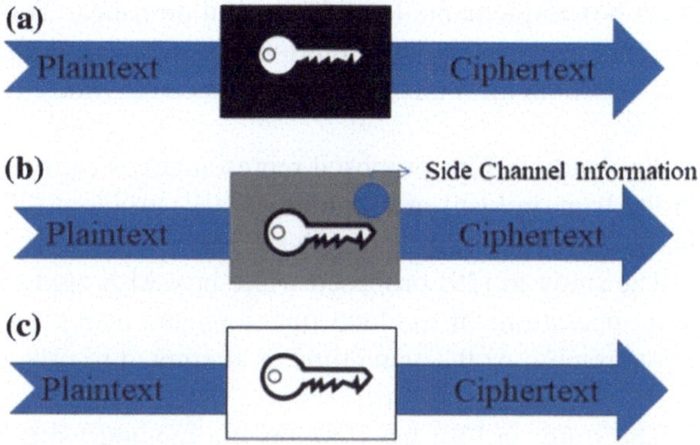

Fig. 5 Cryptology models **a** Black Box Cryptography Model, **b** Gray Box Cryptography Model, **c** White Box Cryptography Model

and important data, can access and edit the algorithm in the system. As shown in Fig. 5c, the system functions fully open.

4 White Box Cryptography (WBC)

White Box Cryptography (WBC) [18] provides a nonbreaking system that ensures security even if the attackers have full access on the system. It aims to keep application on secure software structures while providing security through trusted hardware with Trusted Execution Environment.

The white box cryptography technique aims to protect against the detection and renewal of the key used in the encryption algorithm of the software application even though the attackers have full control over the system.

A secure white box cryptography application aims to establish the system in a structure that prevents to disclose the cryptographic key despite all the details of the algorithm being known on an unreliable platform and the operation of the cryptographic algorithm being clear.

Initial work on white box cryptography started with Chow and friends' encryption policies on white box applications with Advanced Encryption Standard (AES) [19] and white box applications with Data Encryption Standard (DES) [20]. In both constructs, encryption is based on the basic strategy of allocating to loop functions, which is shown as a summary of lookup tables in small size.

The DES algorithms developed by Chow et al. was broken in the complexity of 2^{14} [21], the AES algorithms developed by Chow et al. was broken in the complexity of 2^{22} [22]. On the other hand, the CEJO structure for the design of the white box application using lookup tables has been used in many encryption schemes. Most of the researchers used the CEJO structure to develop and generate attack scenarios on it.

The study in [23] (known as BGE attacks) showed that the secret key can be modified in the complexity of 2^{30}. The study in [24] proposed an attack in the generic white box implementation of Substitution Linear Transformation (SLT) cryptography. The study in [25] has improved BGE attack in the best time complexity. Study [22], based on the study [25], and reduced the time complexity of the BGE attack by O (2^{22}).

The study in [26] proposed representatives equivalent to the same ciphers based on the Isomorphism of Polynomials (IP) problem [27]. But it was broken with the publication of the solution to the IP problem [28].

The study in [29] proposed white box AES application with changing the arithmetic operations of the loop functions and using the binary statement of the AES cryptography. At the same time, it attempted to protect white box AES application by using double password.

The study in [30] has developed a method using Intel's new AES Instructions without using the encryption key or the loop keys stored in memory directly in the decryption code. Thus, this structure is about 2.6 times faster than the lookup table based alternatives is formed. A faster structure has been achieved by using the optimization techniques and the mapped key schedule method.

The study in [31] has created "soft code" keys that translate the bit-string expression into a functional expression. They do this in 4 main steps. In the first step, the randomly generated key is divided into n parts with True Random Number Generator. Each piece is encrypted by passing it through a certain mathematical operation. The mathematical relationship determined in the previous step is held as a C_i instruction and each instruction is transformed into a function that produces the correct output taking the reference part as input. If the entry is incorrect, the original part value cannot be generated, so the entire key will not be accessible. Each piece is converted into successive functions and a software source code chain is created. In the last step, the application code is compiled with "soft-coded" keys and uses software encryption algorithms when necessary.

The study in [32] performs each cycle in three steps using the AES algorithm basically running in 10 cycles. In the first step, line shifting is performed, in the second step, XOR operations are performed, in the third step, adding the loop key, changing the byte, and mixing the columns in the AES algorithm are performed.

The study in [33] used China's block cipher standard the SMS4 algorithm which is used to protect wireless networks because they targeted low memory requirement and low computational cost. Using the $S(x) = I(x * A_1 + C_1) * (A_2 + C_2)$ formula when constructing the S-box of the algorithm. A_1 and A_2 are 8×8 matrices; C_1 and C_2 are 1×8 matrices and include 0 and 1 value. They encrypt the data using the SMS4 encryption algorithm over the resulting S-boxes. They indicated that the size of the lookup tables and matrices is 144.625 KB in their algorithm. By comparing them with other algorithms, they cover less space than the work had done up to that time. They show in comparison to other algorithms that the algorithm is in medium speed and there is no attack is known yet.

Some security notions for white-box cryptography have been studied in [34, 35]. Unbreakability; the first security objective of WBC is given a program to preserve the privacy of the key k embedded in the program. An adversary cannot extract the secret key of a randomly generated white-box implementation. One-Wayness should be difficult to invert the program computations and white-box implementation allows one to encrypt and not enable decryption. Incompressibility is that an adversary cannot produce a functionally equivalent program of significantly smaller size. Traceability; White-box enable to derive several functionality equivalent versions of the same encryption or decryption program. A user can transform its version of the program with the same functionality.

5 Conclusion

Studies show that systems are not only interested in making it safer but also are made to perform the operations quickly. It has become popular to develop the encryption algorithms used to increase the reliability against existing attacks. Studies show that the information that is important in the system is not completely confidential. The most important information for system must be stored securely and must be delivered unaltered and untouched by attackers. Secure storage of personal data should be intended to speed up the operation on the system.

Researchers aimed to create a new cryptographic solution model by adapting the widely used hardware with ensuring the confidentiality and security of storage, processing and flow of personal data.

In white box model, the attacker has full control of the targets execution environment, has complete access to the implementation algorithms with full privilege, can observe dynamic code execution and important data, can reach and alter detailed algorithms in the system. On the other hand, its execution time is less than the other models and application with WBC is faster than the others. If you want to develop secure payment system with white box cryptography model, you must consider some security notations such as onewayness, unbreakability, incompressibility, and traceability. It is understood that, the best cryptography model is white box cryptography to develop faster application.

Acknowledgements This work is also a part of the Ph.D. thesis titled Model Design and Performance Analysis for Secure Storage of Personel Data in Mobile Payment Systems at Istanbul University, Institute of Physical Sciences.

References

1. BAKM (BKM) (2014) Kredi Kartı Kullanım Alışkanlıkları Araştırması 2014, Kart Monitör. http://bkm.com.tr/wp-content/uploads/2015/06/kart_monitor_2014.pdf
2. BAKM (BKM) (31 Mart) Mektupla/ Telefonla Sipariş Ve İnternette Yapılan Kartlı Ödeme İşlemleri Raporu. http://bkm.com.tr/mektupla-telefonla-siparis-ve-e-ticaret-islemleri
3. (2013, 31 Mart). Türkiye'deki Ödeme Sistemlerinin Kırılımı: Alternatif Ödeme Sistemleri ve Detayları. http://www.odemesistemleri.org/
4. Lamport L (1981) Password authentication with insecure communication, (in English). Commun ACM 24(11):770–772

5. Gong LY, Pan JX, Liu BB, Zhao SM (2013) A novel one-time password mutual authentication scheme on sharing renewed finite random sub-passwords, (in English). J Comput Syst Sci 79(1):122–130
6. Huang Y, Huang Z, Zhao HR, Lai XJ (2013) A new one-time password method (in English). In: 2013 International conference on electronic engineering and computer science (EECS 2013), vol 4, pp 32–37
7. Khitrov M (2013) Talking passwords: voice biometrics for data access and security. Biom Technol Today 2013(2):9–11
8. Dhamija R, Perrig A (2000) Deja Vu: a user study using images for authentication (in English). In: Proceedings of the ninth usenix security symposium, Usenix Association, pp 45–58
9. Brostoff S, Sasse MA (2000) Are passfaces more usable than passwords? A field trial investigation (in English). In: People and computers Xiv - usability or else!, pp 405–424
10. Wiedenbeck S, Waters J, Birget J-C, Brodskiy A, Memon N (2005) Authentication using graphical passwords: Effects of tolerance and image choice. In: Proceedings of the 2005 symposium on usable privacy and security, 2005. ACM, pp 1–12
11. Jansen W (2003) Authenticating users on handheld devices. In: Proceedings of the Canadian information technology security symposium, 2003, pp 1–12
12. Jansen W (2004) Authenticating mobile device users through image selection. In: WIT transactions on information and communication technologies, vol 30
13. Vigila SAMC, Muneeswaran K, Antony WTBA (2015) Biometric security system over finite field for mobile applications (in English). IET Inf Secur 9(2):119–126
14. Dandawate YH, Inamdar SR (2015) Fusion based multimodal biometric cryptosystem (in English). In: 2015 International Conference on Industrial Instrumentation and Control (ICIC), pp 1484–1489
15. Faquih A, Kadam P, Saquib Z (2015) Cryptographic techniques for wireless sensor networks: a survey (in English). In: 2015 IEEE bombay section symposium (IBSS)
16. Shankar K, Eswaran P (2016) A new k out of n secret image sharing scheme in visual cryptography (in English). In: Proceedings of the 10th international conference on intelligent systems and control (ISCO'16)
17. Panchal G, Samanta D (2016) Comparable features and same cryptography key generation using biometric fingerprint image (in English). In: Proceedings of the 2016 IEEE 2nd international conference on advances in electrical & electronics, information, communication & bio informatics (IEEE AEEICB-2016), pp 691–695
18. Beunardeau M, Connolly A, Geraud R, Naccache D (2016) White-box cryptography: security in an insecure environment (in English). IEEE Secur Priv 14(5):88–92
19. Chow S, Eisen P, Johnson H, Van Orschot PC (2003) White-box cryptography and an AES implementation (in English). Sel Areas Cryptogr 2595:250–270
20. Chow S, Eisen P, Johnson H, van Oorschot PC (2002) A white-box DES implementation for DRM applications (in English). Digit Rights Manag 2696:1–15
21. Wyseur B, Michiels W, Gorissen P, Preneel B (2007) Cryptanalysis of white-box DES implementations with arbitrary external encodings. In: International workshop on selected areas in cryptography. Springer, Berlin, pp 264–277
22. Lepoint T, Rivain M, De Mulder Y, Roelse P, Preneel B (2014) Two attacks on a white-box AES implementation (in English). In: Selected areas in cryptography - Sac 2013, vol 8282, pp 265–285
23. Billet O, Gilbert H, Ech-Chatbi C (2005) Cryptanalysis of a white box AES implementation (in English). Sel Areas Cryptogr 3357:227–240
24. Michiels W, Gorissen P, Hollmann HDL (2009) Cryptanalysis of a generic class of white-box implementations (in English). Sel Areas Cryptogr 5381:414–428
25. Tolhuizen L (2012) Improved cryptanalysis of an AES implementation. In: Proceedings of the 33rd WIC symposium on information theory in the Benelux, Boekelo, The Netherlands, May 24–25, 2012, 2012: WIC (Werkgemeenschap voor Inform.-en Communicatietheorie)
26. Billet O, Gilbert H (2003) A traceable block cipher (in English). Adv Cryptol Asiacrypt 2894:331–346

27. Patarin J (1996) Hidden fields equations (HFE) and isomorphisms of polynomials (IP): two new families of asymmetric algorithms (in English). In: Advances in Cryptology - Eurocrypt '96, vol 1070, pp 33–48
28. Faugere JC, Perret L (2006) Polynomial equivalence problems: algorithmic and theoretical aspects (in English). In: Advances in cryptology - eurocrypt 2006, proceedings, vol 4004, pp 30–47
29. Karroumi M (2010) Protecting white-box AES with dual ciphers. In: ICISC, 2010, vol 6829. Springer, Berlin, pp 278–291
30. Gueron S (2013) White box AES using Intel's new AES instructions (in English). In: Proceedings of the 2013 10th international conference on information technology: new generations, pp 417–421
31. Azhar I, Ahmed N, Abbasi AG, Kiani A, Shibli A (2014) Keeping secret keys secret in open systems (in English). In: 2014 international conference on open source systems and technologies (ICOSST), pp 100–104
32. Luo R, Lai XJ, You R (2014) A new attempt of white-box AES implementation (in English). In: 2014 International conference on security, pattern analysis, and cybernetics (SPAC), pp 423–429
33. Shi Y, He ZJ (2014) A lightweight white-box symmetric encryption algorithm against node capture for WSNs (in English). In 2014 IEEE wireless communications and networking conference (WCNC), pp 3058–3063
34. Delerablee C, Lepoint T, Paillier P, Rivain M (2014) White-box security notions for symmetric encryption schemes (in English). In: Selected areas in cryptography - SAC 2013, vol 8282, pp 247–264
35. Saxena A, Wyseur B, Preneel B (2009) Towards security notions for white-box cryptography. In: Proceedings of Information security, pp 49–58

User Association for Enhancing Physical Layer Security in Heterogeneous Network

Haji M. Furqan, Guftaar Ahmad Sardar Sidhu and Huseyin Arslan

Abstract In this work, a fair user association scheme is proposed for heterogeneous networks aiming to enhance physical layer security (PLS). More specifically, the user association problem is formulated as a Nash bargaining problem (NBS) with the objective of maximizing sum secrecy rate. Firstly, we develop a two player bargaining scheme for two base stations (BSs) in order to bargain user association for enhancing secrecy rate and then it is extended to multi-player bargaining scheme using hungarian algorithm. Simulation results are presented for proposed scheme and compared with SINR based association to show the effectiveness of our algorithm.

Keywords Heterogeneous network · Physical layer security

1 Introduction

Heterogeneous networks (HetNets) are one of the most promising architecture for 5G and beyond. Due to the concurrent operation of different technologies and various kinds of base stations (BSs), the HetNets are able to achieve energy and spectrum efficient communications [1]. User association or cell association in Hetnets refers to as which base BS and UE should be associated with each other [1]. In [2], authors proposed maximum signal to interference and noise ratio (SINR) based user association scheme. In this scheme most of the users were connected to macro BS because of its high power, due to which small cell deployment becomes inefficient. To overcome the issues of [2], the received power at user's end was artificially increased by adding a bias, however, this approach have interference issue from marco-cells [3].

H. M. Furqan (✉) · H. Arslan
School of Engineering and Natural Sciences, Istanbul Medipol University, 34810 Istanbul, Turkey
e-mail: hamadni@st.medipol.edu.tr

G. A. S. Sidhu
COMSATS Institute of Information Technology, Islamabad, Pakistan

H. Arslan
Department of Electrical Engineering, University of South Florida, Tampa, FL 33620, USA

A. Boyaci et al. (eds.), *International Telecommunications Conference*, Lecture Notes in Electrical Engineering 504,
https://doi.org/10.1007/978-981-13-0408-8_19

In [4], authors proposed user association which was based on maximizing the sum rate of all users using convex optimization, however, fairness among different users could not be guaranteed with the obtained solution.

In order to fairly distribute the resources, Nash bargaining solution (NBS) is one of the most popular approach. In [5], authors proposed a fair resource allocation scheme using NBS for orthogonal frequency division multiple access (OFDMA). In [6], authors proposed an opportunistic user association for multi-service Hetnets. In [7], authors proposed a fair user association scheme for maximizing the data rate based on NBS.

For user association, different metrics have been adopted in the literature, for example, Spectrum efficiency, Energy efficiency, QoS, Fairness, etc.,[2]. However, few of the aforementioned works report user association for enhancing PLS. In traditional cellular networks PLS techniques were designed based on SINR based association [8]. However, such approaches cause the problem related to load unbalancing. This load unbalancing situation causes high power consumption and also degrades the performance, especially for the delay sensitive applications. Hence, load unbalancing should be taken into account while designing PLS techniques [8].

In [9], authors proposed PLS approach for two tier Hetnets based on design of secrecy beamforming. However, the study was not from the perspective of network analysis and they also didn't consider random spatial positions of users and Eve. In [10], the authors proposed an access threshold based secrecy user association. The threshold based algorithm associate the user with the BS that provides the maximum truncated average received signal power beyond a threshold. However, fairness was not discussed in [10].

Motivated by [7–10], in this work we formulate a user association problem for enhancing secrecy rate as a bargaining problem, in which different BSs compete for the serving users. In the first step two player bargaining algorithm, based on NBS, is developed for user association [6]. The two player bargaining algorithm is then extended to multi-player bargaining algorithm by using Hungarian algorithm.

The rest of the paper is organized as follows: The system model is presented in Sects. 2, 3 presents problem formulation, two player bargaining scheme for two BSs, and also the extension of two player bargaining scheme to multi-player bargaining scheme. Simulation results are presented in Sect. 3.4, and the paper is concluded in Sect. 3.5.

2 System Model and Preliminaries

In this paper, we consider a downlink transmission of a 2-tier Hetnet, where the BSs of tier 1 are modeled as macro BSs while the BSs of tier 2 are modeled as pico BSs as shown in Fig. 1. In this model all the BSs share same frequency band such that pico BS are deployed with much low transmit power than macro BS. The proposed solution can also be applicable to multi-tier Hetnets. In this work, we consider M BSs, where mth BS is denoted by BS_m. Without loss of generality, we assume BS_1 is

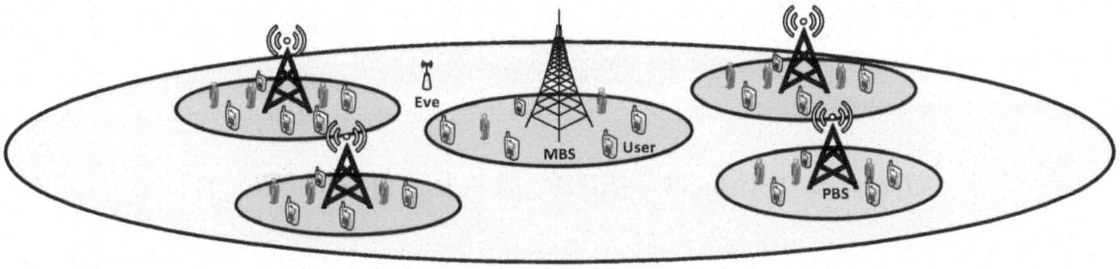

Fig. 1 Basic system model

a macro BS while $\{BS_2, BS_3, \ldots, BS_M\}$ are $M-1$ pico BSs, and there are N user equipments, denoted as UE_n, $(n \in 1, 2, 3, \ldots, N)$. We also consider an Eve that is trying to eavesdrops the communication between BSs and users. It should be noted that the Eve can be considered as an unauthorized user of network which is trying to eavesdrop the communication of authorize UEs and BSs.

To formulate the user association, any element of association matrix, X, is defined as:

$$x_{mn} = \begin{cases} 1, & \text{if } UE_n \text{ is associated with } BS_m, \\ 0, & \text{otherwise.} \end{cases} \quad (1)$$

The downlink SINR at UE_n associated with BS_m is given as:

$$\text{SINR}_{mn} = \frac{P_m g_{mn}}{\sigma^2 + \sum_{i \in 1,2,\ldots,M, i \neq m} P_i g_{in}}, \quad (2)$$

where g_{mn} is the average channel gain between UE_n and BS_m that includes shadowing and pathloss effects, P_m is the transmit power of BS_m and σ^2 is the estimated noise power level. It should be noted that the SINR for association is averaged over the association time and is approximately constant irrespective of dynamic channels.

The downlink SINR at Eve, E, that is trying to eavesdrop the downlink communication of B_m and UE_n is given as:

$$\text{SINR}_{me} = \frac{P_m g_{me}}{\sigma^2 + \sum_{i \in 1,2,\ldots,M, i \neq m} P_i g_{ie}}, \quad (3)$$

where g_{me} is the average channel gain between Eve and BS_m. The secrecy rate (SSR_{mn}) or the utility of user equipment UE_n associated to BS_m in the presence of Eve is given as:

$$SSR_{mn} = \frac{W}{\sum_{n=1}^{N} x_{mn}} \left[\log_2(1 + \text{SINR}_{mn}) - \log_2(1 + \text{SINR}_{me}) \right], \quad (4)$$

where W is the bandwidth. In a particular cell, each user is provided with an independent spectrum of bandwidth $\frac{1}{\sum_{n=1}^{N} x_{mn}}$. The Eq. (4) can be re-written as follows:

$$SSR_{mn} = \frac{1}{\sum_{n=1}^{N} x_{mn}}[SRR_{mn}] \tag{5}$$

$$SRR_{mn} = W[\log_2(1 + \text{SINR}_{mn}) - \log_2(1 + \text{SINR}_{me})] \tag{6}$$

3 Proposed NBS Based User Association

Nash barganing solution(NBS) is a type of cooperative game theory for solving problems related to resource allocation among competing players. The bargaining problem is explained as follow [5]: Assume that we have M players (BSs) that are competing for resources and the minimal payoff of each player m, ($m \in 1, 2, \ldots, M$), is $\mathbf{U}_m^{min}$, where $\mathbf{U}_m^{min} = (U_1^{min}, U_2^{min}, U_3^{min}, \ldots, U_M^{min})$. The set of feasible payoff that will be received by players, if they cooperate, is given by $\mathbf{U} = (U_1, U_2, U_3, \ldots U_M)$. From the concepts of NBS, all of the cooperating players will atleast get their minimum payoff such that $\{U_m \in \mathbf{U}|U_m \geq U_m^{min}, \forall m \in (1, 2, 3, \ldots, M)\}$ is a non empty set and (**U**, **Umin**) is called a M person bargaining problem. We can formulate Nash bargaining (NB) problem as follows:

$$U* = arg \max_{U} \prod_{m=1}^{M} (U_m - U_{min}), \tag{7}$$
$$\text{s.t.} \quad U_m \geq U_{min}.$$

According to [5, 6], there exists an optimal and unique NBS if U_m is upper bounded and unique.

3.1 Problem Formulation

The user association problem for enhancing physical layer security can be modelled as a NB problem among different BSs that are competing for users to connect to them. In other words, the BSs are the players in the bargaining problem and the payoff of the BS_m is U_m, where U_m is defined as follows:

$$U_m = \sum_{n=1}^{N} x_{mn} SSR_{mn} \tag{8}$$

where SSR_{mn} is the utility function of UE_n associated with BS_m as defined in Eq. (4). Putting the value of SSR_{mn} in (8) we get:

$$U_m = \sum_{n=1}^{N} x_{mn} \frac{W}{\sum_{n=1}^{N} x_{mn}} [\log_2(1 + \text{SINR}_{mn}) - \log_2(1 + \text{SINR}_{me})] \tag{9}$$

In this work the goal of optimization is to determine the user association that maximizes the NBS utility function. The optimization problem can be defined as:

$$\begin{aligned} \max_X U = & \prod_{m=1}^{M} (U_m - U_{min}) \\ \text{s.t.} \quad & U_m \geq U_{min} \forall m \\ & x_{mn} = \{0, 1\}, \forall m, n \\ & \sum_{m=0}^{M} x_{mn} = 1, \forall n \end{aligned} \tag{10}$$

To conclude, the bargaining problem of user association for enhancing secrecy rate is defined as follows: Each BS_m has payoff U_m, the optimization goal here is to find X to maximize all U_m simultaneously under the constraints of $U_m \geq U_m^{min}$. In order to find unique, fair and optimal X, the fast and effective method is crucial.

3.2 NBS Based Algorithm for Two-BS Case

In this sub-section, we present two player bargaining algorithm for two BSs ($M = 2$) and then we extend it for multiple BSs in the next sub-section. The two players (BS_1 and BS_2) are any arbitrary part of the network and can be modelled as two pico BSs or a macro BS and a pico. In this work, a simple two band partition is applied for user association. This approach is inspired from [11, 12]. The authors in [12] presented sub-carrier assignment algorithm by two band partition. The authors proved that two band-partition is near optimal for the optimization goal of weighted maximization of users' rate for two-user subcarrier assignment.

The basic algorithm for two-band partition is presented in Fig. 2. In this work the goal of two player bargaining is to maximize the NBS utility $\prod_{m=1}^{M}(U_m - U_{min})$. We relax the constraints $x_{mn} = \{0, 1\}$ to continuous values with $0 \leq x_{mn} \leq 1$. Based on (10), the lagrangian function is as follows:

$$L = \prod_{m=1}^{2} (U_m - U_{min}) + \sum_{n=1}^{N} \lambda_n \sum_{m=1}^{2} (x_{mn} - 1) \tag{11}$$

Step1: Initialization: Initialize the user association such that each BS_m gets U_m^{min} **Set** $U_{max}(0) = U = \prod_{m=1}^{M}(U_m - U_{min})$, **Calculate** $f(SSR_{1n}, SSR_{2n})$
Step2: Sorting of users **Sort** UE_n from largest to smallest based on $f(SSR_{1n}, SSR_{2n})$
Step3: For n=1, ... , N-1 Calculate U(n), where $UE_1, \dots \dots \dots UE_n$ is associated with BS_1 and $UE_{n+1}, \dots \dots \dots UE_N$ is associated with BS_2
Step4:: Choose two band partition $\acute{K} = \arg\max_n \mathbf{U(n)}$ **Set** $U_{max}(i) = U$ **(Ḱ)**
Step5: Update User association **If** $U_{max}(i) \le U_{max}(i-1)$, **Iteration ends** **Else,** Update $f(SSR_{1n}, SSR_{2n})$ based on new partition **Set i=i+1**

Fig. 2 Two band partition algorithm

where λ_n is the Lagrange multiplier. Equation (11) can be re-written as follows:

$$L = \left(\sum_{n=1}^{N} x_{1n} SSR_{1n} - U_{1min}\right)(U_2 - U_{2min}) + \sum_{n=1}^{N} \lambda_n \left\{(x_{1n} - 1) + (x_{2n} - 1)\right\} \tag{12}$$

By applying karush-kuhn-Tunker (KKT) [12] conditions, we have partial derivative of (12) with respect to x_{1n} as follows:

$$\frac{\partial L}{\partial x_{1n}} = \frac{\partial}{\partial x_{1n}}\left(\left(\sum_{n=1}^{N} x_{1n} SSR_{1n} - U_{1min}\right)(U_2 - U_{2min}) + \sum_{n=1}^{N} \lambda_n \{(x_{1n} - 1) + (x_{2n} - 1)\}\right) \tag{13}$$

$$\frac{\partial L}{\partial x_{1n}} = \left(SSR_{1n} + x_{1n}\frac{\partial SSR_{1n}}{\partial x_{1n}})(U_1 - U_{1min}\right) + \lambda_n \tag{14}$$

Similarly, the Partial derivative of (12) with respect to x_{2n} is given as follows:

$$\frac{\partial L}{\partial x_{2n}} = \left(SSR_{2n} + x_{2n}\frac{\partial SSR_{2n}}{\partial x_{2n}}\right)(U_2 - U_{2min}) + \lambda_n \tag{15}$$

Now, based on KKT, we can get:

$$\frac{(SSR_{1n} + x_{1n}\frac{\partial SSR_{1n}}{\partial x_{1n}})}{(U_1 - U_{1min})} = \frac{(SSR_{2n} + x_{2n}\frac{\partial SSR_{2n}}{\partial x_{2n}})}{(U_2 - U_{2min})} \tag{16}$$

From (16), we have:

$$\frac{\frac{SRR_{1n}}{\sum_{n=1}^{N} x_{1n}} - \frac{x_{1n} SRR_{1n}}{(\sum_{n=1}^{N} x_{1n})^2}}{(U_1 - U_{1min})} = \frac{\frac{SRR_{2n}}{\sum_{n=1}^{N} x_{2n}} - \frac{x_{2n} SRR_{2n}}{(\sum_{n=1}^{N} x_{2n})^2}}{(U_2 - U_{2min})} \tag{17}$$

$$\frac{SSR_{1n} - \frac{x_{1n} SSR_{1n}}{(\sum_{n=1}^{N} x_{1n})}}{(U_1 - U_{1min})} = \frac{SSR_{2n} - \frac{x_{2n} SSR_{2n}}{(\sum_{n=1}^{N} x_{2n})}}{(U_2 - U_{2min})} \tag{18}$$

$$\frac{\frac{(\sum_{n=1}^{N} x_{1n}) SSR_{1n} - x_{1n} SSR_{1n}}{(\sum_{n=1}^{N} x_{1n})}}{(U_1 - U_{1min})} = \frac{\frac{(\sum_{n=1}^{N} x_{2n}) SSR_{2n} - x_{2n} SSR_{2n}}{(\sum_{n=1}^{N} x_{2n})}}{(U_2 - U_{2min})} \tag{19}$$

The right and left side of Eq. (19) can be considered as a marginal benefits of UE_n for BS_1 and BS_2, respectively. It should be noted that when UE_n is associated with one BS, Eq. (19) becomes inequality, such that if the right side is greater than the left side the UE_n will be connected to BS_2 and vice versa with BS_1. Finally, we define $f(SSR_{1n}, SSR_{2n})$ as:

$$f(SSR_{1n}, SSR_{2n}) = \frac{\frac{(\sum_{n=1}^{N} x_{1n}) SSR_{1n} - x_{1n} SSR_{1n}}{(\sum_{n=1}^{N} x_{1n})}}{(U_1 - U_{1min})} - \frac{\frac{(\sum_{n=1}^{N} x_{2n}) SSR_{2n} - x_{2n} SSR_{2n}}{(\sum_{n=1}^{N} x_{2n})}}{(U_2 - U_{2min})}, \tag{20}$$

where we define $f(SSR_{1n}, SSR_{2n})$ as a difference of marginal benefits of UE_n for BS_1 and BS_2. Thus, the user association to BS_1 or BS_2 can be decided by checking the value of $f(SSR_{1n}, SSR_{2n})$, such that if the value of $f(SSR_{1n}, SSR_{2n})$ is greater than zero, the user should be associated with BS_1 but if the value is less than zero it should be associated to BS_2. It should be noted that when $f(SSR_{1n}, SSR_{2n}) = 0$ then the user will be associated to both BSs simultaneously due to continuous relaxation of x_{mn} [12]. Note that such user can be associated with any of these two BSs without affecting the performance because the number of users is much greater than the number of base stations. It is clear that $f(SSR_{1n}, SSR_{2n})$ is the monotonic function of n, such that, we sort the indices of UEs to make $f(SSR_{1n}, SSR_{2n})$ decrease in n, then the Eq. (20) will be similar to weighted maximization in [11]. Thus, the two band partition is near optimal to the optimization problem of (10). According to simulations the proposed algorithm converged within three to four rounds and have complexity of $O(N \log_2 N)$ with binary search.

Step1: Initialization: Initially associated all UE_s to BSs
Step2: Grouping Group the BSs into pairs (coalition) by a. Random method b. Hungarian method *If M is odd, create a dummy BS and no BS bargains with this dummy BS
Step3: Bargaining Bargain user association by algorithm in Table 1
Step4: Repetition Go to step 2. until there is no further improvement

Fig. 3 Multi-player barganing algorithm

3.3 NBS Algorithm for Multi-BS Case

In the literature the user association problems related to multiple BSs are mainly solved in a centralized manner, but such approaches have computational complexity which increases with the increase in M and N. Hence, in this work we are using two step iterative algorithm. In this approach, the base stations are first grouped into pairs and then two player bargaining algorithm is applied on each pair as explained in Fig. 2. The basic details of multi-player bargaining is presented in Fig. 3. The grouping of BSs can be solved efficiently by using Hungarian algorithm [13] because it is an assignment problem. The detailed assignment problem formulation is as follows:

Let's suppose that the benefits for the ith BS, that is barganing UEs with the jth BS is defined as b_{ij}, where b_{ij} is the element of matrix b and is defined as follows:

$$b_{ij} = \max_{i,j\in\{1,\ldots,M\}} \big(U(\widetilde{U}_i, \widetilde{U}_j) - U(\widehat{U}_i, \widehat{U}_j), 0\big), \tag{21}$$

where $\widehat{U}_i$ and $\widehat{U}_j$ are the payoff for BS_i and BS_j before barganing while $\widetilde{U}_i$ and $\widetilde{U}_j$ are the payoff of BS_i and BS_j after barganing. The barganing algorithm presented in Fig. 2 calculates $b_{ij}\forall i, j$. We define a matrix h_{ij} whose elements represents the pairing status between two BSs and is given as follows:

$$h_{ij} = \begin{cases} 1, & \text{if } UE_n \text{ is associated with } BS_n, \\ 0, & \text{otherwise.} \end{cases} \tag{22}$$

Then the assignment problem is about how to group the barganing pairs (base stations) such that over all benefits get maximized and it can be formulated as follows:

$$\max_{h} \sum_{i=1}^{M}\sum_{i=1}^{M} h_{ij}b_{ij}, \tag{23}$$

$$\text{s.t.} \sum_{i=1}^{M} h_{ij} = 1, \forall i, \sum_{j=1}^{M} h_{ij} = 1, \forall j, h_{ij} \in 0, 1, \forall i, j$$

The optimization problem in (23) is modified as $\min_h \sum_{i=1}^{M}\sum_{i=1}^{M} -h_{ij}b_{ij}$, in order to take into account the minimization goal of hungarian algorithm [13]. The Hungarian algorithm have complexity of $O(M^3)$. Hence, each round of multi-player barganing has complexity of $O(M^2 N \log_2 N + M^3)$. Based on simulations, it converges within six to seven rounds.

3.4 Simulation Results

This section presents the simulation results to show the effectiveness of our proposed algorithm. In this work, a downlink hetNet is simulated in which there is a central Macro BS and there are several symmetrically located pico BSs along a circle of radius 120m around marco base station.

The users are assumed to be randomly distributed in hetnet area. The path loss between pico base station and UE is give as $140.7 + 36.7 \log 10\mathrm{d}$ (km) while path loss between pico base station and UE is given as $128.1 + 37.6 \log 10\mathrm{d}$ (km) [14]. The basic parameters are presented in Table 1 that are based on [6, 14].

The performance of the proposed algorithm is compared with SINR based user association. Figure 4 shows the average sum secrecy rate versus number of users when 1 pico and 1 macro are simulated. In should be noted from the figure that the sum secrecy rate for proposed association scheme is much better than the SINR based association scheme. Also note that the sum secrecy rate increases with the increase in the number of users for both proposed and SINR based scheme because there is more diversity with the increase in the number of users.

Figure 5 presents user fairness for different number of pico base stations with fixed number of users. We use the widely accepted Jain's fairness index (JFI) [7] and is defined as follows: $FI = \frac{(\sum_{n=1}^{N} SSR_n)^2}{N\sum_{n=1}^{N} SSR_n}$, where SSR_n is the secrecy rate of UE_n when the user association is used and $n = \{1, 2, \ldots, N\}$. It should be noted that the higher value of J indicates more fairness between users. From the Fig. 5, it is clear that the proposed scheme is superior in terms of fairness as compared to the SINR based association. The proposed scheme not only provides better secrecy rate but also improves the fairness among the users which indicates that users can fully access the resources in the proposed NBS solution. Figure 6 shows the threshold versus number of iterations for the proposed NBS scheme convergence by using Hungarian algorithm and random method. The proposed algorithm with the Hungarian algorithm converges in 7–8 rounds while the proposed scheme using random method converges

Table 1 Simulation parameters

Parameter	Value
Bandwidth	10 MHz
Tx power of pico BS	30 dBm
Tx power of macro BS	46 dBm
Noise power	−174 dBm/Hz

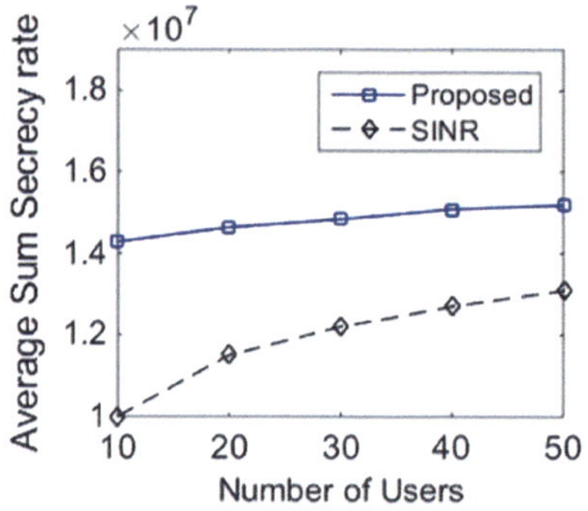

Fig. 4 Average sum secrecy rate of users

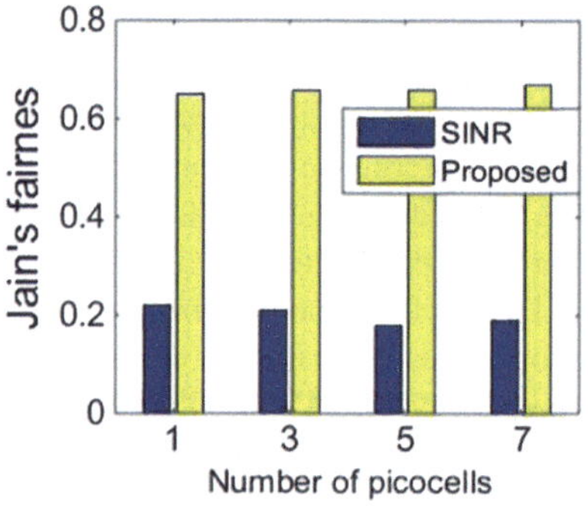

Fig. 5 The Jain's fairness index

much slower. This is due to the fact that the best bargaining pairs can be found out efficiently and fastly by using Hungarian method, which have lower computational cost and lower complexity.

3.5 Conclusion and Future Work

In this work, a cooperative game theory, based on NBS, is applied to develop a fair user association scheme to enhance the physical layer security of downlink hetnets. Firstly, a bargaining scheme for two BSs is presented. And, then scheme for two BSs is extended to multi-player bargaining scheme. In the multi player bargaining scheme, firstly, the BSs are grouped in to pairs by Hungarian algorithm and then in

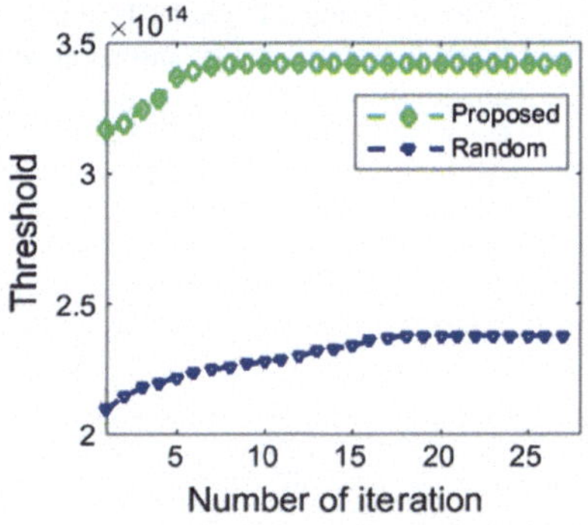

Fig. 6 Convergence

each pair two player bargaining scheme is applied. The two BSs case have complexity of $O(N \log_2 N)$ while multi player case have complexity of $O(M^2 N \log_2 N + M^3)$. Simulation results indicate that the proposed solution can enhance security while providing fairness among users.

Acknowledgements This work has been supported by The Scientific Research Council of Turkey (TUBITAK) under grant No. 114E244.

References

1. Damnjanovic JM, Wei Y, Ji T, Luo T, Vajapeyam M, Yoo T, Song O, Malladi D (2011) A survey on 3GPP heterogeneous networks. IEEE Trans Wirel Commun 8:10–21
2. Liu D, Wang L, Chen Y, Elkashlan M, Wong K, Schober R, Hanzo L (2016) User association in 5G networks: a survey and an outlook. IEEE Commun Surv Tutor 18:1018–1044
3. Ye Q, Rong B, Chen Y, Al-Shalash M, Caramanis C, Andrews JG (2013) User association for load balancing in heterogeneous cellular networks. IEEE Trans Wirel Commun 12:2706–2716
4. Corroy S, Falconetti L, Mathar R (2012) Dynamic cell association for downlink sum rate maximization in multi-cell heterogeneous networks. In: 2012 IEEE international conference on communications (ICC), Ottawa, ON, pp 2457–2461
5. Han Z, Ji Z, Liu K (2005) Fair multiuser channel allocation for OFDMA networks using Nash bargaining solutions and coalitions. IEEE Trans Commun 53(8):1366–1376
6. Liu D, Chen Y, Chai KK, Zhang T, Elkashlan M (2014) Opportunistic user association for multi-service hetnets using nash bargaining solution. IEEE Commun Lett 18(3):463–466
7. Liu D, Chen Y, Chai KK, Zhang T (2014) Nash bargaining solution based user association optimization in HetNets. In: IEEE 11th consumer communications and networking conference (CCNC). Las Vegas, pp 587–592
8. Bojkovic Z, Bakmaz B, Bakmaz M (2016) Security aspects in emerging wireless networks. Int J Commun 1:158–165
9. Lv T, Gao H, Yang S (2015) Secrecy transmit beamforming for heterogeneous networks. IEEE J Sel Areas Commun 33(6):1154–1170
10. Wang HM, Zheng TX, Yuan J, Towsley D, Lee MH (2016) Physical layer security in heterogeneous cellular networks. IEEE Trans Commun 64(3):1204–1219
11. Yu W, Cioffi JM (2002) FDMA capacity of gaussian multiple-access channels with ISI. IEEE Trans Commun 50(1):102–111

12. Ksairi N, Bianchi P, Ciblat P, Hachem W (2010) Resource allocation for downlink cellular OFDMA systemspart I: optimal allocation. IEEE Trans Signal Process 58(2):720–734
13. Papadimitriou CH, Steiglitz K (1998) Combinational optimization: algorithm and complexity. Dover Publications, New York
14. 3GPP (2010) Further advancements for E-UTRA physical layer aspects (tr 36.814)

Dynamic Feature Selection for Spam Detection in Twitter

M. Salih Karakaşlı, Muhammed Ali Aydin, Serhan Yarkan and Ali Boyaci

Abstract Social Networks continue to increase their popularity day by day. With the widespread availability of Internet access, interest of people in social networks has also increased significantly. The fact that, popularity of social media makes it tempting to use social media platforms for bad purposes. Malicious people are attempting to gain unfair profits by using fake accounts and various techniques. Among these initiatives, SPAM is one of the most frequently used methods. Today, SPAM attacks on social networks are increasing and many social network users are exposed to this and similar attacks. To identify SPAM users among billions of social network users, the examination of massive amounts of data requires a challenging large-scale data analysis. In this study, we group similar Twitter users and introduce a dynamic feature selection technique that use different features for each user groups instead of use static feature set and apply machine learning algorithms to classify spam users on Twitter.

Keywords: Social Media · Spam Detection · Feature Selection · Big Data

---

M. S. Karakaşlı · M. A. Aydin
Istanbul University, Istanbul, Turkey
e-mail: muhammed.karakasli@ist.ogr.edu.tr

M. A. Aydin
e-mail: aydinali@istanbul.edu.tr

S. Yarkan (✉)
Center for Applied Research on Informatics Technologies, Istanbul Commerce University, Istanbul, Turkey
e-mail: syarkan@ticaret.edu.tr

A. Boyaci
Istanbul Commerce University, Istanbul, Turkey
e-mail: aboyaci@ticaret.edu.tr

A. Boyaci et al. (eds.), *International Telecommunications Conference*,
Lecture Notes in Electrical Engineering 504,
https://doi.org/10.1007/978-981-13-0408-8_20

1 Introduction

Social networks have become most popular tools in web. Twitter, Instagram, Facebook and LinkedIn are most used social networks. Only Twitter has more than 320 million active users and total user count of top 10 social networks is more than 9 billion [1]. This popularity encourages attackers to use social network.

Social media does not have a trusted environment, so all users are potentially at risk. Even an ordinary person can easily access to the extremely important information about any user via social network [2]. Malicious people are using various techniques to obtain unfair profit by creating fake accounts on social media sites. In this direction, one of the most frequently used method is SPAM. SPAM, although expressing a general concept as meaning, is often associated with unsolicited e-mail [3]. However, SPAM is not just an e-mail restricted concept and it can be defined as unwanted content in the shortest way. Social media is also faced with the threat of spam like search engines and mails. Spammers often use social media sites to share harmful links, send unwanted messages, publish pornographic contents, illegal advertising, defraud the users [4]. Spammers in social networks have the intention of collecting personal information and attacking user profiles that they identify. However, attackers share SPAM content with malware links and expect users to install this software on their computers. Computers hacked in this way, can be used in subsequent spam attacks and DDOS-like network attacks [5]. The fact that users on the social network trust their friends' circles and think that harmful content will not come from them allows harmful content to spread easily [6].

Nowadays, SPAM attacks are increasing in social media, and many social media users are exposed to these and similar SPAM attacks. There is a need to develop effective systems for detecting SPAM accounts and SPAM contents so that social networks can be cleaned and users can have a better experience [8].

In this study, it has been shown that SPAM users operating in Twitter are not a uniform structure, so instead of using a static feature set, a method with dynamic feature selection ability for different users has been proposed. After users are clustered according to a set of properties, the set of sub properties that can be most effective for each cluster is dynamically determined. Finally, machine learning techniques were applied to the dataset using these identified features. A brief overview of different spam detection works and algorithms drawn in Sect. 2. Section 3 explains why we are focusing on this point. Section 4 describe the stages of our methodologies. Experimental results are described in Sect. 5. Concluding and what we want to do next are outlined in Sect. 6.

2 Related Works

The basic methods used to identify SPAM accounts in the social media are focused on content based and graphically based. Content-based methods are based on evaluating information such as the number of followers obtained through social media profiles, the number of people followed, and the sharing frequency through different machine learning techniques. Graphical methods are based on examining the user's relationship to his / her friends and the location on the network, rather than examining the user himself / herself.

URL analysis is a method that helps us to determine if the user account is a SPAM account by analyzing links in their posts. URL shortened services use hash functions to generate much simple and shorter links. This kind of services making it difficult to identify SPAM accounts. There are different ways to detect SPAM accounts using URL analysis methods. In a study [6] to identify harmful URLs shared by social users, two methods were used. These methods are domain name anomaly and social anomaly. While domain name anomaly analyzes the credibility of the keywords in the domain names, social anomaly analyzes the behavior of the users on the social platform. All gathered attributes for identifying harmful contents are divided into groups using Naive Bayes classifier. In another study [9], URL sharing behaviors on Twitter were put into two different perspectives. First one lets us analyze how these links are shared and the other one shows us how these posts were received by other users. In a different study [10] an event based approach is used to prevent low quality content in social media. At the same time shortened URLS on the Twitter were analyzed to identify statistics for SPAM links.

Meaning analysis lets us relate the posts in social media with its content and the tags using natural language processing methods. In this regard a study [11] was performed not only about social media but also web pages. With the help of Natural language processing methods, SPAM comments were tried to be detected using empty spaces, indents and non-ASCII characters. Where social accounts are analyzed, it is possible to get comprehensive personal information. These can contain information like their interests and hobbies. A study [12] tries to take advantage of these information to identify SPAM contents. The main point for this approach is possibility of being SPAM content can differ from user to user.

Feature analysis tries to use differences of social media users' information based on their profiles to isolate real accounts from SPAM accounts. Most important element for this analysis is the effects of attributes that are used to successfully differentiate real accounts and spam accounts. In this regard, the most comprehensive study [13] was performed by Benvenuto. Benvenuto tried to identify most suitable attributes to detect SPAM accounts. As a result, 10 attributes were selected. In another study [14], SPAM videos on YouTube were tried to be identified by using comment count, watch count, like/dislike ratio and videos' categories. Also in a similar study [15], tweeting frequency, follower and following counts and elapsed time since the account creation was analyzed to produce results. According to these results, elapsed time since registration and follower ratings produce similar results, while follower and following counts and tweeting frequency values are much higher for SPAM accounts. Similarly, to analyze SPAM attacks' effectiveness on Top Trends on Twitter, a study [16] was performed. All tweets that target the top trends were collected for 7 days using Twitter API. From 9 million collected tweets, 1500 tweets were selected randomly to train Naive Bayes classifier. These 1500 tweets were analyzed one by one to identify if it is SPAM or not. As a result, 2.8 percent of the tweets that targets top trend found to be a SPAM.

Network location analysis method is one of the most used Graphical-based analysis methods to detect SPAMs. It is used to detect SPAMS for not only social media, but also for, SMS [17] and e-mail services [18]. In a study [19] performed by Bougessa, social network users are likened to nodes, and messaging and sharing between them are likened to node links. Through this definition, it has been tried to determine whether the users are SPAM. To achieve this results, a prestige value was assigned to each user. Hybrid method were used to identify prestige values that were assigned to users.

Neighbor attribute analysis targets SPAM account's network. It analyzes specific attributes of connected users. This methodology bases its results on differences between real user's network and spam user's network.

3 Motivation

We can consider social media as a place that lets users freely share their opinion. Everyone has their unique thoughts and feelings; these differences are expressed in their actions. Just like in real life, these differences are observed in social media platforms where everyone shares their opinions. Just like how we can classify people as unsociable, calm, shy or talkative, bold, brave from their actions in daily life, we can also do similar classifications between social media users. We can't expect personal traits of a social media user to be consistent with real life. People can easily hide their real identity in a social media platform, this allows them to act differently than how they do in real life. But this doesn't change the fact that we can classify social media users by some predefined set of classifications.

Classifying people into categories can also be applied to fake users. Countless fake accounts are created to achieve different goals. These fake accounts need to have specific personas to achieve their goals. There are all kind of fake users: trying to profit from possessing private information or distributing malware by sharing malicious links, getting popular with adult content, illegal ads, marketing products and increasing popularity of people on social media.

Every fake account needs to go through different phases and embrace different characters in each phase before it becomes usable. After a fake account is created it needs to spread and interact with as much account as possible. After this stage is completed, the account goes into the active working phase where it makes movements towards its goal. Active working phase is repeated over some intervals and in between lies the resting phase. In the resting phase, this account needs to act similar to a

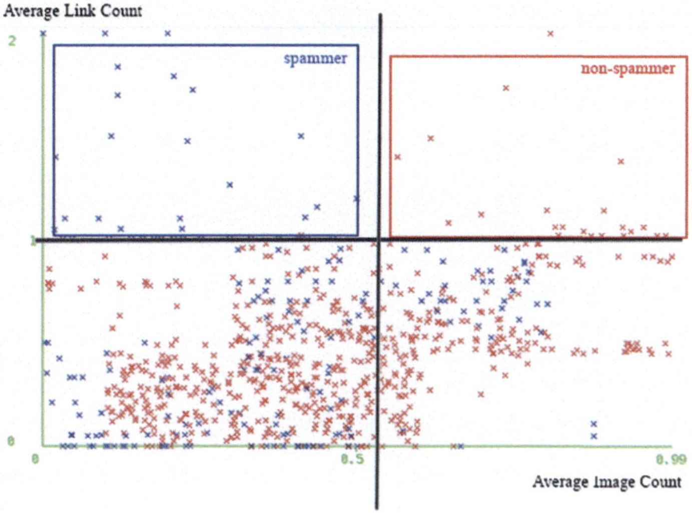

Fig. 1 Characterize of features

real user as much as possible to avoid automated security systems that detect fake accounts.

In this study, we practiced machine learning techniques to detect SPAM users. Features used in machine learning processes are selected dynamically unlike typical studies in this field. In classic methods, most effective features that will be used in machine learning techniques are selected by some kind of preprocessing and applied to the whole dataset. Machine learning techniques are better leveraged by dynamically selecting most effective features specific to an individual user instead of using same features for the whole dataset. It's because the distinguishing effect of every feature is different for each user groups. It's observed that although a feature has a strong distinguishing effect on one user group doesn't have the same effect for other groups. In a study about link sharing rates, it's observed that news accounts and SPAM accounts typically have above average link sharing. When we analyze users that have above average link sharing we observed that SPAM users usually use more images than news sites. If we filter by only image sharing count we can't find SPAM users but if we apply this filter to users that shares more links than other users, image sharing filter becomes more distinctive. This study concludes that every feature has different effects for individual user groups.

Above studies concludes that first, we need to classify each user with predefined parameters and then apply machine learning techniques for each user group. In this study, we use machine learning techniques with the most effective features selected for each user group to spot SPAM users (Fig. 1).

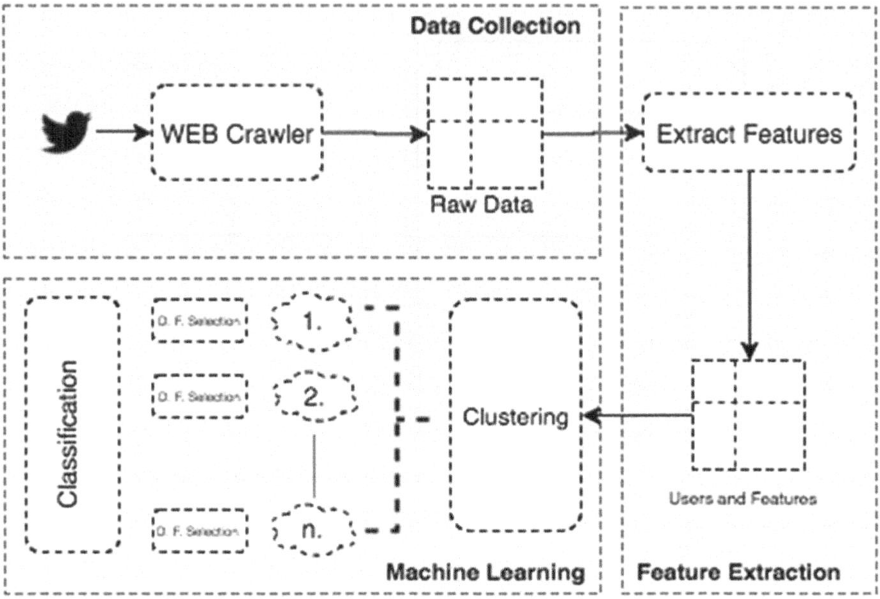

Fig. 2 Flow structure

4 Methodology

4.1 Architecture of Method

The graphic diagram of the proposed method is shown in Fig. 2. As can be seen in diagram, implementation phases of the method basically can be studied in three separate groups.

The first stage is the data collection phase, which allows the collection of the raw data required for the work to be performed. Following the data collection phase is the feature extraction phase, which allows the properties to be obtained from the raw data for use in machine learning techniques. The final stage is the machine learning phase, which is the process of deciding whether the user is SPAM (Fig. 3).

4.2 Data Collection

A JAVA-based CRAWLER software has been developed to collect user information via Twitter. Thanks to this software, user data can be collected without being tied to the API provided by Twitter. The restrictions that the Twitter API imposes on the user are not suitable for use because it prevents data collection intensively. Thanks

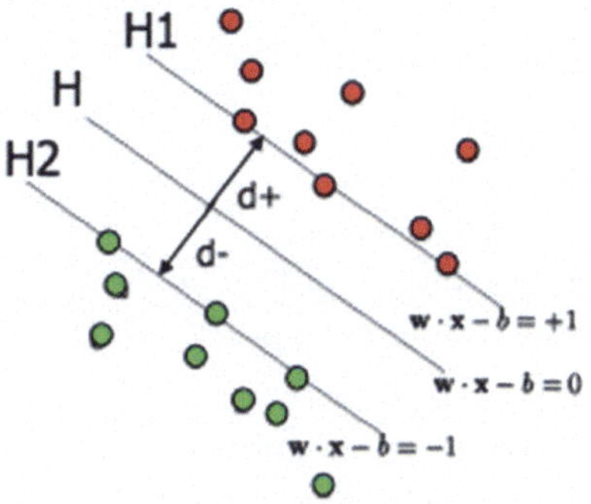

Fig. 3 An example of a two-class problem with SVM

to the software we have developed, data can be collected continuously without being subject to any restrictions.

Before the data collection process starts, users were randomly selected from among users who shared about Twitter's agenda items. The profiles of the selected users were examined one by one to determine whether the users were SPAM. These operations were performed manually by more than one person, and all participants were selected for inclusion in the user list.

This user list has been used as the input of the CRAWLER software. The CRAWLER software collects and records the raw data of the user and the user's shares from the profile pages of all the users in the user list.

In the Data Collection phase, a total of 823 twitter user data were collected. 649 of these users belong to non-spam real users while the remaining 174 belong to spam users. The spam user rate in the entire dataset was set at 21%.

4.3 Feature Extraction

Once the raw user data has been obtained, these data must be extracted from the data in order to be used in machine learning techniques. Property pool was created by making calculations on raw user data. The features found in the property pool are shown in Table 1.

While the features are accommodated in the property pool have been determined, attention has been paid to the most commonly used features in the literature, which do not require very high costs. In this way, it has been tried to establish a decision mechanism with high accuracy by keeping the necessary time and resources required for collecting and extracting features at a minimum level (Table 2).

Table 1 Feature pool

User based features	Content based features
User age	Link count for per tweet
Total tweet count	Average hashtag count
Total followers count	Average mention count
Total followings count	Average favorite count
Tweet count for per age	Average retweet count
Follower count for per age	Retweeted rate
Followings count for per age	Followers count for per tweet
Total common followers count	Average spam keyword
Total common followings count	Average time difference
Average common followers count	Average image count
Average common followings count	

Table 2 Experimental results with dynamic feature selection

Groups	KNN		SVM	
	Correct	Incorrect	Correct	Incorrect
1. Group	271	18	267	22
2. Group	319	8	312	15
3. Group	193	14	192	15
General	783	40	771	52

Table 3 Experimental results without dynamic feature selection

KNN		SVM	
Correct	Incorrect	Correct	Incorrect
721	102	693	130

4.4 Clustering and Dynamic Feature Selection

After extracting features from the raw user data, the X-Means algorithm was applied using the most basic user properties such as User Age, Tweet Count, Followers Count and Followings Count to create groups of similar users. The X-Means algorithm is derived from the K-Means algorithm and is developed to determine the number of sets to be automatically generated according to the Bayesian Information Criteria value [20]. The distribution of the clustered users with the X-Means algorithm is as follows (Table 3).

After grouping similar users with the K-Means algorithm, the best features were determined for each user group. The most important point that distinguishes my thesis from other similar activities is that it uses dynamically different properties rather than a static set of features. By specifying fewer and more effective features for each user group;

- Acceleration of the classification algorithm,
- Reduction of model complexity,
- Increase the accuracy rate,
- Reduce the likelihood of over fitting are provided.

4.5 Machine Learning

The machine learning phase is the last phase in which the target user is made a decision as to whether it is a SPAM account. In this phase, users who are already grouped are tried to be classified with various classification algorithms together with dynamically determined properties for the group they are belonging to. In this study k-NN, SVM classifiers are used, the results are compared with each other for two investigations.

4.5.1 k-Nearest Neighbors

One of the most popular machine learning algorithms, k-Nearest Neighbors, is used extensively for classification operations. It can also be used in regression analysis. The kNN algorithm is basically based on the calculating the distance of an unclassified data to the classified data. After the distances of the data to each other are calculated in the feature space, the unclassified data is assigned to the most appropriate class. The simplified steps of the algorithm are as follows;

- The number of neighbors to be taken into consideration (k) is determined by the user.
- The distances of each point in the training data are calculated with the unclassified data.
- The nearest k neighbors are determined considering this distances.
- The class with the largest number of matches is selected.

4.5.2 Support Vector Machine (SVM)

One of the very effective and simple methods used for classification is SVM. Mathematical formulas originally designed for the classification of two-class linear data have been generalized to classify multi-class and non-linear data. The general working principle of SVM is to estimate the most appropriate decision function that separates classes from each other. In order to do this, it is necessary to define the hyperplane that separates the classes from each other.

More than one hyperplane can be drawn for separating the two classes. However, the SVM's goal is to find the hyperplane that maximizes the distance between its nearest points. In a classification problem with two classes, we assume that the training data consisting of N samples is *{x_i,y_i} for i = 1, 2,3, .., N* with $x_i \in R^d$ d-dimensional space, $y_i \in$ *{-1, +1} label of classes.*

$$g(x) = \begin{cases} x_i w + b \geq +1 \\ x_i w + b \leq -1 \end{cases} \tag{1}$$

where w is the weight vector, b is a constant.

To maximize the spacing between the H1 and H2 hyperplanes, value of || w || needs to be minimized. As a result, the decision function for two class problems is obtained as follows.

$$\frac{1}{2} w^T w + C \sum_{i=1}^{N} \xi_i \tag{2}$$

subject to constraints;

$$y_i \left(w^T \varphi(x_i) + b \right) \geq 1 - \xi_i \text{ and } \xi_i \geq 0, i = 1, \ldots, N \tag{3}$$

where C is the capacity constant. SVM also requires the users to define the kernel function. There are number of kernels that can be used in SVM models such as linear, quadratic, radiak basis, polynomial.

5 Evaluation

To evaluate the method that we have presented, we compared the results obtained with using dynamic feature selection and without dynamic feature selection for same dataset.

When the users were grouped, features with the most distinguishing characteristics were identified. And features have been dynamically selected for each user group.

We used 4 feature to group users;

1. User Age
2. Tweet Count
3. Followings Count
4. Followers Count

For this study, all users are grouped into 3 separate clusters, and the features specified for each cluster are as follows;

The results obtained for each group according to two different machine learning algorithms as follows.

1. Group	2. Group	3. Group
Average Hashtag Count Average Link Count Average Retweet Count Average Common Followers Average Spam Keywords	Followings Count Average Tweet Count for Per User Age Average Followers Count for Per User Age Average Retweet Count Tweet Count	Average Common Followings Average Image Count Average Link Count Followers Count For Per Tweet Average Time Difference

When the KNN algorithm was used, the success rate for the first group was 93.7%, the success rate for the second group was 97.5%, and the success rate for the third group was 93.2%. Overall success rate was 95.1%. When the SVM algorithm was used, the success rate for the first group was 92.3%, the success rate for the second group was 95.4%, and the success rate for the third group was 92.7%. Overall success rate was 93.6%.

According to the obtained results, the KNN algorithm gave slightly better results to SVM. The following results were obtained when the same dataset was not divided into groups and the dynamic properties were not determined and again classified by KNN and SVM algorithms.

According to these results, the overall success rate was 87.6% for KNN algorithm and 82.9% for SVM algorithm.

With the method that we have set out, the successful forecast rate has been increased by approximately 8 points. These positive results have proved that dynamic property selection can be used for spam detection in social media.

6 Conclusions and Future Work

In this study, in order to determine the SPAM accounts in Twitter, special dynamic features were determined for each user groups which grouping according to the similarity of characteristics and classification were performed by feeding machine learning algorithms with these features. Due to the selection of dynamic features in the results obtained, the accuracy rate has increased by 8-10% compared to other studies. With the reduction in size of the feature set, performance enhancement has been achieved for future work on intensive data, accuracy has been increased, model complexity has been reduced, and the possibility of over fitting has been reduced.

Further studies are planned to increase the amount of data and to utilize large data technologies to establish a spam detection mechanism much closer to the truth.

Acknowledgement This work is also a part of the M.Sc. thesis titled Big Data Analysis in Social Media at Istanbul University, Department of Computer Engineering.

References

1. https://www.statista.com/statistics/272014/global-social-networks-ranked-by-number-of-users/
2. Kandasamy KT, Koroth P (2014) An integrated approach to spam classification on Twitter using URL analysis. In: IEEE students' conference on electrical, electronics and computer science
3. Chaitanya KT, Ponnapalli H, Herts D, Pablo J (2012) Analysis and detection of modern spam techniques on social networking sites. In: Third international conference on services in emerging markets
4. Abu-Nimeh S, Chen T, Alzubi O (2011) Malicious and spam posts in online social networks. IEEE Comput Soc 44(9)
5. Amit AA, Reddy N, Yadav S, Gu G, Yang C (2013) CATS: characterizing automation of Twitter spammers, communication systems and networks (COMSNETS). Fifth Int Conf 2013:1–10
6. Chen C, Guan DJ, Su Q (2015) Feature set identification for detecting suspicious URLs using Bayesian classification in social networks. Inf Sci 289:133–147
7. Twitter, 2015, Twitter Kullanımı ve Şirket Verileri, https://about.twitter.com/tr/company, [Ziyaret Tarihi: 27 Ekim 2015].
8. Wank K, Wang Y, Li H, Zhang X (2011). A new approach for detecting spam microblogs based on text and user's social network features. In: Proceedings of the VLDB endowment, vol 4, No 12. Seattle, Washington
9. Cao C, Caverlee J (2014) Behavioral detection of spam URL sharing: posting patterns versus click patterns. In: International conference on advances in social networks analysis and mining
10. Wang D (2014) Analysis and detection of low quality information in social networks, PhD symposium of 30th IEEE international conference on data engineering (ICDE 2014). Chicago, IL, United States
11. Radulescu C, Dinsoreanu M, Potolea R (2014) Identification of spam comments using natural language processing techniques. In: 2014 IEEE 10th international conference on intelligent computer communication and processing
12. Shen H (2014) Leveraging social networks for effective spam filtering. IEEE Trans Comput 63(11)
13. Fabricio B, Magno G, Rodrigues T, Almeida V (2010) Detecting spammers on Twitter, collaboration, electronic messaging, anti- abuse and spam conference (CEAS), vol 6. National Academy Press
14. Rashhid C, Nuriddin M, Mahmud GAN, Rashedur M (2013) A data mining based spam detection system for YouTube. In: Eighth international conference on digital information management, pp. 373–378
15. Sarita Y, Daniel R, Grant S, Danah B (2010) Detecting spam in a twitter network. Microsoft Res First Monday, 15(1)
16. Stafford G, Louis LY (2013) An evaluation of the effect of spam on twitter trending topics. IEEE, New York
17. Zhao Y, Zhaoxiang Z, Yungonh W, Liu J (2012) Robust mobile spamming detection via graph patterns. In: 21st international conference on pattern recognition.
18. Boykin O, Roychowdhury VP (2005) Leveraging social networks to fight spam. Computer (Impact Factor: 1.44) 38:61–68.
19. Mohammed B (2011) An unsupervised approach for identifying spammers in social networks. In: 23rd IEEE international conference on tools with artificial intelligence.
20. Pelleg D, Moore A (2000) X-means: extending K-means with efficient estimation of the number of clusters. ICML.

Phishing Analysis of Websites Using Classification Techniques

Doğukan Aksu, Zeynep Turgut, Serpil Üstebay and Muhammed Ali Aydin

Abstract In today's world, where all records are carried into an electronic environment, cyber security represents a very broad scope, with the primary objective of preventing the loss of financial and/or emotional loss of people, institutions, organizations through the security of data in the digital environment. Today, the most common cyber security threat is phishing attacks. With the phishing attack, the attacker aims to capture the data which are very important for the individuals like identification number, social security number, bank account information, and so on. In this study, using deep learning, it was checked whether the web sites are real or not by using neural networks and support vector machine, decision tree and stacked autoencoders as classification methods. As a result of the study, 86% success rate was reached by using stacked autoencoders which are a part of deep learning techniques.

Keywords Phishing · Stacked autoencoders · Classification · Support vector machine · Decision trees

1 Introduction

With the acceleration of technology, the internet has become an integral part of everyday life. Thanks to the use of Internet technologies, many transactions can be easily done, and many transactions are carried out over the internet. According to

D. Aksu · S. Üstebay · M. A. Aydin (✉)
Computer Engineering Department, Istanbul Kültür University, Istanbul, Turkey
e-mail: aydinali@istanbul.edu.tr

D. Aksu
e-mail: d.aksu@iku.edu.tr

S. Üstebay
e-mail: serpil.ustebay@istanbul.edu.tr

Z. Turgut
Computer Engineering Department, Halic University, Istanbul, Turkey
e-mail: zeynepturgut@gmail.com

A. Boyaci et al. (eds.), *International Telecommunications Conference*,
Lecture Notes in Electrical Engineering 504,
https://doi.org/10.1007/978-981-13-0408-8_21

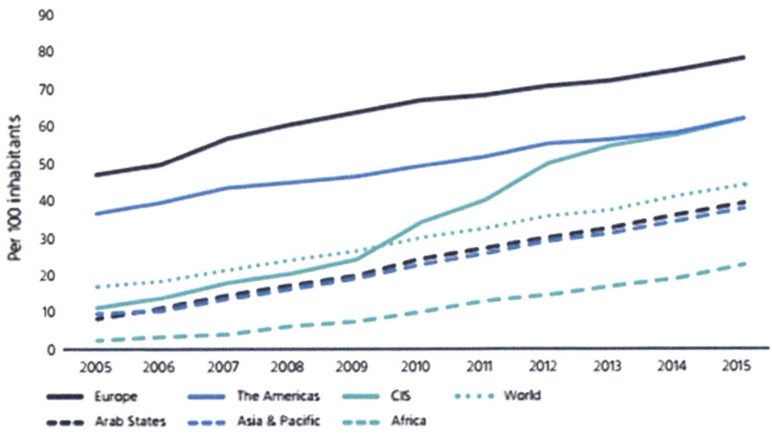

Fig. 1 Internet access rate of continents

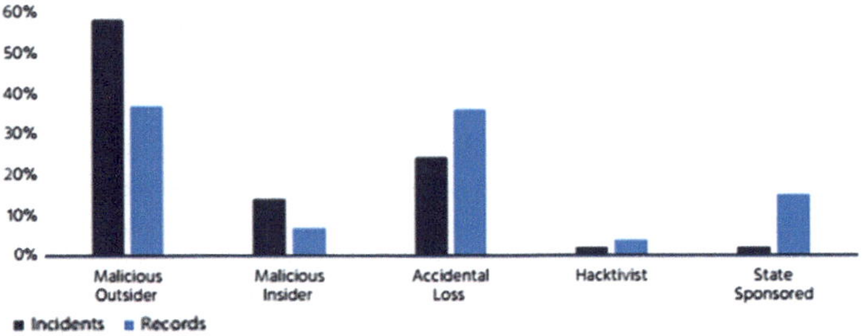

Fig. 2 Sources of security problems on the internet

the report published by International Telecommunication Union (ITU) in 2016 [1], the internet access rates of continents is shown in Fig. 1.

As the internet access rates of the people around the world are increasing, security of the internet has become a priority and the cyber security issue has become popular. Again according to the same report, Fig. 2 [1] shows the sources of security problems on the internet.

Accordingly, it appears that the greatest danger comes from external sources. However, the preventions taken by the users in order to avoid these malicious are shown in Fig. 3 [1].

Phishing attacks are the process by which an attacker seizes information from a user through a fake website similar to the actual website. The user is redirected to the fraudulent website where personal information is requested. If user is not been suspicious of the site and if the user enters his/her information with confidence, the information is compromised by the attacker. The attacker then uses this information for their own personal purposes. Therefore, protection from phishing attacks and awareness of counterfeit web sites will prevent many material and moral losses.

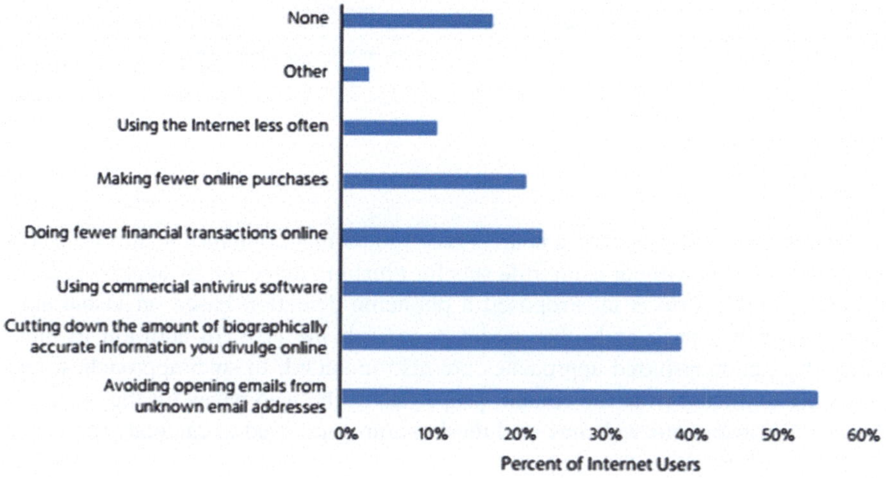

Fig. 3 Preventions taken by the internet users

In this study, fake web sites were identified using deep learning, which is one of the machine learning methods and continues to grow in popularity. In order to detect fraudulent web sites, URLs belonging to web sites that are punctuated by internet are collected and analyzed together with malicious websites. Neural networks, stacked autoencoders, support vector machines and decision trees are compared in order to determine the discriminatory qualities for detecting false websites and to group fake/non-fake websites.

In the phishing analysis, information about previous studies about the subject was given and the developed method was presented in the method part. In the methodology part working principles of neural networks, stacked autoencoders, support vector machines and decision trees are presented. The results are tested with different parameters and the findings are shared in the same part. In the conclusion section, obtained results are shared.

2 Phishing Analysis

Toolboxes based on the analysis of fake web pages are available for the detection of phishing attacks [2, 3]. Blacklist/whitelist-based anti-phishing techniques such as Google's secure browsing API [4] and Mozilla's Firefox [5] are used in related browsers. Ma and his colleagues [6, 7] used online learning and statistical methods to identify malicious websites.

Mohammed et al. [8] have extracted 17 features from web sites to determine whether a URL is harmful, and have manually defined rules based on these features.

phish_id	url	phish_detail_url	submission_time
3053619	http://www.randomactsofjake.com/Blackboard-Login.htm	http://www.phishtank.com/phish_detail.php?phish_id=3053619	2015-03-16T16:42:40+00:00
3053476	http://alturl.com/2q95b	http://www.phishtank.com/phish_detail.php?phish_id=3053476	2015-03-16T15:05:34+00:00
3051521	http://getir.net/yg5a	http://www.phishtank.com/phish_detail.php?phish_id=3051521	2015-03-15T17:28:32+00:00
3050769	http://getir.net/yg4t	http://www.phishtank.com/phish_detail.php?phish_id=3050769	2015-03-15T03:23:55+00:00
3048627	http://scaccom.biz/Calendar/Menu/ccessing.html	http://www.phishtank.com/phish_detail.php?phish_id=3048627	2015-03-13T10:01:28+00:00
3048049	http://www.glamorous.hk/OurTime.com/ourtime.com/www.ourti	http://www.phishtank.com/phish_detail.php?phish_id=3048049	2015-03-13T01:11:08+00:00

Fig. 4 Phishing data set

Nguyen et al. [9] proposed a neuro-fuzzy model that combines neural networks and fuzzy systems without using rule sets for phishing detection by analyzing URL properties, while Eric et al. Proposed a phoneme detection based on visual similarity between a page and other target pages scheme [10]. In addition to these approaches, content-based approaches are also included. In such approaches, features were extracted from the content properties of the web pages for the detection of generally suspicious websites, and these features are used to calculate the visual similarities of the pages [11–13].

In the study of Aksu et al. [14], the features are statically determined and the system is trained with support vector machines which is a machine learning method. The disadvantage of this system is that the qualities are determined statically. For this reason, the system's accuracy gives a high accuracy rate on the websites that is compatible with the qualifications, which causes the accuracy rate to drop on pages that have never been identified.

3 Methodology

In this study, the URL of the website that made 1000 phishing attacks for phishing analysis was taken from PhisTank website [15].The data set contain the relevant information that performs the phishing attack is shown in Fig. In order to reflect the actual sample of the analysis performed, 1000 URLs of web sites belonging to non-malicious web sites which are collected from authors' browsing history have been added to the dataset.

After the dataset was collected, the data set belonging to 1000 malicious and 1000 non-malicious URLs in the dataset was subjected to classification process in order to identify the data with phishing attack. The URLs are converted to ASCII values so that the information contained in the data set are examined in detail and the classification is performed on matrices containing ASCII values.

In the classification process, feedforward Neural networks and stacked automatic encoders from deep learning methods are used. 70, 80 and 90% slices of the existing data set were used for the training process and the rest of the data set are used for test process (Fig. 4).

Table 1 Results of neural network

Method	Training data (%)	Hidden layer	Accuracy (%)
Neural network	70	699	78.50
	80	799	76.75
	90	899	79.50

3.1 Neural Networks

Neural Networks consist of interconnected processing units called neurons. Different types of connections can be made between neurons to achieve the desired result. By using a learning rule defined on neural networks, it is intended to reduce the error to zero [16]. The network weights can be changed for this purpose. In this study, a feedforward neural network is used. In the feedforward neural networks, the information which comes to the network is processed through the input layer, the hidden layer and the output layers respectively and the result is obtained. However, while the data are processed, they do not have connections within the same layer. Results of the neural network method can be seen in Table 1The best accuracy is obtained when 90% of the data used in training

3.2 Stacked Autoencoders

In this study autoencoders are used. An autoencoder is one kind of feedforward neural networks which aims to reconstruct its own inputs. The parameters which are trained in autoencoders have same amount of data in input and output. An autoencoder can be used for dimensionality reduction if the number of hidden units are less than the number of input units. An autoencoder extracts features from their input in two phase: encoding and decoding. The autoencoder is responsible for preprocessing and extracting the features of the successors. It is possible to use more than one autoencoder. If a structure containing more than one autoencoder is used for feature extraction and they are stored and evaluated in a stack [17].

Two autoencoders are used in the implemented deep learning application. It is necessary to reduce the input size in every step by going to the output result of the features from 2000 data in data set.

If the situation in which 70% of the data are used for training is taken into consideration; Autoencoder 1 (AE1) extracts 699 features from 1400 data which come as the input. The Autoencoder 2 (AE2) reduces features to 2 from 699 which come from AE1. The implemented Stacked Autoencoder structure is shown in Fig. 5.

During the coding phase of the stacked autoencoders, the sigmoid transfer function, denoted by f, is used in the coding phase. In the decoding phase calculations done by using linear function in Eq. 2.

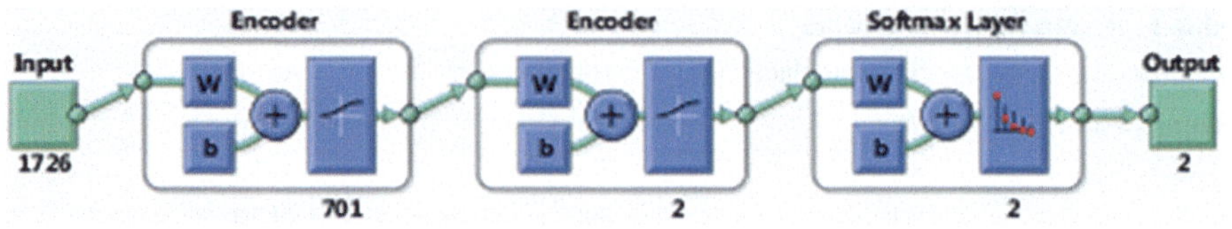

Fig. 5 Structure of stacked autoencoders

Table 2 Results of deep learning

Method	Training data (%)	Inputs of AE1	Inputs of AE2	Accuracy (%)	Accuracy after fine tuning (%)
Deep learning	70	699	2	84	83
	80	799	2	86.50	86.50
	90	899	2	83.50	84

$$f(z) = \frac{1}{1 + e^{-z}} \tag{1}$$

$$f(z) = z \tag{2}$$

Hidden layers' output is calculated according to Eq. 3, where c represents the hidden layer output, w represents the weighting parameter between the input and hidden layer, and b represents the bias vector.

$$c = f(b + w^T x) \tag{3}$$

Table 2 represent the results of the stacked autoencoder method. While 80% of the data used in training, the best accuracy is obtained.

3.3 *Support Vector Machine*

Support Vector Machines (SVM) aims to find the ideal truth that separates the two groups of data in a plane. From the infinite number of lines, both groups are selected to be the farthest line, i.e. the one with the highest margin. Support vector machine aims finding the optimal separating hyper plane. This method reduces the error for unseen patterns [14]. There is no assumption about the distribution of data. In comparison to other methods, the problem of overfitting is rare, adaptation to multidimensional data is easy. It can be adapted to both linear and nonlinear data. As shown in Table 3, 90% of the data is used in training, the best accuracy is obtained through the results of the SVM.

Table 3 Results of support vector machine

Method	Training Data (%)	Accuracy
Support vector machine	70	56.6500
	80	57.4250
	90	60.0500

Table 4 Results of decision tree

Method	Training Data (%)	Accuracy (%)
Decision trees	70	80.6500
	80	80.8250
	90	81.3000

3.4 Decision Trees

Decision trees can be used as a classification method. They show a model in the form of tree structure consisting of decision and leaf nodes which are evaluated as property and target. The decision tree algorithm is developed by dividing the data set into smaller and even smaller pieces. A decision node may contain one or more branches. It can be used to process both numeric and class data and has low computational complexity. Decision tree has three types of nodes, root node, internal nodes and leaf-terminal nodes. In a decision tree, each classes assign to appropriate leaf nodes. The remaining non-terminal nodes are provided to test the conditions of the specific attributes which are utilized to precise the different characteristic of the data. A decision tree is grown recursively by partitioning the training records into successively purer subsets [18]. Among the results given in Table 4, the best one is where the 90% of data is used for training, in which case accuracy is 81.3%.

4 Conclusion

In this study, to classify the phishing attacks on the web sites, feedforward neural networks, stacked autoencoders, support vector machines and decision trees are used. When the classification process is carried out, the data size used for training has been changed and tested. In Fig. 6 accuracy results of the methods are presented. It is seen that in every case stacked autoencoders reached the best accuracy. The highest accuracy is obtained when 80% of data used for training by using stacked autoencoders.

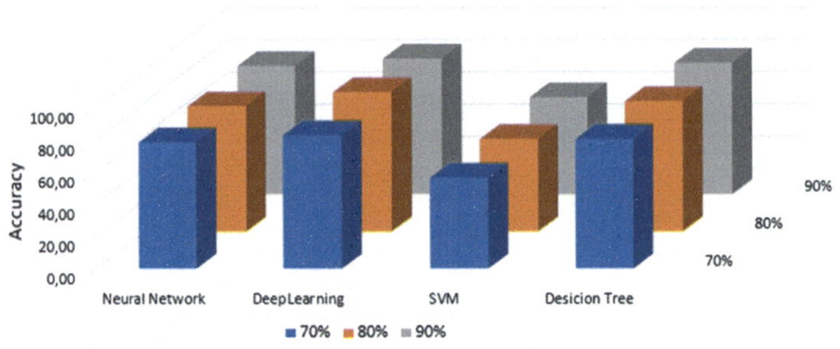

Fig. 6 Accuracy results

References

1. ITU (2016) Global Internet Report 2016
2. C. Inc., Cloudmark Toolbar (2017) http://www.cloudmark.com/desktop/ie-toolbar. Accessed 10 Oct 2017
3. Likarish P, Dunbar D, Hansen TE, Hourcade JP (2008) B-APT : Bayesian Anti-Phishing Toolbar, 2008, pp 1745–1749
4. Google Developers (2011) Google Safe Browsing. http://code.google.com/intl/fr/apis/safebrowsing/. Accessed 01 Jan 2011
5. M. Foundation (2011) Phishing Protection. http://www.mozilla.com/en-US/firefox/phishingprotection/. Accessed 01 Jan 2011
6. Ma J, Saul LK, Savage S, Voelker GM (2009) Beyond blacklists : learning to detect malicious web sites from suspicious URLs. In: World Wide Web Internet and web information systems, pp 1245–1253
7. Ma J, Saul LK, Savage S, Voelker GM (2009) Identifying suspicious URLs : an application of large-scale online learning. In: International conference on machine learning, 2009, pp 681–688
8. Mohammad RM (2012) An assessment of features related to phishing websites using an automated technique, pp 492–497
9. Nguyen LAT, Nguyen HK, To BL (2016) An efficient approach based on neuro-fuzzy for phishing detection. J Autom Control Eng 4(2):159–165
10. Medvet E, Kruegel C (2017) Visual-similarity-based phishing detection
11. Cao Y, Han W, Le Y (2008) Anti-phishing based on automated individual white-list. In: Proceedings of the 4th ACM workshop on Digital identity management—DIM '08, 2008, p 51
12. Liu W, Deng X, Huang G, Fu AY (2006) An antiphishing strategy based on visual similarity assessment. IEEE Internet Comput 10(2):58–65
13. Dunlop M, Groat S, Shelly D (2010) GoldPhish: Using images for content-based phishing analysis. In: 5th International Conference on Internet Monitoring Protocols ICIMP 2010, pp 123–128
14. Dogukan A, Abdullah A, Ali AM (2017) Detecting phishing websites using support vector machine algorithm. In: Proc. 2nd World conference on technology, innovation and entrepreneurship
15. OpenDNS (2017) PhisTank. https://www.phishtank.com
16. Akgündoğdu A (2003) Bulanık-Yapay Sinir Ağları ile Biyomedikal Görüntü İşlemesi. İstanbul Universitesi
17. Karabulut EM (2016) Investigation of deep learning approaches for biomedical data classification. Cukurova universitesi
18. Kumar V, Tan PN, Steinbach M (2014) Introduction to data mining. First. Pearson

Part VI
Signal Processing

Ear Biometric Recognition in Unconstrained Conditions

Amir Benzaoui and Abdelhani Boukrouche

Abstract Recognizing identity from morphological shape of the human ear using one sample image per person in training-set (i.e. only one model of the individual to be identified is registered in the database and available for the task of identification), with insufficient and incomplete training data, dealing with strong person-specificity can be very challenging. In addition, most encountered testing-images in real world applications are not in high quality due to their acquisitions in difficult conditions (ex, video-surveillance) which cause more challenges like: rotated images or images with low resolution. In continuation to our previous works on ear recognition, we present in this paper an experimental and comparative study on the effects of rotation and scaling of ear images using only one sample image per person in training-set which are considered as problems largely encountered in real world applications. Several local color texture descriptors are tested and compared under several color spaces. Support Vector Machine (SVM) is used as a classifier. We experiment with USTB-1 ear database. The experiments show very acceptable and interesting results in comparison to those reported in literature.

1 Introduction

Over the last few decades, the explosion of computers and communication networks has significantly increased the need for identification of individuals. Traditionally, there are two common methods of authentication: the first method is based on what is known (ex, PIN code, password, etc.), while the second is based on possession

A. Benzaoui (✉)
Department of Electrical Engineering, Faculty of Sciences and Applied Sciences, University of Akly Mohand Oulhadj, Bouira, Algeria
e-mail: amirbenzaoui@gmail.com

A. Boukrouche
Laboratory of Inverse Problems, Modeling, Information and Systems (PI:MIS) Department of Electronics and Telecommunications, Faculty of Sciences and Technology, University of May 08th 1945, Guelma, Algeria

A. Boyaci et al. (eds.), *International Telecommunications Conference*,
Lecture Notes in Electrical Engineering 504,
https://doi.org/10.1007/978-981-13-0408-8_22

(ex, badge, identity card, smart card, etc.). However, these methods pose serious problems of reliability. Each of them suffers from weaknesses that can seriously degrade their usefulness. Indeed, passwords can be forgotten or decrypted via specific software, and badges can be lost or even stolen. Since the recent terrorist attacks and threats in many countries, reliable identification of people has become a major problem for security reasons. Biometrics is an emerging solution to this problem; its objective is to determine an irrefutable identity. Contrary to what we know or what we possess, biometrics is based on what we are, and also avoids duplication, theft, forgetting or loss [1]. Biometrics is defined as the science served to verify or determine the identity of an individual on the basis of his chemical characteristics (ex, DNA, odor, etc.), behavioral characteristics (ex, gait, electronic signature, voice, etc.), or morphological characteristics (ex, fingerprint, hand/finger geometry, retina, iris, face, etc.) [2–4]. It is a scientific method used to identify individuals and to control access to safe areas (ex, Networks) [5]. The advantage of identification based on this technique; is that each individual has its own physical characteristics that cannot be changed, lost or stolen.

During several years, the human ear was used as a mean of identification in forensic investigations. The human ear has a wealth of information located on a curved 3D surface; this wealth of information has attracted the attention of legal scientists [6]. The outer morphology of the ear is relatively stable over a period of time which is acceptable for biometric applications. Several studies on the external stability of the ear (i.e. the auricle) suggest that the shape of the ear matures very early while its expansion continues with a very low rate [7]. Ear images can be acquired simultaneously with facial images and used together to improve significantly the accuracy of recognition. It is also possible to use the ear and the face as a complementary part of information (multi-biometrics), especially in the applications of monitoring, tracking or continuous personal authentication [8]. The principal advantages related to the use of the 2D ear images as a biometric modality reside in the stability against the variations in facial expression, the acceptability by people, and the convenience of secretly acquiring images for surveillance applications. Over the past decade, research has evolved slightly to develop automated ear identification technologies. However, significant efforts are still required to improve ear detection, segmentation, and identification in order to make a deployment in surveillance and in other commercial applications.

As in most biometric applications, variations caused on appearance by unconstrained conditions are always considered as open challenges for ear recognition. Among these challenges, we quote: *rotation* and *scaling*. Following our previous works on ear recognition [9–13], we present in this paper an experimental and comparative study on the effects of rotation and scaling of ear images which are considered as problems largely encountered in real world applications. The rest of the paper is organized as follows: in the next section, we describe some related work. In Sect. 3, our proposed approach is presented. In Sect. 4, we present our experimental results by applying the proposed approach on USTB-1 database. Finally, a conclusion related to this work is given in Sect. 5.

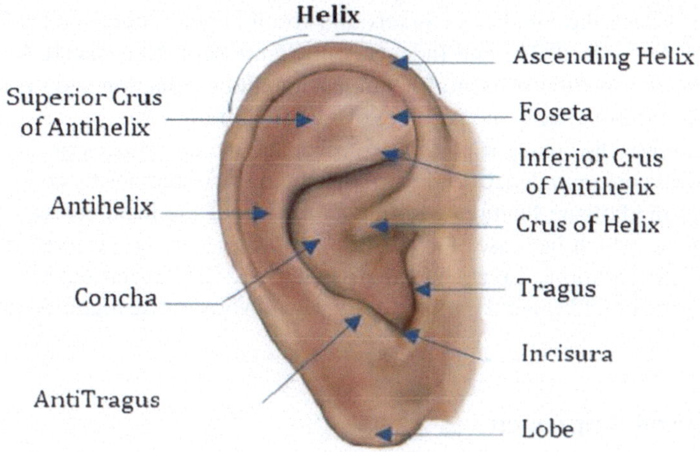

Fig. 1 Anatomy of the human external ear [6]

2 Related Work

The possibility of identifying persons by using the external form of the ear was discovered for the first time by the French criminologist [14] and confirmed by the American police officer [15]. The latter proposed the first automatic system of ear identification, based on only seven characteristics. The detailed structure of the ear (Fig. 1) is not only unique, but also stable, as the appearance of the ear does not change during the life of the human being. In addition, the acquisition of the ear images does not necessarily require the cooperation of the person, it is considered by most researchers as a non-intrusive modality.

Actually, most of biometric systems based on the ear modality consist in extracting a set of features and comparing these features with those models stored in the database. Based on this principle, previous research into the problem of ear recognition can be divided into three classes: local approaches, global approaches, and hybrid approaches.

The algorithms based on local approaches seek to find specific landmark points in the image, and then extract the characteristics around these specific points. Local approaches are more robust than holistic approaches in the unconstrained environments. Despite the capacity of these algorithms to tackle the problem of variations in pose and size, they have a difficulty in finding a precise and exact method to locate the specific points. The approaches which have attracted most interest in this class are: the *Force Field Transform* [16], the *Scale Invariant Feature Transform* (SIFT) [17], the *Speeded Up Robust Features* (SURF) [18], etc.

Global approaches are based on pixel information; all pixels of the image are treated as a single vector and the total number of pixels represents the size of the vector. Most methods in this approach use another space of representation (sub-space) in order to reduce the number of pixels and to eliminate redundancies. Principal Component Analysis (PCA) and Independent Component Analysis (ICA) [19] are the most popular methods used for the dimensionalities reduction and the selection of the useful information in the problem of ear recognition.

Hybrid approaches make it possible to associate the advantages of the methods based on local correspondences and the statistical transformations by combining the detection of the holistic features with the extraction of the local characteristics of its appearance, which increases the stability of the recognition performance in the unconstrained conditions. Lu et al. [20], Yuan and Mu [21], Jeges and Mate [22], as well as Zhang et al. [23] are the most well-known hybrid ear recognition methods.

3 Proposed Approach

In our previous works: Benzaoui et al. [9–11], we have tested and compared three recent gray local texture descriptors, namely: Local Binary Pattern (LBP) [24], Local Phase Quantization (LPQ) [25], and Binarized Statistical Image Features (BSIF) [26]. These methods can easily derive an effective feature model which combines the global form of the analyzed object and the local texture of its appearance in a single feature vector. In fact, surprising and very interesting rank-1 recognition rates (96% with USTB-1 database), in comparison to other related work, were emerged with BSIF descriptor using only a single referential ear image per person in the training-set.

Following our previous works on ear recognition based on local texture descriptors, we have proposed, in Benzaoui et al. [11], to use anatomical and embryological information about the human ear in order to find the autonomous components and the locations where large inter-individual variations can be detected. Embryology is particularly relevant to our approach as it provides information on the possible changes that can be observed in the external structure of the ear. In fact, the elimination of the semi rigid-information gave very interesting improvements especially when using only a single image per person in the training-set (an improvement from 96 to 96.27% with USTB-1 database).

In Benzaoui et al. [12], we have proposed to exploit the color information of the local texture descriptors (LBP, LPQ, and BSIF). For example, the Color Local Binary Patterns (LBP) descriptor can be used to extract the joint color-texture information from the ear images. In this descriptor, the LBP histograms are extracted from the individual image bands. Subsequently, these histograms are concatenated to form the final descriptor, as shown in Fig. 2.

In continuation to this last work based on the color information, we propose in this paper to test and compare the effectiveness of several color descriptors (LBP, LPQ, and BSIF) with several color spaces (RGB, HSV, and YCbCr) in the unconstrained conditions, namely: *rotation* and *scaling*. In fact, we test the robustness of our biometric system against several changes in rotation and scaling: we randomly rotate

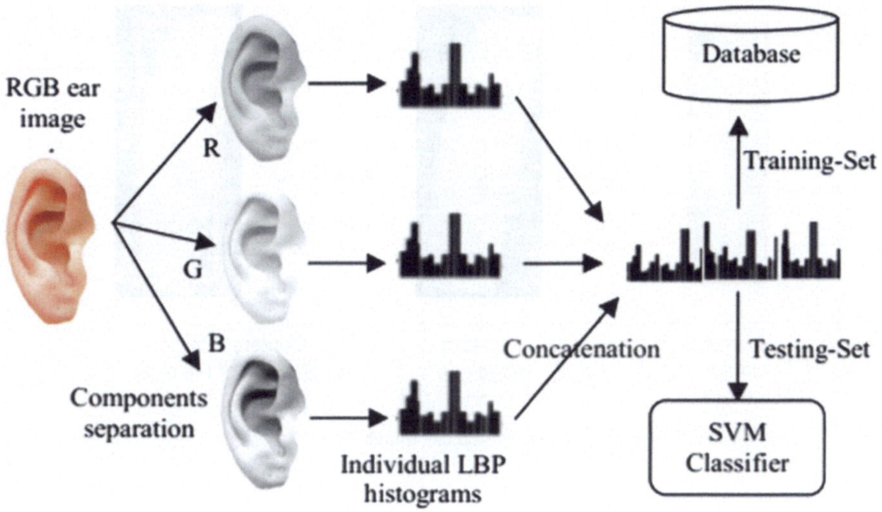

Fig. 2 General diagram of the proposed system

each test image by an angle θ in the interval $[-10$ deg, $+10$ deg$]$ and we randomly resize the test image by an enlargement or reduction factor γ in the interval [0.6, 1.4].

4 Experimental Results

To evaluate the performance of the proposed system in the unconstrained conditions, we carried out a series of experiments using three local texture descriptors, namely: LBP, LPQ, and BSIF, as well as three color spaces, namely: RGB, HSV, and YCbCr, in order to extract discriminating features from a set of ear images from the USTB-1 [27] database. The database and the protocol used to evaluate its performances are discussed in the following sub-section.

4.1 Dataset and Settings

The first version of the USTB database [27] consists of ear images collected by UST Beijing University. The images of this database were taken under different conditions of lighting and noise. It contains 185 images of 60 subjects, with at least three images of each subject. This database is delivered with automatically normalized and cropped ear images of 80×150 pixels in size.

The experimental protocol uses only one ear image per person in the training-set and the remaining ear images of the same person are used in the testing-set.

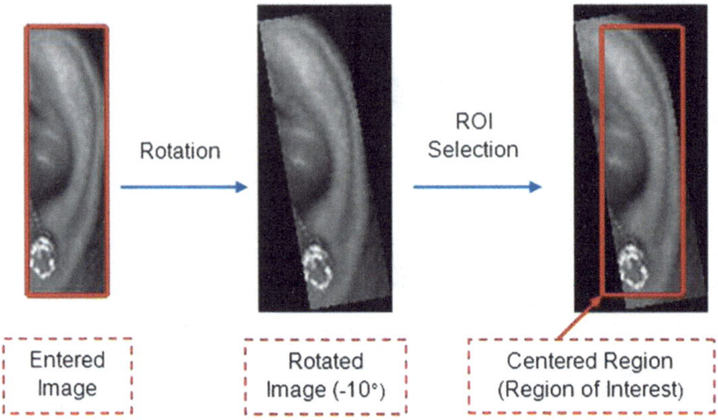

Fig. 3 Example of ROI extraction from rotated ear image

As the majority of the subjects in the databases have three images, we carried out three permutations and reported the average rank-1 recognition rate. We used this protocol because in real world applications (e.g., ID cards or passports); only one image per person is used in the training phase i.e. only one model of the individual to be identified is registered in the database and available for the task of identification.

As the size of the tested images will be changed by the effect of rotation or scaling, the size of the feature vector will also be automatically changed. For example with rotation, we selected the centered region (or the region of interest (ROI)) from the rotated image, as shown in Fig. 3. This ROI must have the same size as the entered image to generate a feature vector that has the same size as the vectors stored in the database.

4.2 *Experiments*

In the first experiment, we tested the effects of rotation and scaling of the three local texture descriptors namely (LBP, LPQ, and BSIF) on the gray level space in addition to the color spaces (RGB, HSV, and YCbCr). The obtained results were recorded in the Tables 1 and 2. Each descriptor is characterized by its own parameters: the LBP has two parameters (P,R) representing the sampling points (P) on a circle of radius (R), the LPQ is defined by the radius of the filter (P), while the BSIF depends on the filter size l and the length n of the bit string. In our previous works [9–11], the performance of each descriptor with all possible instances of parameters were measured, tested, and compared in order to find the best texture descriptor and the parameters that give the best recognition performance. A detailed discussion of these tests and experiments can be found in [9–11].

We can see and demonstrate from Tables 1 and 2 that:

Table 1 Best rank-1 recognition rates of different color descriptors applied on the gray level and color spaces using a single ear image per person in the training-set under different angle of rotation θ

		−10°	−05°	0°	+05°	+10°
LBP	Gray	1.6	15.2	48.8	9.6	1.6
	RGB	1.6	14.4	44.8	8.8	1.6
	HSV	1.6	5.6	34.4	8.8	2.4
	YCbCr	4	12	42.4	11.2	4
LPQ	Gray	8	13.6	31.2	17.6	6.4
	RGB	12.8	17.6	30.4	17.6	8.8
	HSV	1.6	5.6	28	4.8	1.6
	YCbCr	6.4	9.6	32	13.6	4.8
BSIF	Gray	76.8	92	96	93.6	73.6
	RGB	**82.76**	**92**	***96.53***	**94.4**	**86**
	HSV	59.2	78.4	93.33	80	54.4
	YCbCr	67.2	83.2	88.8	84.8	58.4

Table 2 Best rank-1 recognition rates of different color descriptors applied on the gray level and color spaces using a single ear image per person in the training-set under different enlargement/reduction factor γ

		0.6%	0.8%	0%	1.2%	1.4%
LBP	Gray	10.4	17.6	43.2	6.4	4
	RGB	11.2	20	44.8	6.4	4
	HSV	4.8	7.2	34.4	3.2	1.6
	YCbCr	5.6	15.2	42.4	8	8
LPQ	Gray	11.4	20	30.2	24	16.8
	RGB	14.2	20.8	31.4	28.8	20.8
	HSV	6.4	7.2	28	4	4
	YCbCr	6.4	16.8	32	12.8	8
BSIF	Gray	79.2	93.6	96	94.4	91.2
	RGB	**86.8**	**93.6**	***96.53***	**95.2**	**92**
	HSV	55.2	83.2	93.33	80.8	53.6
	YCbCr	60	86.4	88.8	84	70.4

- The results of the BSIF descriptor outperform all other descriptors (LBP and LPQ) in terms of recognition rates.
- The results of the gray level exceed the results of the HSV and YCbCr color spaces, while the RGB color space presents more improvements and very interesting results in terms of recognition in comparison to the gray level space or the other color spaces.
- For all approaches, the recognition performance degrades proportionally with the increase in the absolute value of the angle θ or in the changes of the factor γ. How-

ever for all tested instances of the parameters θ or γ, the results obtained with the configuration (BSIF-RGB) appear very acceptable and interesting in comparison to those reported with the rest of configurations and to some results presented in the literature, which demonstrates the powerful and the robustness of the use of the color information in the unconstrained conditions, namely: *rotation* and *scaling* variations.

- It should be noted that most existing approaches in the literature employ two ear (sample) images per person in the training-set, but in real world applications (e.g., ID cards or passports), only one image per person is used in the training phase i.e. only one model of the individual to be identified is registered in the database and available for the task of identification. Our study demonstrates that this approach based on the color information (the application of BSIF descriptor on the color space RGB) is reliable and can be implemented in real world applications.

5 Conclusion

In this paper, we have studied the effects of rotation and scaling in the problem of ear biometric recognition by using only one sample image in training-set. In fact, several color texture descriptors under several color spaces are tested and compared in order to find the optimal parameters that give the best performance. The results obtained with the configuration (BSIF-RGB) presented very acceptable and interesting performance in comparison to those reported with the rest of configurations and to some results presented in the literature. Our study demonstrates that this approach based on the color information is reliable and can be implemented in real world applications.

References

1. O'Gorman L (2003) Comparing passwords, tokens, and biometrics for user authentication. Proc IEEE 91(12):2021–2040
2. Jain AK, Ross AA, Nandakumar K (eds) (2011) Introduction to biometrics. Springer Science+Business Media, New York
3. Pato JN, Millet LI (eds) (2009) Biometric recognition: challenges and opportunities. The National Academic Press, Washington, D.C.
4. Benzaoui A (2014) Face analysis, description, and recognition using improved local binary patterns in one dimensional space. J Control Eng Appl Inf (CEAI) 16(4):52–60
5. Wayman JL, Jain AK, Maltoni D, Maio D (eds) (2005) Biometric systems: technology, design and performance evaluation. Springer, London
6. Arbab-Zavar B, Nixon MS (2011) On guided model-based analysis for ear biometrics. Comput Vis Image Understand (Elsevier) 115(04):487–502
7. Ross A, Abaza A (2011) Human ear recognition. IEEE Comput Biometric Compendium 44(11):79–81
8. Huang Z, Liu Y, Yang M, Chen L (2013) A robust f ace and ear based multimodal biometric system using sparse representation. Pattern Recognit (Elsevier) 46(08):2156–2168

9. Benzaoui A, Hadid A, Boukrouche A (2014) Ear biometric recognition using local texture descriptors. J Electron Imaging (JEI-SPIE) 23(5):053008
10. Benzaoui A, Hezil N, Boukrouche A (2015) Identit y recognition based on the external shape of the human ear. In: Proceedings of the international IEEE conference on applied research in computer science and engineering (ICAR)
11. Benzaoui A, Adjabi I, Boukrouche A (2016) Person identification based on ear morphology. In: Proceed ings of the International IEEE conference on advances aspects of software engineering (ICAASE)
12. Benzaoui A, Adjabi I, Boukrouche A (2017) Experiments and improvements of ear recognition based on local texture descriptors. Opt Eng (OE-SPIE). 56(4):043109
13. Benzaoui A, Boukrouche A (2017) Ear recognition using local color texture descriptors from one sample image per person. In: Proceedings of the international IEEE conference on control, decision, and information technologies (CODIT)
14. Bertillon A (1980) La photographie Judiciaire, avec un appendice sur la classification anthropométrique. Technical Report, Gauthier-Villars, Paris
15. Iannarelli A (1989) Ear identification. Forensic Identification Series, Paramount Publishing Company, Fremont, California
16. Abdel-Mottaleb M, Zhou J (2006) Human ear recognition from face profile images. In: Proceedings of the international conference on biometrics (ICB). Lecture notes in computer science, vol 3822. Springer, pp 786–792
17. Bustard JD, Nixon MS (2010) Toward unconstrained ear recognition from two-dimensional images. IEEE Trans Syst Man Cybern (Part A: Syst Hum) Special Issue on Recent Advances in Biometrics 40(3):486–494
18. Prakash S, Gupta P (2013) An Efficient ear recognition technique invariant to illumination and pose. Telecommun Syst (Springer) 52(3):1435–1448
19. Zhang H, Mu Z (2008) Compound structure classifier system for ear recognition. In: Proceedings of the IEEE international conference on automation and logistics (ICAL), pp 2306–2309
20. Lu L, Xiaoxun Z, Youdong Z, Yunde J (2006) Ear Recognition based on Statistical Shape Model. In: Proceedings of the 1st international IEEE conference on innovative computing, information, and control (ICICIC), vol 03, pp 353–356
21. Yuan L, Mu ZC (2007) Ear recognition based on 2D images. In: Proceedings of the 1st international IEEE conference on biometrics: theory, applications, and systems (BTAS), pp 1–5
22. Jeges E, Mate L (2007) Model-based human ear localization and feature extraction. Int J Intell Comput Med Sci Image Process (Taylor & Francis). 01(2):101–112
23. Zhan B, Mu Z, Zeng H, Luo S (2014) Robust ear recognition via nonnegative sparse representation of Gabor orientation information. Sci World J (Hindawi)
24. Benzaoui A, Boukrouche A (2013) Face recognition using 1DLBP texture analysis. In: Proceedings of the 5th international IARIA conference on computational technologies and applications (Future Computing), pp 14–19
25. Ojansivu V, Heikkil J (2008) Blur insensitive texture classification using local phase quantization. In: Proceedings of the 3rd international conference on image and signal (ICSIP). Lecture Notes in Computer Science (Springer), pp 236–243
26. Kannala J, Rahtu E (2012) BSIF: Binarized statistic al image features. In: Proceedings of the international IEEE conference on pattern recognition (ICPR), pp 1363–1366
27. Mu Z (2009) USTB Ear Image Database, Beijing, China. http://www1.ustb.edu.cn/resb/en/index.htm

Identification of Non-Conforming Cordless Phone Signals in Licensed Bands

Selen Geçgel, Mehmet Akif Durmaz, Hakan Alakoca, Güneş Karabulut Kurt and Cem Ayyıldız

Abstract Identification of interference signals is critical in telecommunication systems increasing the need for automated signal identification. As deep convolutional neural networks demonstrate significant achievements in pattern recognition problems, it can be inferred that deep learning methods will give successful results in the field of wireless communications, especially about identification of signal. This paper investigates a new approach of signal identification based on deep convolutional neural network with the convolution architecture for feature extraction (CAFFE) framework. Authors provide the identification for the types of interference signals based on non-conforming digital enhanced cordless telecommunications (DECT) devices. For training NVIDIA-DIGITS, the NVIDIA Deep Learning GPU Training System, is used. The classification accuracy of the system under additive white Gaussian noise and Rayleigh fading channel conditions is observed to be high despite low signal to noise ratio values.

Keywords Deep learning · Signal identification · Classification · Rayleigh · AWGN

S. Geçgel (✉) · M. A. Durmaz · H. Alakoca · G. K. Kurt
Istanbul Technical University, Wireless Communication Research Laboratory (WCRL), Istanbul, Turkey
e-mail: gecgel16@itu.edu.tr

M. A. Durmaz
e-mail: durmazm@itu.edu.tr

H. Alakoca
e-mail: alakoca@itu.edu.tr

G. K. Kurt
e-mail: gkurt@itu.edu.tr

C. Ayyıldız
Turkcell Technology Research and Development Laboratory, Turkcell İletişim Hizmetleri A.Ş., Istanbul, Turkey
e-mail: cem.ayyildiz@gmail.com

A. Boyaci et al. (eds.), *International Telecommunications Conference*, Lecture Notes in Electrical Engineering 504,
https://doi.org/10.1007/978-981-13-0408-8_23

1 Introduction

Identification of signals in wireless environments consisting of noise, interference with multiple non-stationary signals has a major importance in communication systems. Network service providers use several frequency bands in the radio spectrum for their mobile phone networks. Use of digital enhanced cordless telecommunications (DECT) 6.0 phones are incompatible with systems in some countries since they cause significant interference in the bands of the universal mobile telecommunications system (UMTS) licensed services. DECT phones use the same or adjacent frequency band for up and down link channels. Furthermore most DECT devices keep RF transmission active therefore inter-systems interference occurs between 3G mobile communication systems and DECT. 30 relevant scenarios have been elaborated in [1].

Current approaches generally focus on machine learning and statistical methods which consist pre-processing, feature extraction and classification sections. When the amount of data and the number of dimensions increase, longer processing times and reduced identification performance can be observed. Using deep learning methods can improve the performance and make it more independent from human control. In [2], the comparison the performance deep autoencoder networks and machine learning technique in the field of communication signal recognition is given, and the deep learning method's performance is shown to be superior. In [3], a new technique for modulation format identification in digital coherent receivers by applying deep neural network (DNN) based on pattern recognition is preferred and the experimental results demonstrate good identification accuracies for three commonly-used modulation formats despite having no prior optical signal to noise ratio (OSNR) information. In [4], automatic cognitive radio waveform recognition system which can identify 8 kinds of waveforms (BPSK, LFM, Costas codes, Frank code and T1-T4) with a success rate exceeding 93.7% is developed by using convolutional neural network (CNN) classifier and Choi-Williams time frequency distribution. Deep learning can help learn data representations with multiple levels of abstraction according to the mathematical model which is composed of multiple processing layers [5].

There are various studies for signal identification and classification. Detection of a signal using cyclostationary classifiers are covered work in [6]. Studies [7, 8] target detection of modulation type at receiver node by using cyclostationary classifiers. In [9] spectral correlation function of received RF signals is used for signal estimation. A real time signal identification application is indicated in [10], where orthogonal frequency-division multiple access (OFDMA) based signal identification process is handled by using cyclic autocorrelation function. However, cyclostationary based processes require high speed robust digital signal processing processors.

Time-frequency analysis allows simultaneous description of a signal in time and frequency, so that the temporal evolution of the signal spectrum can be analyzed [11]. There are many methods but due to its computational simplicity the short-time Fourier transform (STFT) is more resourceful than most of adaptive approximations. The most well-known approach in this regard is the spectrogram. In our paper, in order to obtain an image from radio signal data, the spectrogram of the data is

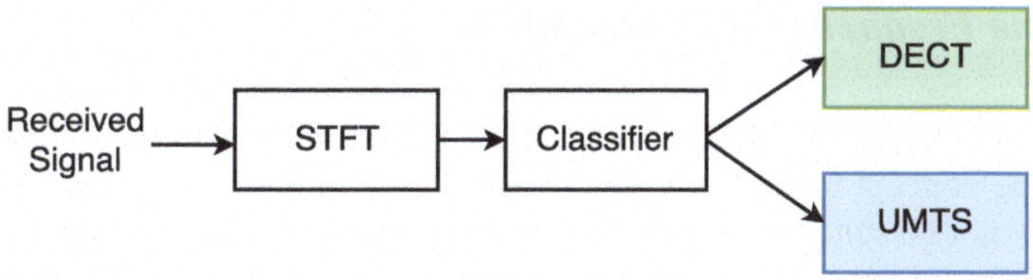

Fig. 1 Block diagram of the system

calculated to represent the DECT and UMTS signals as an image. Since classification methods constitute an essential component of signal identification, here we propose a system that brings together deep convolutional neural networks and signal processing methods for the detection of DECT signals. Main block diagram of the system is given in Fig. 1. We can summarize our contributions as follows:

1. Based on an actual problem encountered by network service providers, identification of DECT signals in 3G bands is investigated.
2. An identification scheme based on STFT and convolutional deep neural network is proposed.
3. The performance of the proposed approach is shown via simulation results.

2 System Model

2.1 Signal Model

We use 4-QAM modulation for DECT signal and BPSK modulation for code-division multiple access (CDMA). These systems were tested in two different scenarios. First, time domain signals transmits with additive white Gaussian noise (AWGN) and received signal $y(n)$ are given as

$$y(n) = s(n) + w(n), \tag{1}$$

where $s(n)$ is transmitted signal and $w(n)$ is white Gaussian noise component. We then included Rayleigh fading channel to obtain a more realistic environment for the second scenario. Received signal in the presence of fading is changed as

$$y(n) = s(n) * h(n) + w(n), \tag{2}$$

where, $h(n)$ is the impulse response of the wireless channel and $*$ denotes convolution operator.

2.2 Time Frequency Representation

The spectrogram has been the most widely used tool for the analysis of time-varying spectra. The concept behind it is simple and powerful. Discrete Fourier transform (DFT) may not be suitable because DFT extracts spectral content of the signal over time but does not provide the spectral content at a particular time. Evaluating after partitioning the signal to small portions, taking short-time Fourier transform of signal is quite useful, and calculated as

$$y(n) = \sum_{n=0}^{N-1} y(n)h_w(n)m(k,n). \tag{3}$$

The function $h_w(n)$ is the window, $m(k,n)$ is DFT kernel and $y(k,n)$ denotes the short-time Fourier transform of the signal, where

$$m(k,n) = e^{-j2\pi kn/N}. \tag{4}$$

The spectrogram is [12]

$$P(n) = \sum_{k=0}^{N-1} \mid y(n) \mid^2, \tag{5}$$

which can be considered as energy density.

3 Classifier Design and Test Results

3.1 NVIDIA - DIGITS

Deep Learning GPU Training System (DIGITS) is a system for training deep-learning systems developed by NVIDIA. It provides an easy-to-use interface simple to start building an image-recognition system with a web-based front end. Designing, training and visualizing deep neural networks for image classification, segmentation and object detection are simplified through the use of DIGITS. Also in case of insufficient data, it is possible to download pre-trained models such as AlexNet, GoogLeNet and LeNet from the DIGITS Model Store or to use different frameworks like the convolution architecture for feature extraction (CAFFE), Torch and Tensorflow.

3.2 Dataset

The simulated dataset consists of 6 sub-datasets and each of them has 1000 images in PNG format. Dataset is formed under different noisy environmental conditions or SNR values. One sub-dataset is decomposed into 3 parts: a training set of 704 images, a validation set of 96 images and a test set of 200 images. All images are generated

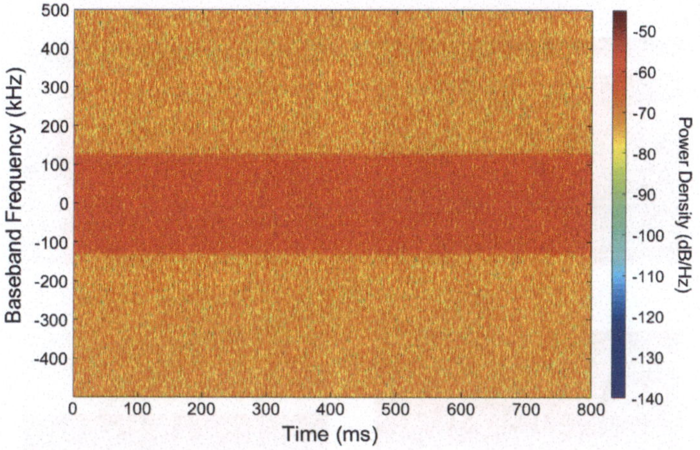

Fig. 2 A spectrogram example for DECT signal

with simulations of the DECT and UMTS signals. Each signal contains 8×10^6 symbols. We use an 8-tap random channel for Rayleigh fading channel simulations. Simulations are run in MATLAB and its spectrogram is used for visualization of the data. Sampling frequency is selected as 100 MHz and window size is 128. Axis labels and values are not included in images but an example for the outputs is shown in Fig. 2. Down-converted frequency is centered between 500 and −500 kHz. Colorbar represents power density, which is limited between −45 and −140 dB. The size of an image is 875×656 pixels. An image has 2 dimensions also its each row is an RGB vector that defines one color and specifies the intensity of red, green, and blue. We aim to train on these RGB values of pixels which shows that energy density of signals. Example images from each sub-dataset can be seen at Figs. 3 and 4.

3.3 Deep CNN Architecture - GoogLeNet

GoogLeNet uses Inception modules instead of many pool and convolution layers, because stacking these layers or using a large number of filters can cause overfitting. In one inception module, all these operations are performed in parallel. The inception model can be seen at Fig. 5, which has one input and output and the layers between them. We use one 3×3 convolution, one 5×5 convolution, one pooling and four 1×1 convolutions. All the convolutions, including those inside the Inception modules, use rectified linear activation. 1×1 convolutions are used to compute reductions before the expensive 3×3 and 5×5 convolutions. Besides being used as reductions, they also include the use of rectified linear activation making them dual-purpose [13]. The complete architecture is a 22 layer CNN, with 9 inception modules. Full connected layers were not utilized. There was preferred an average pool in this way reduced the number of parameters. Detailed structure of GoogLeNet architecture can simply be thought of as a network in the network.

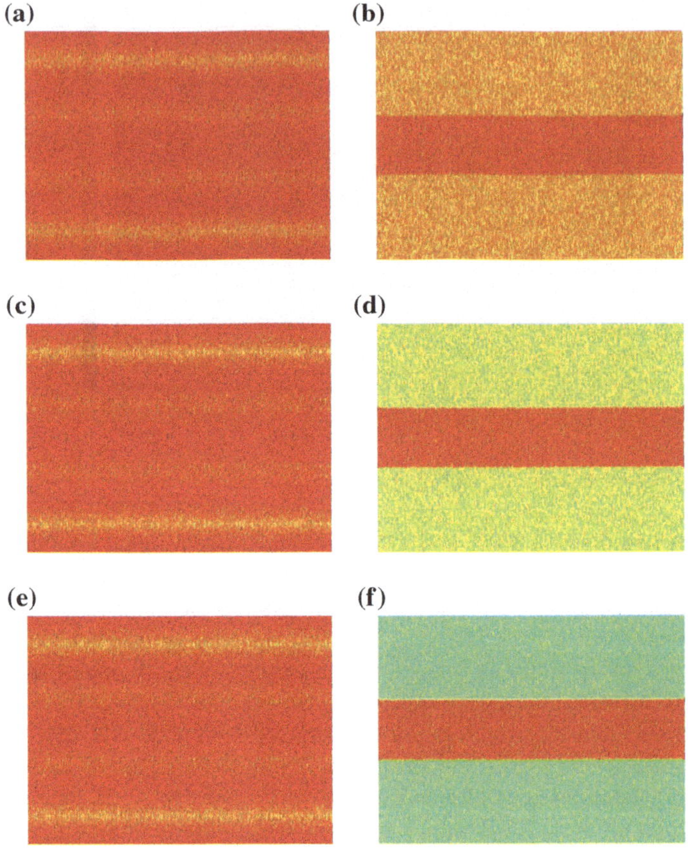

Fig. 3 AWGN channel examples: **a** CDMA with 10 dB SNR, **b** DECT with 10 dB SNR, **c** CDMA with 20 dB SNR, **d** DECT with 20 dB SNR, **e** CDMA with 30 dB SNR, **f** DECT with 30 dB SNR

The network was designed with computational efficiency and practicality in mind, so that inference can be run on individual devices including even those with limited computational resources, especially with a low-memory footprint [13].

3.4 Proposed Classification Model

In this study, identification of the corruptive effect of non-conforming DECT signals are aimed by using above mentioned methods and tools. Firstly, DECT and UMTS signals are created during this processes under different SNR values and different channel models to observe classification accuracy. Signals are processed according to time selected time freq. distri. tech., the spectrogram because of its simplicity.

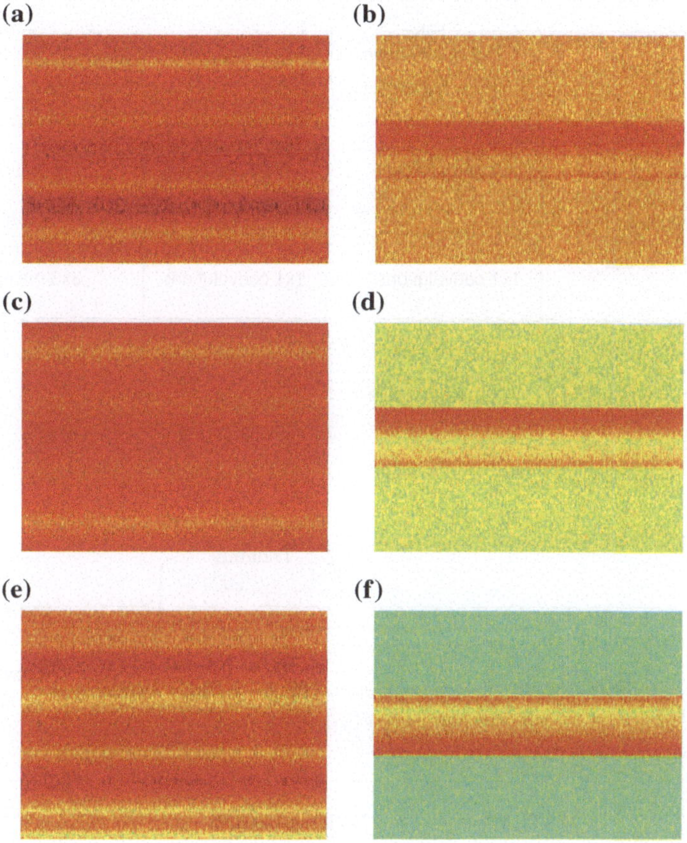

Fig. 4 Rayleigh fading channel examples: **a** CDMA with 10 dB SNR, **b** DECT with 10 dB SNR, **c** CDMA with 20 dB SNR, **d** DECT with 20 dB SNR, **e** CDMA with 30 dB SNR, **f** DECT with 30 dB SNR

Dataset obtained by rendering imported to the NVIDIA-DIGITS and in GoogLeNet trained. Our proposed scheme can be seen in the aggregate in Fig. 6.

3.5 Results

Based on the above described system model with different SNR and channel types about the identification of non-conforming DECT signals, the designed classifier achieved approximately 100% success. Test accuracy values calculated over test datasets which have consist of 1200 images totally. For each dataset results is listed in Table 1.

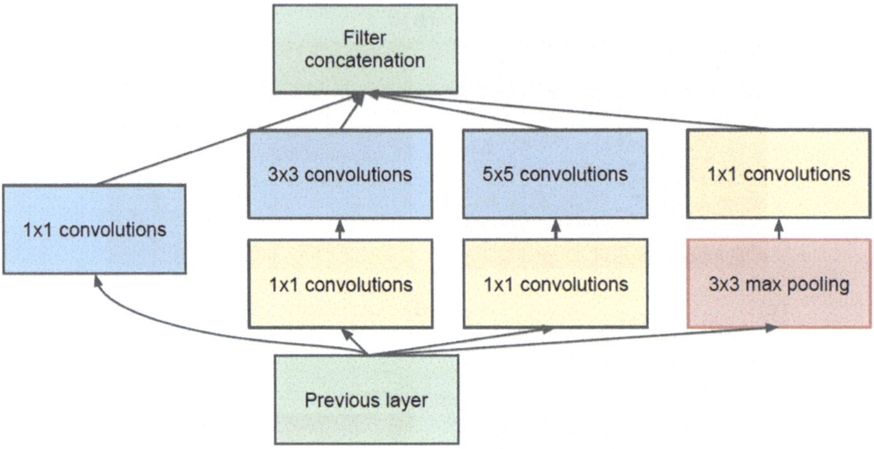

Fig. 5 Full inception model [13]

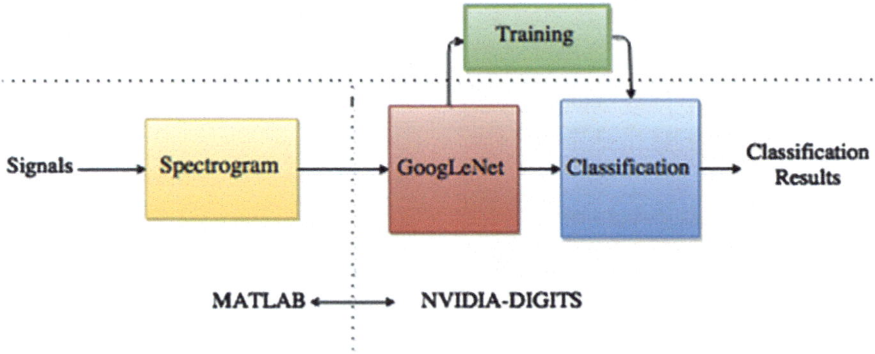

Fig. 6 Proposed classification structure

4 Conclusion

The problem of identifying of DECT signals which interfere to the UMTS frequency bands is investigated. A high accuracy approach by using the deep CNN is proposed. Classification results demonstrate that by using deep learning though spectrogram based images can provide accurate classification results in low SNR conditions, even in the presence of fading channels.

Table 1 Accuracy results of the system over test datas

Channel type	SNR	Test image	Accuracy
AWGN	10	DECT	%100
		CDMA	%100
	20	DECT	%100
		CDMA	%100
	30	DECT	%100
		CDMA	%100
AWGN + Rayleigh Fading	10	DECT	%100
		CDMA	%99.97
	20	DECT	%100
		CDMA	%100
	30	DECT	%100
		CDMA	%100

References

1. European conference of postal and telecommunications administrations (1999). Adjacent band compatibility between UMTS and other services in the 2 GHz band
2. Li J, Qi L, Lin Y (2016) Research on modulation identification of digital signals based on deep learning (2016). In: 2016 IEEE international conference on electronic information and communication technology (ICEICT), pp 402–405
3. Khan FN, Zhong K, Al-Arashi WH, Yu C, Lu C, Lau APT (2016) Modulation format identification in coherent receivers using deep machine learning. IEEE Photonics Tech Lett 28(17):1886–1889
4. Zhang M, Diao M, Guo L (2017) Convolutional neural networks for automatic cognitive radio waveform recognition. IEEE Access 5:11074–11082
5. Lecun Y, Bengio Y, Hinton G (2015) Deep learning. Nature 521(7553):436–444
6. Gardner WA (1994) Cyclostationarity in communications and signal processing. Technical report, Statistical Signal Processing Inc Yountville Ca
7. Fehske A, Gaeddert J, Reed JH (2005) A new approach to signal classification using spectral correlation and neural networks. In: International symposium on new frontiers in dynamic spectrum access networks (DySPAN), pp 144–150
8. Davy Manuel, Gretton Arthur, Doucet Arnaud, Rayner Peter JW (2002) Optimized support vector machines for nonstationary signal classification. IEEE Signal Process Lett 9(12):442–445
9. Zhou R, Li X, Yang TC, Liu Z, Wu Z (2012) Real-time cyclostationary analysis for cognitive radio via software defined radio. In: Global communications conference (GLOBECOM), pp 1495–1500
10. Bahadir Tuğrel H, Alakoca H, Tekbiyik K, Karabulut Kurt G, Ayyildiz C (2016) OFDMA system identification using cyclic autocorrelation function: a software defined radio testbed. In: Signal processing and communication systems (ICSPCS), pp 1–7
11. Lopez-Risueno G, Grajal J, Sanz-Osorio A (2005) Digital channelized receiver based on time-frequency analysis for signal interception. IEEE Trans. Aerosp Electron Syst 41(3):879–898
12. Cohen L (1989) Time-frequency distributions-a review. Proc IEEE 77(7):941–981

13. Szegedy C, Liu W, Jia Y, Sermanet P, Reed S, Anguelov D, Erhan D, Vanhoucke V, Rabinovich A (2015) Going deeper with convolutions. In: Conference on computer vision and pattern recognition (CVPR), pp 1–9

Fast Frequency Estimation of a Complex Sinusoidal When the Phase Is Already Known

Ahmet Serbes

Abstract This paper describes a fast and precise algorithm for the frequency estimation of a single complex sinusoidal in noisy environments. The method interpolates on the shifted samples of its DFT. Computer simulations show that the proposed method outperforms existing popular DFT-interpolation based frequency estimation algorithms. The proposed method can be successfully employed in noisy environments, where the phase of the complex sinusoidal is already known.

Keywords Frequency estimation · Complex sinusoidals · Cramer-Rao lower bound

1 Introduction

Estimating frequency of a sinusoidal signal is one of the fundamental problems in signal processing. The solution to this problem is of interest to many areas including radar, sonar, communications, audio, biomedical, and seismic signal processing. Among all of the existing frequency estimation methods, the most accurate one is the maximum likelihood estimation (MLE) method. However, MLE may be computationally demanding.

Sinusoidal frequency estimation methods can be roughly divided into two: noniterative and iterative methods. Iterative methods generally rely on optimizing a cost function. Being accurate, they are usually inefficient in terms of the computational cost. The celebrated MLE method is an example of iterative methods.

This paper describes a non-iterative method for estimating the frequency of a single complex sinusoidal signal buried in additive white Gaussian noise. The method is non-iterative and interpolates on the samples of the DFT coefficients. Existing DFT-interpolation based algorithms [1–6] can be used for sinusoidal signals whose phase and frequency are simultaneously unknown. However, the proposed algorithm

A. Serbes (✉)
Yıldız Technical University, Esenler İstanbul, Turkey
e-mail: aserbe@yildiz.edu.tr

A. Boyaci et al. (eds.), *International Telecommunications Conference*,
Lecture Notes in Electrical Engineering 504,
https://doi.org/10.1007/978-981-13-0408-8_24

is only for signals whose phase is already known. Although the requirement that the phase should be known in prior might seem to be a disadvantage, simulation results show that our algorithm outperforms the existing DFT interpolation based methods.

The paper is organized as follows. Problem definition and background information is given in Sect. 2. Section 3 introduces and studies the proposed algorithm. Numerical computer simulations are examined Sect. 4. Finally, Sect. 5 concludes the paper by discussing advantages and disadvantages of the proposed algorithm and the required computational cost.

2 Problem Definition

A discrete, single complex sinusoidal signal under additive white Gaussian noise is modeled by

$$s[n] = A\ \exp\left(j\frac{2\pi}{N} + f_0 n + j\phi\right) + w[n], \tag{1}$$

where A and f_0 are the real unknown amplitude and frequency, respectively. Here, ϕ is the known phase parameter. $w[n]$ are assumed to be zero-mean, white, circular symmetric, complex, and additive white Gaussian noise process with variance σ_w^2.

The frequency parameter can be decomposed into two additive parts as

$$f_0 = k_p + \delta, \tag{2}$$

Where k_p and $|\delta| < \frac{1}{2}$ are the integer and the fractional frequencies, respectively. The 2 goal of this paper is to estimate the frequency f_0 fast and accurately, when the phase ϕ is known a priori. The Cramér-Rao lower bound (CRLB) on the performance of the estimator, in the case of known phase is given by [7] as

$$f_0 = k_p + \delta, \tag{3}$$

where N is the total number of samples in the signal and $SNR = A^2 / \sigma_w^2$ is the signal-to-noise ratio. Without losing generality, from now on we set the phase to zero. Even when the phase is not zero, it can always be compensated by multiplying the signal with $\exp(-j\phi)$, which eventually sets the phase to zero. When the phase is set to zero, the DFT of the signal in (1) at the index $(k_p + k)$ can be derived easily as

$$S[k_p + k] = A\ \exp\left(j\pi\frac{N-1}{N}(\delta - k)\right)\frac{\sin(\pi(\delta - k))}{\sin\left(\frac{\pi(\delta-k)}{N}\right)}. \tag{4}$$

The above equality shows that the DFT of the signal reaches its maximum at the index k_p, i.e., when $k = 0$ for integer k. When $\delta = 0$, the DFT is complex conjugate symmetric around the index k_p. For other values of δ, the symmetricity is not preserved. All of the DFT-interpolating algorithms employ this property of symmetricity to estimate the fractional frequency component. The inverse also holds. That is, if the DFT is not symmetric, the residual frequency δ is non-zero. Consequently, the residual frequency is a function of the change in the symmetricity.

The next section introduces and discusses the proposed frequency estimation algorithm.

3 Proposed Method for Estimating the Frequency

Our proposed method first evaluates N-point DFT of the signal to estimate the integer frequency k_p. The integer frequency is estimated by determining the index at which the DFT reaches its peak by employing

$$\hat{k}_p = \arg\max_k \left| \sum_{n=0}^{N-1} s[n] \exp\left(-j\frac{2\pi}{N}nk\right) \right|^2. \tag{5}$$

where we assume that $\hat{k}_p = k_p$ for lower values of *SNR*. Then, we propose to estimate the residual frequency by

$$\hat{\delta} = \frac{1}{\pi} \arctan\left\{ \frac{\mathrm{Re}\{S[k_p + \frac{1}{2}] - S[k_p + \frac{1}{2}]\}}{\mathrm{Im}\{S[k_p - \frac{1}{2}] - S[k_p - \frac{1}{2}]\}} \right\}, \tag{6}$$

where Re$\{\cdot\}$ and Im$\{\cdot\}$ represents real and imaginary parts, respectively. $S[k_p \pm \frac{1}{2}]$ are the DFT of $s[n]$ evaluated at the indexes $(k_p \pm \frac{1}{2})$. Evaluating $S[k_p \pm \frac{1}{2}]$ is obtained by multiplying s[n] by exp is obtained by multiplying s [n] by exp $\left(\mp\frac{j\pi}{N}\right)$ and then taking the DFT.

Using (4), it can be shown that

$$\begin{aligned}
&\mathrm{Re}\left\{ S\left[k_p + \frac{1}{2}\right] - S\left[k_p - \frac{1}{2}\right] \right\} \\
&= A\,\mathrm{Re}\left\{ e^{\left(j\pi\frac{N-1}{N}(\delta-\frac{1}{2})\right)} \frac{\sin(\pi(\delta-\frac{1}{2}))}{\sin\left(\frac{\pi(\delta-\frac{1}{2})}{N}\right)} - e^{\left(j\pi\frac{N-1}{N}(\delta-\frac{1}{2})\right)} \frac{\sin(\pi(\delta+\frac{1}{2}))}{\sin\left(\frac{\pi(\delta+\frac{1}{2})}{N}\right)} \right\} \\
&= -A\cos(\pi\delta) \left\{ \frac{\cos(\pi\frac{N-1}{N}(\delta-\frac{1}{2}))}{\sin\left(\frac{\pi(\delta-\frac{1}{2})}{N}\right)} + \frac{\cos(\pi\frac{N-1}{N}(\delta+\frac{1}{2}))}{\sin\left(\frac{\pi(\delta+\frac{1}{2})}{N}\right)} \right\}.
\end{aligned} \tag{7}$$

Similarly,

$$\mathrm{Im}\left\{S\left[k_p-\frac{1}{2}\right]-S\left[k_p-\frac{1}{2}\right]\right\}=$$

$$A\times\mathrm{Im}\left\{\mathrm{e}^{\left(j\pi\frac{N-1}{N}(\delta-\frac{1}{2})\right)}\frac{-\cos(\pi\delta)}{\sin\left(\frac{\pi(\delta-\frac{1}{2})}{N}\right)}-\mathrm{e}^{\left(j\pi\frac{N-1}{N}(\delta+\frac{1}{2})\right)}\frac{-\cos(\pi\delta)}{\sin\left(\frac{\pi(\delta+\frac{1}{2})}{N}\right)}\right\}$$

$$=-A\cos(\pi\delta)\left\{\frac{\sin\left(\pi\frac{N-1}{N}(\delta-\frac{1}{2})\right)}{\sin\left(\frac{\pi(\delta-\frac{1}{2})}{N}\right)}+\frac{\sin\left(\pi\frac{N-1}{N}(\delta+\frac{1}{2})\right)}{\sin\left(\frac{\pi(\delta+\frac{1}{2})}{N}\right)}\right\}. \tag{8}$$

Therefore,

$$\frac{\mathrm{Re}\left\{S\left[k_p+\frac{1}{2}\right]-S\left[k_p-\frac{1}{2}\right]\right\}}{\mathrm{Im}\left\{S\left[k_p-\frac{1}{2}\right]-S\left[k_p+\frac{1}{2}\right]\right\}}$$

$$=\frac{\cos\left(\pi\frac{N-1}{N}(\delta-\frac{1}{2})\right)\sin\left(\frac{\pi(\delta+\frac{1}{2})}{N}\right)+\cos\left(\pi\frac{N-1}{N}(\delta+\frac{1}{2})\right)\sin\left(\frac{\pi(\delta-\frac{1}{2})}{N}\right)}{\sin\left(\pi\frac{N-1}{N}(\delta-\frac{1}{2})\right)\sin\left(\frac{\pi(\delta+\frac{1}{2})}{N}\right)-\sin\left(\pi\frac{N-1}{N}(\delta+\frac{1}{2})\right)\sin\left(\frac{\pi(\delta-\frac{1}{2})}{N}\right)}. \tag{9}$$

Employing trigonometric product-to-sum identities $\sin(\alpha)\cos(\beta)=\frac{1}{2}[\sin(\alpha-\beta)+\sin(\alpha+\beta)]$ and $\sin(\alpha)\cos(\beta)=\frac{1}{2}[\cos(\alpha-\beta)-\cos(\alpha+\beta)]$ in the above yields

$$\frac{\mathrm{Re}\left\{S\left[k_p+\frac{1}{2}\right]-S\left[k_p-\frac{1}{2}\right]\right\}}{\mathrm{Im}\left\{S\left[k_p-\frac{1}{2}\right]-S\left[k_p+\frac{1}{2}\right]\right\}}=-\frac{\cos\left(\pi(\delta-\frac{1}{N})\right)-\cos\left(\pi(\delta+\frac{1}{N})\right)}{\sin\left(\pi(\delta-\frac{1}{N})\right)-\sin\left(\pi(\delta+\frac{1}{N})\right)} \tag{10}$$

Applying trigonometric sum-to-product formulas in the above provides,

$$\frac{\mathrm{Re}\left\{S\left[k_p+\frac{1}{2}\right]-S\left[k_p-\frac{1}{2}\right]\right\}}{\mathrm{Im}\left\{S\left[k_p-\frac{1}{2}\right]-S\left[k_p-\frac{1}{2}\right]\right\}}=\tan(\pi\delta). \tag{11}$$

Therefore, (6) is satisfied in the noiseless case. As it can be seen through (6), the proposed method leverages the impaired symmetricity introduced by the residual fre-quency component. The reason that we only employ two samples of the shifted samples is that most of the energy of the DFT is concentrated around k_p. Moreover, (11) shows that the proposed method is unbiased, which is a preferred circumstance. After estimating the residual component, estimate of the frequency can be found by

$$\hat{f}_0=\hat{k}_p+\hat{\delta} \tag{12}$$

Next section discusses performance of the proposed method by comparing its performance with various existing methods. Computer simulations show that the proposed method outperforms other existing methods.

Table 1 Existing DFT-Based Frequency Estimation Algorithms

Method	Estimator
Jacobsen [1]	$\hat{\delta} = \mathrm{Re}\left\{\frac{S[k_p-1]-S[k_p+1]}{2S[k]-S[k-1]-S[k_p+1]}\right\}$
Candan [2]	$\hat{\delta} = \frac{\tan(\pi/N)}{\pi/N} Re\left\{\frac{S[k_p-1]-S[k_p+1]}{2S[k_p]-S[k_p-1]-S[k_p+1]}\right\}$
Quinn [4]	$\alpha_1 = \mathrm{Re}\left\{\frac{R[k_p-1]}{R[k_p]}\right\}, \quad \alpha_2 = \mathrm{Re}\left\{\frac{R[k_p+1]}{R[k_p]}\right\}$ $\delta_1 = \alpha_1/(1-\alpha_1), \quad \delta_2 = \alpha_2/(1-\alpha_2)$ If $\delta_1 > 0$ and $\delta_2 > 0$, then $\hat{\delta} = \delta_2$. Else $\hat{\delta} = \delta_1$
MacLeod [5]	$d = \frac{\mathrm{Re}\{S[k_p-1]S^*[k_p]-S[k_p+1]S^*[k_p]\}}{\mathrm{Re}\left\{2\lvert S[k_p]\rvert^2+S[k_p+1]S^*[k_p]+S[k_p-1]S^*[k_p]\right\}}$ $\hat{\delta} = \left(\sqrt{1+8d^2}-1\right)/(4d)$
Aboutanious [6]	$\hat{\delta} = \frac{1}{2}\mathrm{Re}\left\{\frac{S\left[k_p+\frac{1}{2}\right]+S\left[k_p-\frac{1}{2}\right]}{S\left[k_p+\frac{1}{2}\right]-S\left[k_p-\frac{1}{2}\right]}\right\}$
Proposed Algorithm	$\hat{\delta} = \frac{1}{\pi}\arctan\left\{\frac{\mathrm{Re}\left\{S\left[k_p+\frac{1}{2}\right]-S\left[k_p-\frac{1}{2}\right]\right\}}{\mathrm{Im}\left\{S\left[k_p-\frac{1}{2}\right]-S\left[k_p+\frac{1}{2}\right]\right\}}\right\}$

4 Computer Simulations and Results

This section compares the estimation performance of the proposed algorithm with some existing methods in the literature. Table 1 summarizes existing DFT-based interpolation algorithms that are widely used in the literature for estimating residual frequency of a single complex sinusoidal along with our proposed algorithm.

The performance criteria for comparison is the mean square error (MSE) defined by

$$MSE = \frac{1}{M}\sum_{i=0}^{M}\left(f_0 - \hat{f}_0\right)^2, \tag{13}$$

where f_0 is the actual normalized frequency and $\hat{f}_0$ is the estimated frequency, which is the sum of the estimated integer frequency component and the estimated residual frequency, given by (12). M shows the number of Monte-Carlo simulations.

Throughout all the simulations we took the number of Monte-Carlo simulations as $M = 10^4$. Length of the signal is taken as $N = 64$.

Figure 1a plots the performance comparison of the proposed estimator with the other existing estimators for $\delta = 0.25$. As it is clear from the figure, the proposed estimator shows a better performance compared to the others. Moreover, since our estimator is unbiased, the performance follows the CRLB by a constant ratio. All the other estimator performances shown in Fig. 1a suffer bias effect, which becomes apparent when SNR increases. We also see that Candan's algorithm performs the second and Quinn's algorithm performs the least.

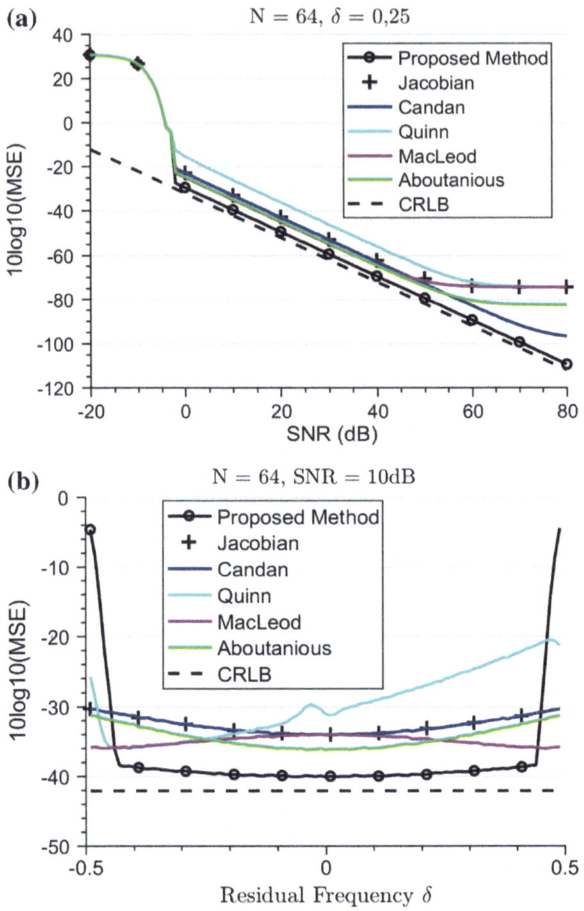

Fig. 1 Performance comparison of the proposed method with the existing methods for a signal length of N = 64. In **a** variation of MSE with respect to SNR is shown when the residual shift $\delta = 0.25$ and in **b** performance of the estimator with respect to the residual frequency under constant SNR are shown

Figure 1b compares the performance with respect to the residual frequency δ under constant *SNR* at 10 dB. We see that the performance of the proposed estimator is nearly flat and better than the existing other algorithms except for when δ near ± 0.5. Our simulations show that the proposed algorithm shows a significantly better performance, compared to the other algorithms at other SNR values.

Consequently, the proposed algorithm outperforms other algorithms that are known in the literature for signals whose phase is already known.

5 Discussions and Conclusions

5.1 Discussions

Even though the proposed method seems to exhibit the best performance amongst the existing methods, its biggest disadvantage is that the method only performs well when the phase is known a priori. This means that the proposed method is only applicable to the signals whose phase is already known. Therefore, it may not be suitable for many problems, where the phase is not known, restricting its applicability. However, the other existing algorithms show similar performance when the phase of the relevant signal is not known. Even so, the our method proposes an important framework for in the future studies for unbiased estimator designs.

5.2 Computational Cost

Computation of our algorithm requires two DFT operations. The first DFT is for evaluating the integer part of the frequency k_p in (5). Computing $S\left[k_p + \frac{1}{2}\right]$ and $S\left[k_p - \frac{1}{2}\right]$ requires the signal to be multiplied with $exp\left(\pm j\frac{\pi}{N}n\right)$ first, and then transformed by taking its DFT. However, for time-critical applications $S\left[k_p \pm \frac{1}{2}\right]$ can be calculated by using the celebrated Goertzel algorithm [8] which requires $\mathcal{O}(N)$ complex multiplications and additions instead of DFT, which requires $\mathcal{O}(N \log N)$ computational cost. Therefore, the total computational complexity for our algorithm is in the order of $(N \log N)$.

5.3 Conclusions

In this paper, we have presented a novel way to estimate frequency of a complex single sinusoidal under noisy environments, when the phase of the signal is known. The proposed estimator is shown to perform better than the existing algorithms in the literature.

References

1. Jacobsen E, Kootsookos P (2007) Fast, accurate frequemcy estimators. IEEE Signal Proc. Mag. 24(3):123–125
2. Candan C (2007) A method for fine frequency estimation from three DFT samples. IEEE Signal Proc Lett 18(6):351–354
3. Quinn BG (1997) Estimation of frequency, amplitude, and phase from the DFT of a time series. IEEE Trans Signal Proc 45(3):814–817

4. Quinn BG (1994) Estimating frequency by interpolation using Fourier coefficients. IEEE Trans Signal Proc 42(5):1264–1268
5. Macleod M (1998) Fast nearly ML estimation of the parameters of real or complex single tones or resolved multiple tones. IEEE Trans Signal Proc 46(1):141–148
6. Aboutanious E, Mulgrew B (1998) Iterative frequency estimation by interpolation on Fourier coefficient'. IEEE Trans Signal Proc 53(4):1237–1248
7. Kay SM (1993) Fundamentals of statistical signal processing: estimation theory. Prentince-Hall, New Jersey
8. Oppenheim AV, Shafer RW, Buck JR (1999) Discrete-time signal processing, 2nd edn. Prentince-Hall, New Jersey

Rapidly Varying Sparse Channel Tracking with Hybrid Kalman-OMP Algorithm

Ayşe Betül Büyükşar, Habib Şenol, Serhat Erküçük and Hakan Ali Çırpan

Abstract It is expected from future communication standards that channel estimation algorithms should be able to operate over very fast varying frequency selective channel models. Therefore, in this study autoregressive (AR) modeled fast varying channel has been considered and tracked with Kalman filter over one orthogonal frequency division multiplexing (OFDM) symbol. Channel sparsity is exploited which decreases the complexity requirements of the Kalman algorithm. Since Kalman filter is not directly applicable to sparse channels, orthogonal matching pursuit (OMP) algorithm is modified for AR modeled sparse signal estimation. Also, by using windows, sparsity detection errors have been decreased. The simulation results showed that sparse fast varying channel can be tracked with the proposed hybrid Kalman-OMP algorithm and windowing method offers improved MSE results.

Keywords OFDM · Fast time-varying channel · Autoregressive model Kalman · OMP · Sparse channel tracking

A. B. Büyükşar (✉) · H. A. Çırpan
Istanbul Technical University, Istanbul, Turkey
e-mail: buyuksar@itu.edu.tr

H. A. Çırpan
e-mail: cirpanh@itu.edu.tr

H. Şenol · S. Erküçük
Kadir Has University, Istanbul, Turkey
e-mail: hsenol@khas.edu.tr

S. Erküçük
e-mail: serkucuk@khas.edu.tr

A. Boyaci et al. (eds.), *International Telecommunications Conference*,
Lecture Notes in Electrical Engineering 504,
https://doi.org/10.1007/978-981-13-0408-8_25

1 Introduction

Channel estimation is an important part of the communication system since the channel deteriorates the transmitted signal in both time and frequency domains. There are numerous channel estimation algorithms for different channel models in the literature. Future wireless communication standard requirements direct researchers to study on fast time varying frequency selective channel estimation algorithms.

Spectral efficiency and handling frequency selectiveness of the channel are important parameters for wireless communication system design. Therefore, Orthogonal Frequency Division Multiplexing (OFDM) has been accepted by many communication standards and also is considered for future ones. Nonetheless, OFDM systems are very sensitive to synchronization errors and not suitable for fast varying channel environments. Through the cyclic prefix length (T_{CP}), inter-symbol-interference (ISI) effect can be canceled. However, fast time variation of the channel causes interference and it should be tracked over time to equalize these effects at the receiver side. In literature many channel estimation algorithms are proposed for OFDM systems.

The fast variation of the channel represented with Basis Expansion model (BEM) and the variation of BEM coefficients from symbol-to-symbol determined as an autoregressive (AR) process in [1]. Also, Kalman based iterative channel estimation algorithm simulated for pilot assisted OFDM systems. Complexity is reduced using subspace tracking with Kalman filter for OFDM channel estimation algorithm in [2].

It is stated in [3] that over complete dictionaries offer sparser signal representation which improves the performance of compressed sensing (CS) algorithm. It is stated that underwater acoustic channels are appropriate for sparse channel estimations. The sparse channel models and implementation of CS algorithm to wireless communication system are presented in [4]. It is proposed that increasing the number of delay grids results in better resolution in [5]. Performance results show that complexity can be decreased with proposed algorithm without any performance loss. In contrast to the grid resolution based approach, [6] proposes gridless sparse channel estimation with atomic norm minimization for pilot aided OFDM systems.

Future communication systems are expected to be durable very fast varying channels. Therefore, the importance of the rapidly varying channel estimation algorithms has been increased. The sparse channel represented with discrete orthonormal Karhunen–Loeve basis expansion model (DKL-BEM) and OMP algorithm are used to estimate the channel in [7].

In this study, we focus on rapidly varying sparse channel tracking for data-aided OFDM systems. The main motivation of the using Kalman filter for this problem is that sparse multipath channel coefficients are modeled as an AR process and Kalman filter effectively track the signal which is not sparse since it is not applicable to sparse signal tracking. Therefore, orthogonal matching pursuit algorithm (OMP) is used to estimate initial sparse channel coefficients [8]. The proposed method considers the same channel model in [8] and solve the channel estimation problem with Kalman filters. Also, dividing into the full OFDM block to subwindows improves the channel estimation performance.

The remainder of the paper is organized as follows. Section 2 presents the system model of an OFDM based wireless mobile communication system over rapidly varying sparse channels and describes the main parameters of the considered channel model. Section 3 proposes channel estimation algorithm based on Kalman Filtering method and OMP based initialization algorithm. Section 4 presents performance results of the proposed hybrid Kalman-OMP channel tracking algorithm. Finally, Sect. 5 concludes the paper.

2 System Model

OFDM system has been considered with N subcarriers which transmit only pilot symbols. The transmitted OFDM signal is

$$s(t) = \frac{1}{N} \sum_{k=-N/2}^{N/2-1} d[k]\, e^{j\frac{2\pi tk}{T_{sym}}}, \tag{1}$$

where T_{sym} expresses the symbol duration of OFDM signal and $d[k]$'s are the digitally modulated data symbols. Sampling duration of the signal is $T_s = \frac{T_{sym}}{N}$.

The received signal is the convolution of channel impulse response (CIR) and transmitted signal since it is assumed that the length of the channel is finite. Mathematical expression of the CIR is

$$h(t, \tau) = \sum_{l=0}^{L-1} \alpha^{(l)}(t)\delta(\tau - \tau^{(l)}(t)) \tag{2}$$

where L is the number of multipaths, $\alpha^{(l)}(t)$ and $\tau^{(l)}$ represent the fast varying multipath coefficients and delay times, respectively. It is considered that channel has sparse multipath delays which are uniformly distributed over cyclic prefix duration, T_{CP}. This way inter-symbol-interference (ISI) could be canceled when the maximum delay spread is less than the CP duration. However, because of the fast variation of the channel coefficients, inter-carrier-interference (ICI) occurs which is hard to equalize. The received signal is

$$\begin{aligned} r(t) &= s(t) * h(t, \tau) + w(t) \\ r(t) &= \sum_{l=0}^{L-1} \alpha^{(l)}(t)s(t - \tau^{(l)}) + w(t) \end{aligned} \tag{3}$$

$$r(t) = \frac{1}{N} \sum_{l=0}^{L-1} \sum_{k=-\frac{N}{2}}^{\frac{N}{2}-1} d[k] \alpha_l(t) e^{\frac{j2\pi k(t-\tau^l)}{T_{sym}}} + w(t) \tag{4}$$

and it is assumed that $\tau^{(l)}$ is an integer multiple of T_s i.e. $\eta^{(l)} = \frac{\tau^{(l)}}{T_s}$, $w(t)$ is the complex Additive White Gaussian Noise (AWGN). Discrete received signal can be written as

$$r[n] = \frac{1}{N} \sum_{k=-\frac{N}{2}}^{\frac{N}{2}-1} \sum_{l=0}^{L-1} d[k] e^{\frac{j2\pi k}{N}(n-\eta^{(l)})} \alpha^{(l)}[n] + w[n]. \tag{5}$$

In matrix form it can be written as,

$$\begin{aligned} r[n] &= \psi^H[n]\, \boldsymbol{D}\, \boldsymbol{\Phi}^{(l)}\, \boldsymbol{\alpha}[n] + w[n] \\ &= \boldsymbol{C}^H[n] \boldsymbol{\alpha}[n] + w[n] \end{aligned} \tag{6}$$

$$\boldsymbol{c}^H[n] = \boldsymbol{\mathcal{F}}^{-1}[n]\, \boldsymbol{D}\, \boldsymbol{\Phi} \tag{7}$$

$$\boldsymbol{\mathcal{F}}^{-1}[n] \stackrel{\text{def}}{=} \frac{1}{N} \left[e^{\frac{j2\pi}{N}(-N/2)n}, \ldots, e^{\frac{j2\pi}{N}(N/2-1)n} \right]$$

$$\boldsymbol{\Phi} \stackrel{\text{def}}{=} \left[\phi^{(0)}, \phi^{(1)}, \ldots, \phi^{(N_{cp}-1)} \right]$$

$$\phi^{(l)} \stackrel{\text{def}}{=} \left[e^{-\frac{j2\pi}{N}(-N/2)\eta^{(l)}}, \ldots, e^{-\frac{j2\pi}{N}(N/2-1)\eta^{(l)}} \right]^T$$

$$\boldsymbol{D} \stackrel{\text{def}}{=} diag \left\{ d[k]|_{k=-N/2}^{N/2-1} \right\}$$

where $\boldsymbol{\alpha}[n]_{N_{cp}\times 1}$ is the L-sparse channel coefficient vector. a^T is the transpose of a and $diag\{a\}$ represents the matrix with diagonal elements of a. $L \ll N_{cp}$ implies sparseness of the channel.

The objective of this study is to track the $\alpha^{(l)}[n]$ multipath channel coefficients when the delay positions are constant over one OFDM symbol. Even for an all pilot case; there are N received signal samples and N pilot symbols. However, LN unknown channel coefficients and L multipath delay positions have to be estimated from these received signal samples.

Multipath channel coefficients change from sample to sample through combining with the AR process which is defined below

$$\alpha^{(l)}[n] = a\, \alpha^{(l)}[n-1] + b^{(l)}\, u^{(l)}[n]. \tag{8}$$

Here $\alpha^{(l)}[n]$ is the lth path channel coefficient at time instant n. $u^{(l)}[n]$ is the driving channel noise which is distributed as complex Gaussian with unit variance. AR process coefficients a and $b^{(l)}$ can be found by using autocorrelation function of channel coefficients. It is straightforward that each path obeys Jakes' spectrum with the following autocorrelation function

$$\begin{aligned}\rho^{(l)}[n-n'] &= E\{\alpha^{(l)}[n]\alpha^{(l)}[n']^*\} \\ &= \Omega^{(l)} J_0\left(2\pi f_D T_s(n-n')\right).\end{aligned} \tag{9}$$

When $\eta^{(l)}$ is given, the lth path normalized channel power defined as

$$\Omega^{(l)}|\eta^{(l)} = \frac{\lambda}{(1-e^{-\lambda})L} e^{-\lambda \frac{\eta^{(l)}}{N_{cp}}}, \qquad \lambda \geq 0, \tag{10}$$

where λ is multipath power decaying constant and f_D is the maximum Doppler frequency. For general OFDM systems it is accepted that $\lambda = 1$ [7]. $J_0(a)$ represents the zeroth order Bessel function of the a. By applying Yule-Walker equation and with the help of Jakes' autocorrelation function in (9), AR coefficients a and $b^{(l)}$ can be found as

$$\begin{aligned}a &= J_0\left(2\pi f_D T_s\right) \\ b^{(l)} &= \sqrt{\Omega^{(l)}\left(1 - J_0^2\left(2\pi f_D T_s\right)\right)}.\end{aligned} \tag{11}$$

3 Channel Estimation

While normalized delays of the multipath channel are known, then (6) and (8) can be used as observation and state equations, respectively, in classical Kalman filtering. In that case, Eqs. (12)–(16) can be calculated and multipath channel coefficients can be tracked. Thus when a sparse CIR is considered, initially channel delay positions should be obtained. Kalman filtering steps can be adapted to sparse CIR tracking as in [9].

Prediction step:

$$\hat{\boldsymbol{\alpha}}[n \mid n-1] = \boldsymbol{A}\, \hat{\boldsymbol{\alpha}}[n-1 \mid n-1] \tag{12}$$

Error covariance prediction:

$$\boldsymbol{M}[n \mid n-1] = \boldsymbol{A}\, \boldsymbol{M}[n-1 \mid n-1]\, \boldsymbol{A}^H + \boldsymbol{B}\, \boldsymbol{Q}_u\, \boldsymbol{B}^H \tag{13}$$

Kalman gain:

$$\boldsymbol{K}[n] = \boldsymbol{M}[n \mid n-1]\boldsymbol{C}[n]\left(\boldsymbol{C}^H[n]\boldsymbol{M}[n \mid n-1]\boldsymbol{C}[n] + \boldsymbol{C}_w[n]\right)^{-1} \tag{14}$$

Correction step:

$$\hat{\boldsymbol{\alpha}}[n \mid n] = \boldsymbol{K}[n]\left(r[n] - \boldsymbol{C}^H[n]\hat{\boldsymbol{\alpha}}[n \mid n-1]\right) + \hat{\boldsymbol{\alpha}}[n \mid n-1] \tag{15}$$

Error covariance update:

$$\boldsymbol{M}[n \mid n] = \left(\boldsymbol{I} - \boldsymbol{K}[n]\,\boldsymbol{C}^{H}[n]\right) \boldsymbol{M}[n \mid n-1]. \tag{16}$$

3.1 OMP Algorithm for AR Modeled Sparse Signal

CS algorithms offer the estimation of sparse signal from limited number of observations. The main consideration is that these algorithms are derived for sparse signals, which are constant when the observations have been obtained. However, for rapidly varying channels, CIR variations can be modeled with the AR process. Therefore, it is preferred to express each multipath channel coefficient depending on initial estimates as

$$\alpha^{(l)}[n] = a^{n+1}\alpha^{(l)}[-1] + \sum_{k=0}^{n} a^{k} b^{(l)} u[n-k]. \tag{17}$$

This way received signal samples can be expressed using (17) and OMP algorithm can be applied [10]. This representation makes it possible to estimate sparse initial channel coefficients with the OMP algorithm. The mathematical representation is

$$\begin{aligned}
\boldsymbol{r} &= \boldsymbol{D}\boldsymbol{\alpha}[-1] + \boldsymbol{v} \\
\boldsymbol{D} &= \boldsymbol{C}^{H} \odot (\boldsymbol{a} \otimes \mathbf{1}_{1\times N_{cp}}) \\
\boldsymbol{r} &\overset{\text{def}}{=} [r[0],\; r[1],\; \ldots,\; r[M-1]]^{T} \\
\boldsymbol{C}^{H} &\overset{\text{def}}{=} [\boldsymbol{c}[0],\; \boldsymbol{c}[1],\; \ldots,\; \boldsymbol{c}[M-1]]^{H} \\
\boldsymbol{a} &\overset{\text{def}}{=} \left[a^{1}, a^{2}, \ldots, a^{M}\right]^{T},
\end{aligned} \tag{18}$$

where $\boldsymbol{D}$ is the dictionary matrix with size $M \times N_{cp}$ and $\boldsymbol{v}$ represents noise, which includes AWGN and driving noise.

It should be noted that, using all observations causes more complexity despite no additional performance contribution. Therefore, using $M < N$ observations to find initial estimates of channel coefficients and path delays is adequate.

3.2 Windowing

Detection of one or more delay positions wrongly affects the multipath tracking performance adversely when the OMP algorithm is initially run for one OFDM symbol. Therefore, dividing into the one OFDM symbol to windows improves the performance of the sparse channel tracking algorithm. The main consideration is the transition between windows since the estimated normalized delay position can be different for each window. $H[n,k]$ is the channel frequency response (CFR) at the nth time sample and it is defined as

$$H[n,k] = \sum_{l=0}^{L-1} e^{-\frac{j2\pi}{N}k\eta^{(l)}} \alpha^{(l)}[n]. \tag{19}$$

The time sample at the end of each window is p and estimated delay position of the next window is independently calculated. However, for correctly determined delay positions $H[p,k]$ should be the same for adjacent windows. In the notation, l is used for past window delay positions and $\tilde{l}$ is used for the next window, resulting in

$$\begin{aligned} H[p,k] &= \phi^{(l)}\alpha[p] = \phi^{(\tilde{l})}\tilde{\alpha}[p] \\ \tilde{\alpha}[p] &= \underbrace{\left(\phi^{(\tilde{l})H}\phi^{(\tilde{l})}\right)^{-1}\phi^{(\tilde{l})H}\phi^{(l)}}_{T}\alpha[p] \\ \tilde{M} &= TMT^H \end{aligned} \tag{20}$$

where $\tilde{\alpha}$ expresses that α belongs to the new window and T is the transition matrix.

Proposed algorithm takes Doppler frequency and SNR value in addition to the received signal as an input to estimate channel coefficients. The proposed algorithm starts with the OMP algorithm. After the initialization part, Kalman filtering is applied for each window. Then, OMP algorithm is used for next window to estimate and update the delay positions. Using the transition matrix, initial channel coefficients and error covariance matrix are found. This process is repeated for all windows, and the estimated channel frequency response is obtained from (19).

In the following section, performance results will be carried through simulations.

4 Simulation Results

Estimated discrete channel frequency response $\hat{H}[n,k]$ can be found by substituting $\hat{\alpha}^{(l)}[n]$ and $\hat{\eta}^{(l)}$ into (19). Performances are evaluated based on the mean square error (MSE), which is defined as

$$MSE = \frac{1}{N^2}E\left[\sum_{n=0}^{N-1}\sum_{k=-\frac{N}{2}}^{\frac{N}{2}-1}\left|\mathbf{H}[n,k]-\hat{\mathbf{H}}[n,k]\right|^2\right]. \tag{21}$$

Performance of the proposed algorithm will be assessed considering initialization and channel coefficients tracking. The simulation parameters are selected as $N = 1024$, $N_{CP} = 64$.

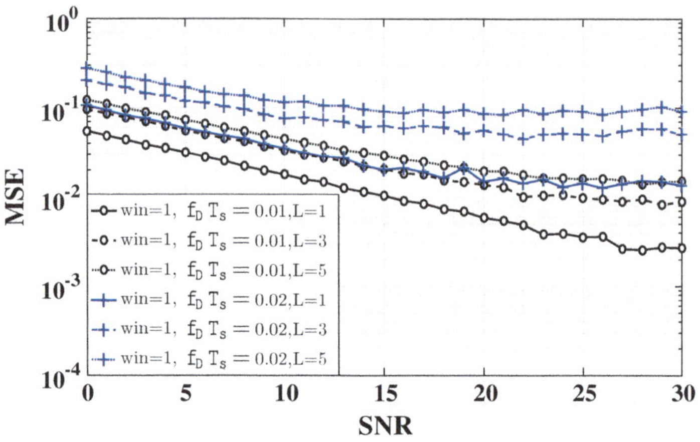

Fig. 1 OMP Kalman MSE results without windowing

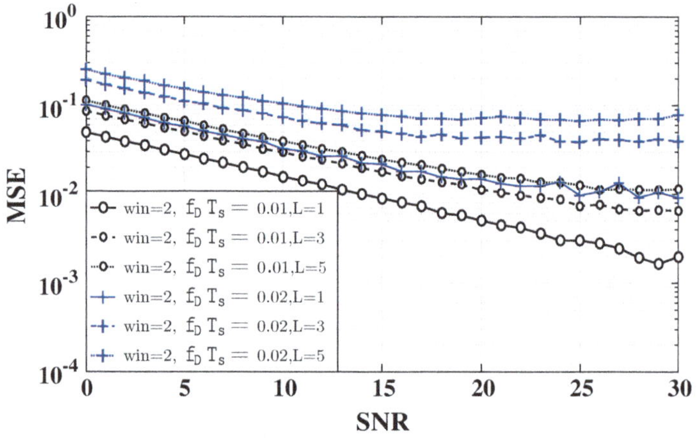

Fig. 2 OMP Kalman MSE results using 2 windows

When OFDM symbol is not divided into subwindows, performance results of hybrid Kalman-OMP algorithm is showed in Fig. 1. It is obvious that Doppler frequency adversely affects the channel estimation performance. It can be inferred that proposed algorithm face with error floor at high SNR values. To show the effect of applying window in the proposed approach, additional simulations are carried, and corresponding results are presented in Fig. 2. It is observed that adding windowing in the proposed technique achieves performance improvement. Incrementation of the number of window decreases the length of window. Therefore OMP algorithm run with less observation which adversely affects the detection performance. Also it should be noted that there is an optimum value of window for a specified num-

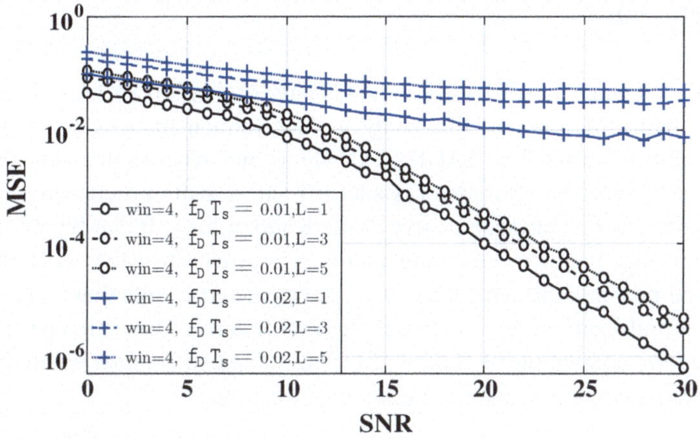

Fig. 3 OMP Kalman MSE results using 4 windows

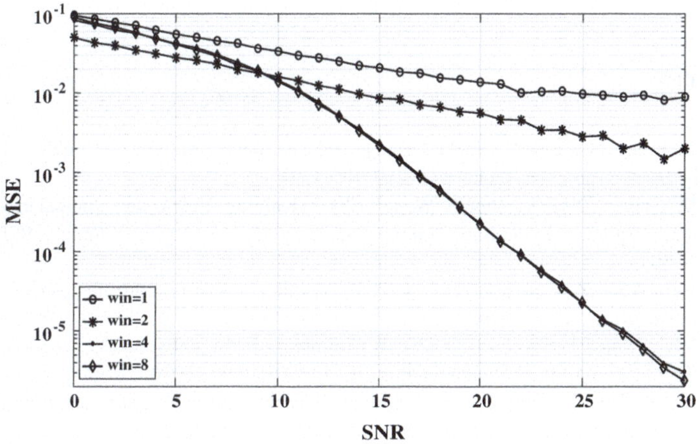

Fig. 4 OMP Kalman MSE results when $f_D T_s = 0.01$ and $L = 3$

ber of OFDM subcarrier. In this study using 4 window found as optimum when the complexity and performance results considered. It is observed that considerable improvement is not achieved when the $f_D T_s = 0.02$. However, high performance improvement obtained for less Doppler values. The performance improvement of the hybrid Kalman-OMP algorithm depend on window number can be seen in Figs. 3 and 4.

5 Conclusion

Very rapidly varying sparse channel model is only studied in [8], best of our knowledge. The main difference of this study is that Kalman filtering is used to track channel variation rather than SAGE-MAP algorithm. Also to decrease the sparse detection error, windowing method is adopted and transition between windows is realized using error covariance matrix of the Kalman filter. It can be inferred from simulation results that hybrid tracking approach improves the MSE with complexity trade off and prevents the error floor. It is also expected results that there is a trade off between multipath number and tracking performance. The main contribution of the study is the adaptation of Kalman filtering to sparse channel estimation since Kalman filter is not applicable to sparse signal tracking.

Acknowledgements This study is supported by The Scientific and Technological Research Council of Turkey (TUBITAK) under project no. 114E298.

References

1. Wan P, McGuire M, Dong X (2011) Near-optimal channel estimation for OFDM in fast-fading channels. IEEE Trans Veh Technol 60(8):3780–3791
2. Huang M, Chen X, Xiao L, Zhou S, Wang J (2007) Kalman-filter-based channel estimation for orthogonal frequency-division multiplexing systems in time-varying channels. IET Commun 1(4):795–801
3. Berger CR, Wang Z, Huang J, Zhou S (2010) Application of compressive sensing to sparse channel estimation. IEEE Commun Mag 48(11):164–174
4. Bajwa WU, Haupt J, Sayeed AM (2010) Compressed channel sensing: a new approach to estimating sparse multipath channels. Proc IEEE 98(6):1058–1076
5. Hu D, Wang X, He L (2013) A New Sparse Channel Estimation and Tracking Method for Time-Varying OFDM Systems. Proceedings of the IEEE 62(9):4648–4653
6. Pejoski S, Kafedziski V (2015) Estimation of sparse time dispersive channels in pilot aided OFDM using atomic norm. Proc IEEE 4(4):397–400
7. Senol H (2015) Joint channel estimation and symbol detection for OFDM systems in rapidly time-varying sparse multipath channels. Wirel Pers Commun 82(3):1161–1178 (Published Online: 14 January 2015)
8. Buyuksar AB, Senol H, Erkucuk S, Cirpan HA (2016) Data-aided autoregressive sparse channel tracking for OFDM systems. In: Thirteenth international symposium on wireless communication systems (ISWCS'16), Poznan, Poland
9. Kay SM (1993) Fundamentals of statistical signal processing volume I: estimation theory. Prentice Hall, New Jersey
10. Elad M (2010) Sparse and redundant representations from theory to applications in signal and image processing. Springer Science Business Media, New York, pp 35–76

Prediction of Physical Activity Times Using Deep Learning Method

Gokhan Ozogur, Mehmet Ali Erturk and Muhammed Ali Aydin

Abstract Sedentary life style causes some serious health problems. In order to minimize these problems, it is recommended to do physical activities regularly. Even though it is possible to track activity level, making physical activity a habit is not easy. In this study, we aimed to predict the times when people will be stationary in terms of physical activity such as sitting or sleeping. Historical physical activity data of each individual is used to generate a model in order to estimate the percentage of being stationary within the next period of time for each individual. In this way, it will be reasonable to suggest a more suitable time for physical activity.

Keywords Deep learning · Recurrent neural networks · Data analytics

1 Introduction

As a result of improvements in technology, people are required to spend much less effort to do a job. There is a higher demand for knowledge worker instead of blue-collar worker in a digital age. These requirements make people spend much more time sitting in front of a computer screen and less time for doing physical activities. Not spending enough time for physical activities is definitely not good for our body. Especially sitting long time in a sedentary situation without giving any break causes some critical health problems. In order to decrease these risks, it is suggested to give short breaks every hour. However tracking break times is not easy in a stressful working environment. There are some devices and applications which can measure your activity level using accelerometers and notify you to give a break and move.

G. Ozogur (✉)
R and D Software Design Department, Arcelik A.S, Istanbul, Turkey
e-mail: gokhan.ozogur@arcelik.com

M. A. Erturk · M. A. Aydin
Computer Engineering Department, Istanbul University, Istanbul, Turkey
e-mail: mehmetali.erturk@istanbul.edu.tr

M. A. Aydin
e-mail: aydinali@istanbul.edu.tr

A. Boyaci et al. (eds.), *International Telecommunications Conference*,
Lecture Notes in Electrical Engineering 504,
https://doi.org/10.1007/978-981-13-0408-8_26

Watching television, using computer, driving car and reading can be given as examples of sedentary behaviors. Giving short breaks in between these sedentary times can be handled using activity level trackers. However this is not enough for a healthy life style. It is recommended to do physical activities at least for 30 min at one session every day. These trackers should be more sophisticated to learn your daily schedule since it may not be possible to give a break for 30-min run in the middle of a business meeting. In order to find a suitable time for physical activities, daily rhythms of people can be analyzed. In our study, we aimed to predict the times when people will be stationary in terms of physical activity such as sitting or sleeping. This prediction can be used to suggest people more suitable times for physical activity according to routines. Using these suggestions, people can reach to a health life style easily.

The organization of our paper is as follows: Sect. 2 presents the related works done in this field. In Sect. 3; data set, models and performance results are explained. Discussions and conclusions about the results are done in Sects. 4 and 5 respectively.

2 Related Works

It is possible to obtain information about health status of individuals from physical activity data of them. One of the determinants of healthy life style is sleep quality. There exist a study in literature about prediction of sleep quality using physical activity data. In that study, sleep quality is derived by comparison of actual and required sleep time [1]. Required sleep time of individuals is calculated on linear regression and deep learning models using physical activity data collected in day-time. Sleep quality information can be used as a performance measure to motivate individuals for healthy life.

Another performance measure is physical activity level of individuals. There exist an application called SmartFit that calculates daily activity times using step counts [2]. In that application, users are motivated to do physical activity using gamification technique. In that technique, non-game related problems are solved using game design elements [3]. In SmartFit application, users gain points according to their physical activity performance. It motivates users for the purpose of gaining physical activity habit. However it doesn't notify users against long stationary durations.

There exist a study which propose a system design to prevent sedentary behaviors of individuals [4]. In that notification system, stationary durations are tracked in real-time by Fitbit device. Users receive a notification if they do not perform required physical activity in a given time. After a short period of time, users receive secondary notification if they do not stop stationary activity. This design is good for reminding individuals against sedentary behavior. However it does not consider availability of the individuals for physical activity in recommended time duration. It would be better to predict more suitable time frames according to routines of each individual.

A study is done about prediction of activity levels of individuals using physical activity data [5]. In that study, physical activity level is calculated as a ratio of sedentary time to given time period. Activity level prediction of next time period is calculated using autoregressive model. In our study, we used deep learning model for the same data set used in that study.

3 Methodology

We built our models on an assumption of physical activity level of individuals in next time period is related with previous activity levels which are calculated from historical physical activity data. Ratio of sedentary time for given time period is calculated using physical activity data, and given as inputs to our model. As an output of the model, prediction of activity level for next time period is obtained. For example, hourly activity levels of an individual can be used for prediction of activity level in next hour. Historical physical activity data is grouped by given time period and sedentary ratio is calculated for each group. Each ratio information is handled as an input for the model. Then data is separated for training and test sets for each individual. We used half of the data of an individual for training and the other half for the test. We used different models for each time period. For example, if we choose 1 h as a time period, then we should train 24 models for each hour. We can easily select model according to group number due to the fact that data is grouped by time period. When training of the model is finished with all training data, we test model using test data. Finally error for predictions is calculated for each individual. Flow diagram of our methodology is shown in Fig. 1.

Performance of our model is measured by mean squared error (MSE) criterion. Errors between prediction outputs of our model and actual results are calculated for each individual. MSE is calculated using the Formula 1 where $\hat{Y}$ is a vector of predictions and Y is a vector of actual results.

$$MSE = \frac{1}{n}\sum_{i=1}^{n}(\hat{Y}_i - Y_i)^2 \quad (1)$$

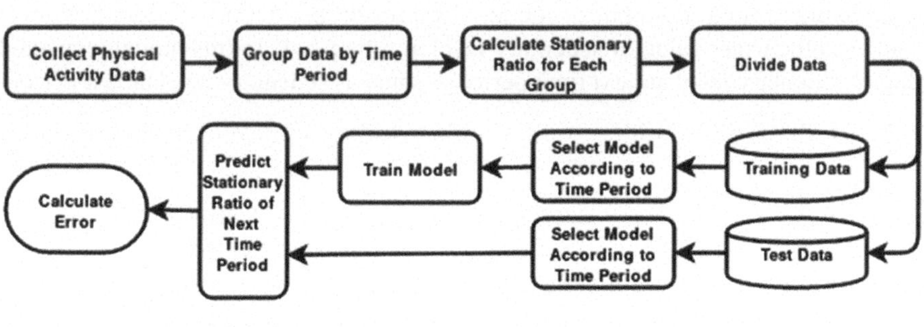

Fig. 1 Flow diagram of deep learning model

3.1 Data Set

In order to measure performance of our model, we used StudentLife dataset [6]. This dataset contains many sensor data collected via smart phones of 49 students of Dartmouth College for 10-week period. Physical activity data in this dataset is a big collection of accelerometer measurements of every 2 s. In order to increase battery life of smart phones, they actively collected data for 1 min and sleep for 3 min. Each measurement is convert to a predicted activity state and shared with time-stamp data. A sample from physical activity data is shown in Table 1 where activity states are labeled with numbers ($Stationary : 0$, $Walking : 1$, $Running : 2$, $Unknown : 3$).

We have done some pre-calculations for using StudentLife data in our model. First of all, activity data is grouped by given time period for each individual. Then, activity times in each time period are calculated using activity state labels. After that, sedentary ratio is calculated using the ratio of stationary labeled data to the other data. For example, if there 1200 measurements labeled as stationary out of 1800 total measurements in a given time period, then the sedentary ratio is 0.67 for that period. If there is no data for given time period, then we assumed that sedentary ratio for that time period is 1 which means the person is fully sedentary.

3.2 Models

In this study, we built a deep learning model in order to predict physical activity level next time periods. Performance of the model is calculated using MSE criterion. Moreover, we obtained results using another model in order to compare results of deep learning model. The benchmark model is called reference model. We built the reference model with one simple rule. If the performance of deep learning model is not better than reference model, then deep learning model is not efficient and reasonable.

Reference Model There is only one simple rule in reference model. Input of this model is activity level at the last time period. Output of the model is 1 time unit lagged value of the only input of model. As a result, prediction of sedentary level at time t is the exact value of sedentary level at time $t - 1$. In this model, expected value of sedentary level at next time period is same as sedentary level at last period.

Table 1 A sample from physical activity data

Timestamp	Activity
1364383644	0 ($Stationary$)
1364383649	1 ($Walking$)
1364383652	0 ($Stationary$)
1364383654	1 ($Walking$)
1364383657	1 ($Walking$)

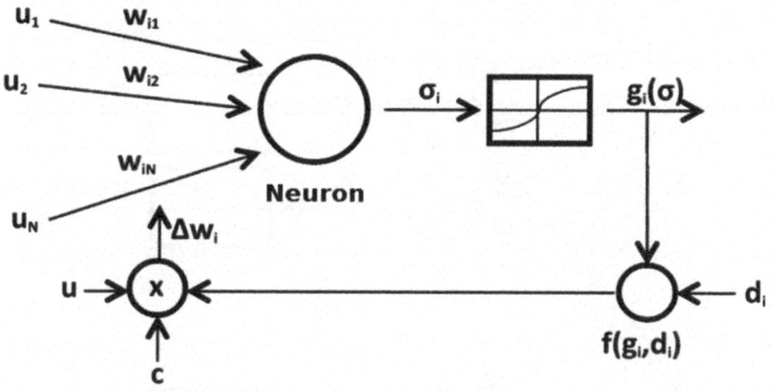

Fig. 2 Artificial neural network model with a single neuron

This is a fast and simple method since there is not any calculation. Therefore, it is expected to obtain not so good result using this model. If the performance of deep learning model is not better than reference model, then deep learning model is using more resource for nothing.

Deep Learning Model In artificial neural networks, calculation is done via neurons. Each input of the neuron is multiplied with related weight coefficient in the neuron. Result of the multiplication is the net effect of the neuron. Net effect got processed by activation function, so that output of the neuron is obtained. In supervised learning methods, weight coefficients are updated to minimize error by processing output of the neuron and desired result. Artificial neural network model with single neuron is shown in Fig. 2 where u is input, w is weight coefficient, σ is net effect, g is output, d is desired result, and c is constant term. In a typical artificial neural network, there are more than one neuron in one layer. In that case, all of the inputs are used in every neuron with related weight coefficient. Artificial neural network model with many neurons in one layer is shown in Fig. 3.

First neuron layer which receives inputs of the model is input layer. Last neuron layer which gives output of the model is output layer. Other layers between input and output layer are hidden layers. Neurons in hidden layer receive inputs from output of neurons in previous layer. It is possible to define deep learning model as artificial neural network model with more than one hidden layer.

In our model, we used recurrent neural network model which is a sub-set of artificial neural network. It is allowed to create backward connections in recurrent neural networks. This makes it possible to connect output of a neuron to input of itself. In this way, output of the neuron depends on previous inputs. We used backward connections in input layer, hidden layers and output layer of our model. Using recurring, we processed not only current input but also previous inputs in input layer. In this way, we had dependency between inputs. We also make a connection from output layer to input layer. This makes our model more stable since previous predictions are handled as an extra feature in input layer. It is possible to think recurrent connections as delay units. Previous outputs of a neuron are handled as inputs for another neuron

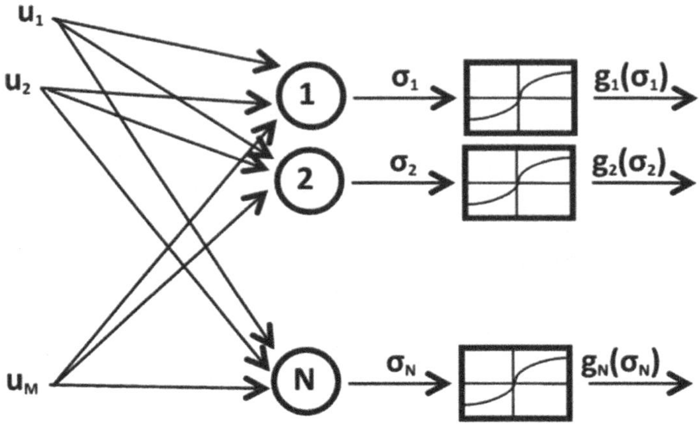

Fig. 3 Artificial neural network model with many neurons in one layer

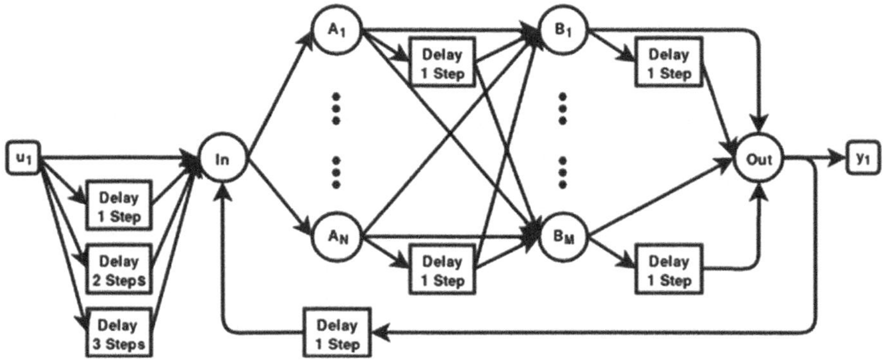

Fig. 4 Recurrent neural network model

with given delay in time. Our model is shown in Fig. 4. We used stationary ratio for current time as a source of information which is showed u_1 in the figure. Moreover inputs of previous 3 days and output of previous day are also handled as input.

3.3 Performance Results

We performed calculations on MATLAB using an open source recurrent neural network toolbox called pyrenn [7]. In that toolbox, it is possible to easily create a network with any number of neurons and delay in each layer. $Levenberg-Marquardt$ algorithm and $Broyden-Fletcher-Goldfarb-Shanno$ algorithm are implemented in that toolbox. We used both algorithms for training of the model and took the average of the results for prediction. We performed training and test calculations on a laptop

Table 2 Top-10 performance results of deep learning model

ID	Deep learning	Reference
44	0.0005341	0.0014425
3	0.0007497	0.0023081
10	0.0010663	0.0080118
23	0.0011769	0.0028271
53	0.0011906	0.0031130
50	0.0012732	0.0013073
15	0.0013299	0.0022649
20	0.0014446	0.0040859
39	0.0016788	0.0005074
18	0.0017248	0.0054256

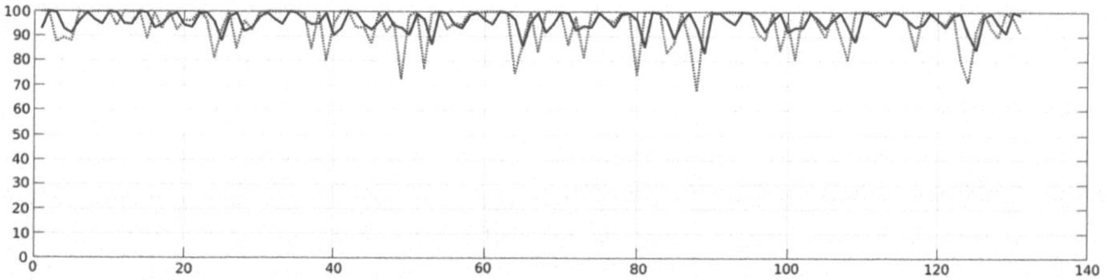

Fig. 5 6-h time period predictions (blue) of deep learning model and actual data (red dashed) for User-58

which has 2.50 GHz dual-core Intel Core i5 CPU and 8 GB RAM. Training of the model takes less than 7 s and testing of the model takes much less than 1 s for an individual.

Reference and deep learning models are tested for 6-h time period data of 49 individuals. Deep learning model is performed better than reference model for 40 individuals out of 49. Top-10 performance results of deep learning model are shown in Table 2 with the results of reference model.

In order to display a sample output of deep learning model, predictions and actual data for $User - 58$ are shown in Fig. 5. Predictions of the model are marked with blue line and actual data is marked in red dashed line.

4 Discussion

Performance results show that predictions of deep learning model is better than reference model for most of the individuals. When we investigate the reason of the performance differences among individuals, we see that data set is not complete for all individuals. There are long period of times with small differences in activity level for some individuals. This indicates that those individuals are not carried on their mobile devices during their activity or they do not move often.

We would like to compare our performance results with the study of He and Agu which uses the same data set for physical activity level prediction [5]. However, our assumption for no data periods in pre-calculation contradicts with their calculations. They take time periods with no data as zero which means an individual is physically active for all of the duration. This causes some individuals move for days without resting. Since our data set is changed after pre-calculation, it is not reasonable to compare the result with that study.

Our performance result also indicates that it is reasonable to use deep learning model with regards to reference model. Training of the model for an individual finishes less than 7 s and testing is much more faster. In real world scenario, it is possible to instantaneous predictions after collecting enough data for training.

5 Conclusion

In our study, we aimed to show that deep learning model can be used for prediction of physical activity levels. Our performance results show that there is not any obstacle to use deep learning model for suggestion of more suitable times for physical activity according to routines of people. They can get a habit of doing physical activity on regular basis and reach to a health life style using these suggestions.

In future works, we would like to develop a mobile application for collecting data and building our own data set. Moreover we would like to make suggestion using mobile application to people about their physical activity level in order to reach a health life. Performance of the model on a mobile device in real time will also be investigated.

Acknowledgements This work is also a part of the M.Sc. thesis titled **Design of a Mobile and Cloud Software for Analysis of Health Data** at Istanbul University, Institute of Physical Sciences.

References

1. Sathyanarayana A, Joty S, Fernandez-Luque L, Ofli F, Srivastava J, Elmagarmid A, Arora T, Taheri S (2016) Sleep quality prediction from wearable data using deep learning. JMIR mHealth and uHealth 4(4): e125
2. Sarwar A, Mukhtar H, Maqbool M, Belaid D (2015) Smartfit: a step count based mobile application for engagement in physical activities. Int J Adv Comput Sci Appl (IJACSA) 6(8):271–278
3. Deterding S, Dixon D, Khaled R, Nacke L (2011) From game design elements to gamefulness: defining gamification. In: Proceedings of the 15th international academic MindTrek conference: envisioning future media environments. ACM, pp 9–15
4. Pina LR, Ramirez E, Griswold WG (2012) Fitbit+: a behavior-based intervention system to reduce sedentary behavior. In: 2012 6th international conference on pervasive computing technologies for healthcare (PervasiveHealth). IEEE, pp 175–178

5. He Q, Agu EO (2016) Towards sedentary lifestyle prevention: an autoregressive model for predicting sedentary behaviors. In: 2016 10th international symposium on medical information and communication technology (ISMICT). IEEE, pp 1–5
6. Wang R, Chen F, Chen Z, Li T, Harari G, Tignor S, Zhou X, Ben-Zeev D, Campbell AT (2014) Studentlife: assessing mental health, academic performance and behavioral trends of college students using smartphones. In: Proceedings of the 2014 ACM international joint conference on pervasive and ubiquitous computing. ACM, pp 3–14
7. Atabay D (2016) pyrenn: first release. https://doi.org/10.5281/zenodo.45022

Author Index

A. Boyaci et al. (eds.), *International Telecommunications Conference*,
Lecture Notes in Electrical Engineering 504,
https://doi.org/10.1007/978-981-13-0408-8

The manufacturer's authorised representative in the EU is Springer Nature Customer Service Centre GmbH, Europaplatz 3, 69115 Heidelberg, Germany. If you have any concerns regarding our products, please contact ProductSafety@springernature.com

Printed and bound by CPI Group (UK) Ltd, Croydon, CR0 4YY
30/09/2026
02225446-0010